Radioecology and the Restoration of Radioactive-Contaminated Sites

edited by

Felix F. Luykx
International Union of Radioecology,
Tervuren, Belgium

and

Martin J. Frissel
International Union of Radioecology,
Heelsum, Belgium

Kluwer Academic Publishers

Dordrecht / Boston / London

Published in cooperation with NATO Scientific Affairs Division

Proceedings of the NATO Advanced Study Institute on
Radio-Active Contaminated Site Restoration
Zarechny, Sverdlovsk, Russia
June 19–28, 1995

A C.I.P. Catalogue record for this book is available from the Library of Congress.

ISBN-13: 978-94-010-6620-4   e-ISBN-13: 978-94-009-0301-2
DOI: 10.1007/978-94-009-0301-2

Published by Kluwer Academic Publishers,
P.O. Box 17, 3300 AA Dordrecht, The Netherlands.

Kluwer Academic Publishers incorporates the publishing programmes of
D. Reidel, Martinus Nijhoff, Dr W. Junk and MTP Press.

Sold and distributed in the U.S.A. and Canada
by Kluwer Academic Publishers,
101 Philip Drive, Norwell, MA 02061, U.S.A.

In all other countries, sold and distributed
by Kluwer Academic Publishers Group,
P.O. Box 322, 3300 AH Dordrecht, The Netherlands.

Printed on acid-free paper

Table of Contents

PREFACE

Since World War II all over the world large scale nuclear facilities have been built and operated, both for civil and defense purposes. Many of these have meanwhile ceased their activities and have been decommissioned.

Some of the sites on which these facilities were located are heavily contaminated with radioactive substances and require intervention to mitigate the radiological consequences for man and nature.

Correct and efficient intervention will only be possible when the behaviour of the radionuclides in the terrestrial environment is sufficiently well known. The study of this behaviour is commonly called radioecology.

Radioecologists have often difficulties in studying radioactivity transfer in agricultural lands and semi-natural ecosystems, because of the complexity and diversity of these environments. In order to have a correct understanding of the phenomena occuring, they must know the mechanisms determining the behaviour of radionuclides in different ecosystems.

To contribute to the understanding of the behaviour of radionuclides in a contaminated environment, a NATO Advanced Science Institute (ASI) on this subject was organized by the International Union of Radioecology (IUR) in June 1995 in Zarechny, Urals, Russia. Over 60 students from 22 countries, professionally involved in radioecology and nuclear site restoration, participated in this course.

In this ASI all factors were analyzed affecting the behaviour of radionuclides, as they move from their point of release through the environment and then enter the tissues of biota living in the ecosystems, in particular plants and animals which are consumed by humans.

The course was held in a region which is heavily contaminated due to radioactive discharges into the environment during nuclear weapon fabrication in the 1950's and 1960's and due to a severe accidental release following an explosion of a rad-waste tank in 1957. This allowed in situ training of the students and to give them experimental evidence of much of the theories presented in the lectures.

The present book is based on the lectures given in the above course. Its main emphasis is on specific radioecologic problems in severely contaminated areas in the former Soviet Union, i.e. the East-Urals Radioactive Trail from the Kysthym accident of 1957, the rivers Techa-Isert-Tobol-Irtish-Ob contaminated from discharges at the MAYAK nuclear fuel reprocessing plant in the late 1940's and the 30 km zone around Chernobyl.

Ecological systems, which have been examined in these areas, include soils, arable and pasture land, forests, lakes and rivers.

Special attention is given to the effects of radiation on natural ecosystems including trees, soil-dwelling organisms and aquatic organisms, and to the possible synergistic effects of radiation and other contaminants on natural populations and ecosystems.

A substantial part is also devoted to short term, medium term and long term countermeasures to mitigate the consequences of radioactive contamination.

While preparing the publication of the present book, the editors have maintained as much as possible the original papers, in order to respect the personal views of the authors. The transcription of cyrillic names of persons and locations was kept as written by the different authors.

ACKNOWLEDGEMENTS

Financial support for organizing this ASI and to publish this book was provided by NATO, to which the International Union of Radioecology is very grateful. The editors thank all the lecturers in the course and the contributors to this book. A special thank is addressed to Dr. A. Trapeznikov and his colleagues of the Institute of Plants and Animals Ecology of the Urals Division of the Russian Academy of Sciences in Ekaterinburg, who made all efforts to make this ASI a success. The editors are also very grateful to the SCK-CEN, Mol, to Dr. H. Vandenhove of the SCK-CEN Mol and Dr. G. Gerber, former Head of Division of the European Commission, for assisting in editing the manuscripts and to Mr. J. Bonnyns for preparing the lay-out of the book.

ORIGIN AND AIM OF RADIOECOLOGY

ARRIGO A. CIGNA
President of the International Union of Radioecology
Fraz Tuffo
I-14023 Cocconato AT, Italy

RERM NATURA SACRA SUA NON SEMEL TRADIT .
INITIATOS NOS CREDIMUS IN VESTIBULO EIUS
HAEREMUS . ILLA ARCANA NON PROMISQUE NEC
OMNIBUS PATENT . REDUCTA ET INTERIORE
SACRARIO CLAUSA SUNT EX QUIBUS ALIUD HAEC
AETAS ALIUD QUAE POST NOS SUBIBIT ASPICIET .

L. A. SENECA *NATURALES QUAESTIONES* Liber VII 30

Nature does not disclose its mystery all at once; we consider
ourselves experts while we are stil standing at the entrance;
such mysteries are not displayed by chance or to everyone:
they are kept inside a safe place, some of them will be
understood by us, some by our descendants.

1. Introduction

The term ecology was firstly introduced by Reiter in 1865 and used again in the following year by E. Haeckel [1] to define the study of the relationships of the living organisms both among themselves and with the abiotic environment. This new discipline developed rather slowly particularly on account of the difficulty to carry out observations without interfering with the organisms observed.

The biological effects of X-rays on living organisms were detected before their discovery by Röntgen was reported in the public press on January 4, 1896. As a matter of fact, in the last months of 1895 Grubbé, a manufacturer of Crookes tubes, used his equipment to study the fluorescence of chemicals and noticed an injury on his left hand. It is interesting to recall here that Grubbé was really a pioneer both in the therapeutic use of ionizing radiations and in health physics practice [2].

Following the suggestions of professor Gilman of the Hahnemann Medical College of Chicago, on January 29 and 30, 1896, he treated, respectively, a patient for carcinoma of the breast and a patient with an ulcerative lupus with his Crookes tube and used lead as a shield to protect the rest of the body [3].

A great number of studies both on therapeutic use and acute effects of ionizing radiation was carried out in the same year 1896 in different laboratories all over the world. Papers on damage to mammals were published by Capranica in 1896 and 1897 [4,5] who reported the results of his studies on the irradiation of mice by X-rays. At the same time also Tarkhanov [6] reported some data on the behaviour of terrestrial and aquatic organisms when irradiated by X-rays.

1

F. F. Luykx and M. J. Frissel (eds.), Radioecology and the Restoration of
Radioactive-Contaminated Sites, 1–15.
© 1996 *Kluwer Academic Publishers.*

In 1899 the antibiotic effect of high doses of the "Becquerel rays" was firstly investigated by Pacinotti & Porcelli [7] who found that bacteria were killed "independently by the presence of oxygen". Curie & Bequerel [8] described some acute effects on the skin after having quoted the findings of Walkoff and Giesel published in 1900.

Other biological effects of the ionizing radiations due to radium were quoted by Sklodowska Curie in 1904 [9]: Giesel found in 1899 that plants' leaves turn yellow and die when irradiated; Himstedt & Nagel confirmed in 1901 that the development of bacterial colonies is slowed down or stopped but this action is "not very strong"; Danysz discovered in 1903 an acute effect in bone marrow and brain of animals which are paralyzed after "one hour" of irradiation and "generally die in a few days".

2. Early evolution of radioecological research

In the first decades of this century the studies became more and more specialized. Some pioneers as Polikarpov, [10] published the results of their studies introducing sometimes quite new concepts.

Gajewskaja [11] reported in 1923 the possibility of a biological effect on *Artemia salina* due to the irradiation by a supposed high concentration of radium in salt lakes near to Sevastopol.

Vernadsky [12] developed around 1929 the concept of the biosphere including the role of radioactivity in it; in particular he studied two species of *Lemna* and introduced the ratio of radioelement concentrations between an organism and water, which was called later in the literature "Concentration Factor" [13].

Stoklasa & Penklava [14] studied a few years later rather in depth the biological effects of radionuclides and described an hormethical effect because "a weak irradiation by radium in plants increases cells reproduction and particularly their growth; the contrary is achieved if the irradiation is severe, i.e. the cells reproduction is stopped, and the growth of the plant is slowed down and its death follows".

It is worthwhile to point out a mistake by Deleage [15] who attributes the creation of the term "radioecology" to Vernadsky in 1935. Vernadsky published a book in 1935: "Les problèmes de la radiogéologie" reported erroneously by Deleage [15] as "Les problèmes de la radioécologie". The origin of the word "radioecology" is therefore unknown.

3. The effects of irradiation

After the 2nd World War, radioecology grew as a field of endeavour which encompasses the relationships between ionizing radiation or radioactive substances and the environment.

During the 1950's the study of the former relationship (i.e. the effects of the ionizing radiation on the environment) received closer attention because negative effects were expected as a result of irradiation after the explosion of a nuclear weapon or a severe criticality accident.

Different kinds of sources were used such as a reactor only partially shielded, thus creating a gamma and neutron radiation field in the midst of a natural area with a variety of habitats, or a "gamma field", i.e. a facility with a radionuclide source (^{60}Co or ^{137}Cs) surrounded by plots in circular array where plants were grown.

The main objectives considered at that time, i.e. 1959, exhaustively described by Platt & Mohrbacher [16], were:

1. To determine what increases in radioactivity in the environment could be attributed to reactor operations and what can be attributed to other sources such as fallout.

2. To establish permissible doses of radiation in nature in terms of the ultimate effects on mankind and also to establish bases for the intelligent use of emergency measures in case of radiation disaster, especially in limited areas.

3. To develop a general body of knowledge and philosophy useful in the evaluation of other pressing problems of the atomic age, such as low level waste disposal and environmental monitoring programmes which are associated with widespread military and civilians uses of radiation sources.

4. To study the ecological effects, in a very broad sense, of continuous and discontinuous gradients of neutron and gamma radiation on selected natural environments over short and long periods of time under natural and controlled conditions. This involves two fundamental approaches to radiation ecology. The first considers the effects on the individual species in terms of their ecological, genetical, physiological, morphological and other inter and intra specific characteristics, which appear throughout their life cycle and under the varying conditions in which they exist. The second is concerned with ecological effects on communities of organisms, especially on those characteristics which are unique to the community rather than to the individuals of which the community is composed. In another sense, the study attempts to include basic aspects of the whole spectrum of edaphic, climatic and biotic investigations.

4. The detection of radioactive fallout

Following the detonation of the first experimental nuclear weapon at Alamogordo, New Mexico (USA) on July 16, 1945, radioactive contamination was found in strawboard material used by Eastman Kodak Co. for packaging X-ray films, which showed fogged spots after about two weeks exposure. This paper board was produced on 6 August 1945, in a mill at Vincennes, Indiana, at approximately 1,000 miles from the point of detonation.

Webb, of the Kodak Research Laboratories, studied immediately this new and unusual contamination [17]. He found that it was mainly due to a beta emitter and that the energy and half-life were 0.6 MeV and 30 days, respectively, corresponding closely to ^{141}Ce which is one of the more abundant fission products. It must be emphasized that, on account of the classification of any data on nuclear weapons, such results were published only four years later after they were obtained. Also the studies carried out by Eisenbud & Harley [18, 19] were classified "secret" for a while before the authors were allowed to disseminate the information without restriction.

On July 20, 1946 Garrigue in France identified an unknown radioactive contamination and attributed it to the nuclear explosion in Bikini on July 1st of that year; he continued for the next ten years to measure and study the fallout produced by experimental nuclear explosions [20-39].

Such kind of measurements was carried out also by other scientists in various countries: USA, Stefanizzi, 1950 [40], Meinke, 1951[41], Eisenbud & Harley, 1953 [19]; Italy, Cigna, 1951 [42,43]; Santomauro & Cigna, 1953 [44]; France, Abribat et al., 1952-55 [45-47], Lecolazet & Hée, 1952 [48], Tanaevsky et al., 1955 [49,50], Jeanno et al., 1959 [51]; Canada, Rose & Katzman, 1952 [52]; Germany, Haxel & Schumann, 1953 [53]; Japan, Nishiwaki, 1955 [54] and Hungaria, Szalay & Berényi, 1955 [55]. A review of the work carried out in this period was published by Gedeonov [56].

The results of these studies on radioactive fallout suggested that not only direct irradiation could be relevant for delivering a dose to persons and organisms but also contamination with radionuclides could play an importante role.

To collect useful information on the way the dust from nuclear bomb explosions was transported into the atmosphere, the rates of surface deposition and the kinetics of ^{90}Sr uptake by plants, animals and humans, a project code-named "Sunshine", started in 1953. The first 2 characters of "Sunshine" refer to Strontium Units, but it is said that the name was related to the excellent weather the scientists involved enjoyed in California during the discussions [18]. The project confirmed that ^{90}Sr was the critical radionuclide in fallout but a number of other radionuclides started also to be investigated.

5. The diffusion of radionuclides into the environment

On account of the great sensitivity of the radioactivity measurements, negligible amounts of radionuclides could be easily identified and measured in different environmental compartments without any slight interference with the metabolism of living organisms. Many processes and phenomena could then be detected and studied. Ecology took advantage from such studies and its growth in a few years was probably greater than in the whole of the previous century.

Up to the very beginning of the 1960's, the compliance with the maximum permissible concentrations established for drinking water, in any kind of environmental water was considered to be a condition sufficient to guarantee the compliance with the intake limits for individuals. But at that time the evidence of some important concentration processes appeared clearly.

E.g., in a fresh water mollusc (Unio sp.) gathered in the Lake Maggiore (North Italy) in 1960 some ^{54}Mn was measured. Initially it was attributed to some unnoticed discharge of this radionuclide from the nearby Nuclear Centre of Ispra, but it was easily shown that the ^{54}Mn was an activation product released by nuclear weapons in fallout. Because of its very low concentration it could be detected in the lake water only when rather large samples were processed. The high value of the concentration factor in the mollusc (ranging between 100,000 and 400,000) made its presence quite evident in the mollusc samples [57, 58].

The first application of this discovery was the use of molluscs as an indicator of the presence of ^{54}Mn in the environment. Soon it became clear that, because of the existence of concentration processes in the environment, the limit of intake for individuals could be exceeded, even if the concentration of a radionuclide in an environmental water would be kept within the drinking water limit.

As a result a great interest in the determination of concentration factors in organisms spread widely in many laboratories. A large number of values were available in a few years time. However, the initial enthusiasm for the concentration factors, considered as a powerful tool in radiation protection research, decreased rapidly because such values were strongly influenced by many unknown parameters and therefore they were assumed to be hardly useful on account of their uncertainty.

The concentration factors went out of fashion in a short time but their role could not be filled in by anything else. Therefore, the scientists engaged in radiation protection had to accept a sort of a compromise. Concentration factors cannot be determined with the same precision as Planck's constant or Avogadro's number; however, if only their order of magnitude is taken into consideration, the information provided by them is quite sufficient to identify important radioecological processes. In addition, it must be emphasized that diffusion processes are often not at equilibrium and, therefore, the effective value of the concentration factors is also time dependent.

Further, it appeared that the transfer of radionuclides from the environment to man could be better evaluated and monitored through the definition of some "critical" quantitiy: a critical group, a critical radionuclide, a critical pathway, etc. It has to be realised that the term "critical" does not imply any idea of danger; it describes a factor which enables decisions to be taken at the level of public health protection. In the same framework another relevant quantity was proposed and used: the "Limiting Radiological Capacity" for a given radionuclide [67,68]. It is defined as the maximum amount of a radionuclide which can be discharged into a particular sector of the environment in such a way that the resulting dose commitment received via a critical pathway does not exceed the annual dose limit for individual members of the public.

A considerable amount of knowledge on the behaviour of radionuclides in the environment and many different models describing such behaviour were available at the time of the Chernobyl accident. The accident provided an opportunity to test and validate these concepts.

Undoubtedly, in the post-Chernobyl situation radioecology is in a better position because the description of the environment is presently much closer to reality and its conclusions much more reliable. But, as it is usual in science development, new problems appeared and new questions were asked.

While agro-ecosystems were investigated rather extensively in the past, new unexplored fields appeared in the wide domain of natural and semi-natural environments. Obviously some information on phenomena occurring in the latter environments were already available some tens of years ago (e.g., the relatively high concentrations of radionuclides in mushrooms). But because of the rather small number of radionuclides detectable during the fallout period, attention was focused on the main pathways of food intake by man.

After the Chernobyl accident also some other fission products, less abundant or with a shorter half-life, were easily measured. At the same time, it became clearly evident that the turnover rate of radionuclides in natural and semi-natural environments was quite different from that found in cultivated environments, not only in view of the different values of the parameters involved, but also due to different mechanisms in each process. A first paper on the Khyshtym accident, which happened on 27 September 1957, was presented for publication by Agre & Korogodin on 11 October 1958 and appeared in 1960 [69], but no reference neither to the place nor to other features of the accident was given. For a rather long time this accident could not be studied exhaustively; but since a few years they are accessible to scientists and represent a potential source of invaluable information.

The knowledge of environmental behaviour of radionuclides is presently adequate for radiological assessment studies. Unfortunately, there are other obstacles in radioecology and environmental radioactivity because of a widespread reduction of facilities where these studies can be carried out. Often, short-sighted politicians and some peculiar conditions may make the situation worse in particular countries. Therefore, every effort should be made to avoid reaching an irreversible degradation of research possibilities, rather than the development of new research.

6. Man and environment

Today, the word "environment" is often coupled with "quality" and it is treated as a substance or condition independent of the beings present in that environment. Of course, there are legitimate reasons for protecting our environment beyond the immediate concerns for human health. The necessity for maintaining an ecological balance is certainly a legitimate concern, as is the preservation of aesthetic values; most people prefer to have clear skies and beautiful vistas [62].

According to some experiments carried out more than thirty years ago, an indication that damaging effects could occur during the development of fish eggs was reported, already with ^{90}Sr-^{90}Y concentrations around 0.1 Bq/l by Polikarpov & Ivanov [63,64]. For some time, these results were considered to be the proof that some organisms could have a sensitivity to ionizing radiation greater than the sensitivity of human individuals.

It is, therefore, important to recall here that such results were never supported by other laboratories and, later on, it was ascertained that the consequences observed on the development of fish eggs was not caused by ionizing radiation but by other experimental conditions [65]. A rather wide review of studies on irradiation of eggs of marine, brackish and freshwater fishes of various taxonomic groups by Polikarpov [66] confirmed that harmful effects were detectable only at concentrations of radionuclides higher than those compatible with the individual limits recommended by ICRP [67-69].

The primary aim of radiation protection is to provide an appropriate standard of protection for each individual of the human population. Man is considered as the most critical element in the general ecosystem and, on the other hand, the protection of the environment is assured when a whole species is not endangered or an imbalance between species is not created. Then, on the basis of the results reported before, it is rather unlikely that the environment could be someway negatively affected if the general principles of radiation protection are applied. That leaves only human health as the justification for environmental radiation control in most cases [62].

Nevertheless, on account of the fact that, in few instances (e.g. when the pathways leading to human irradiation are exceedingly long), it could be possible to reach values of radionuclide contamination not acceptable for the organisms other than man; in these cases the dose limits for such organisms (at least for the principal radionuclides) should be determined.

The relationship between man and environment must also be considered the other way round, i.e. the influence of the environment on man; e.g. the possible effects of high background areas. Epidemiological investigations should be carried out to enlighten the detriment resulting from radon exposure. This is particularly important because radon contributes to about half of the natural background and therefore some relevant information could be obtained on the real effects of low doses.

7. The natural radioactive background

"Every falling raindrop and snowflake carries some radioactive matter to the earth, while every leaf and blade of grass is covered with an invisible film of radioactive material" These words written by Rutherford in 1905 [70] more than 80 years ago, give a fairly realistic view of the environment from the point of view of ionizing radiation.

The assessment of radiation doses from natural background underwent some changes in recent years on account of a better knowledge of the levels and the role played by some radionuclides. A major change was due to a more realistic evaluation of the contribution of ^{222}Rn. As a consequence, the natural background, recently evaluated, is definitely higher than the older one: it must be stressed that only the natural component of the background is taken into account here. The increase is only due to a different approach in calculation and not to an increase of artificial or technologically enhanced sources.

It is interesting to have a comparison of the various values reported from 1962 to 1993 by UNSCEAR, [71-76]. The older data were recalculated as effective dose equivalents assuming that the gonad dose-rate corresponds to that of organs or tissues not explicitly indicated. These data are shown in Table 1. From 1972 to 1982 the estimated values nearly doubled; of course such an increase cannot be applied to any local value but it gives an idea of the differences one can expect when old and recent data are considered.

TABLE 1. Estimated pro capite annual effective dose equivalents from natural sources in areas of
normal background (UNSCEAR, [71-76])

Source of irradiation	Total annual effective dose equivalent (µSv)					
	1962	1972	1977	1982	1988	1993
EXTERNAL IRRADIATION						
Cosmic radiation (incl. neutrons)	500	284	284	301	355	380
Terrestrial radiation	500	440	320	350	410	460
INTERNAL IRRADIATION						
Cosmogenic radionuclides (^{7}Be,^{14}C)	8	7	8	15	15	10
Primordial radionuclides						
^{222}Rn and decay products	34	124	885	930	1,220	1,200
Others (^{40}K, ^{87}Rb, ^{238}U, ^{234}U, ^{230}Th,						
^{226}Ra, ^{232}Th, ^{228}Ra and decay products)	215	214	325	396	381	300
TOTAL	1,257	1,069	1,822	1,992	2,381	2,350

It is interesting to remind here that the consequences of the natural background, in some
cases when the local dose was very high, have been perceived even when the real causes
were obviously not detected. ICRP [69] reported that radiogenic lung cancer is the oldest type of radiation-induced malignancy known. It was recorded as early as in the 15th-
16th century; according to Agricola in 1556 [78]: "Of the illnesses, some affect the
joints, others attack the lungs, some the eyes, and finally some are fatal to men". Among
the miners in the Schneeberger-Jachymov region in Erzgebirge the disease was known
as the "Schneeberger Krankheit" and it was diagnosed as lung cancer in 1879 by Härting
& Hesse [79]. Its possible association with radon was suggested about 40 years later
when the high radon levels in mines of that region were discovered and the real cause of
this disease was recognized in the 1950s only, when the first attempts of lung dosimetry
were made [80, 81].
Recently Cigna [43] found another quotation which is rather older. Titus Lucretius Carus,
a Latin poet, was born around 95 B.C. (perhaps in Pompei, near Naples) and committed
suicide about 44 years later after a love potion drove him mad.
He wrote a poem entitled " De rerum natura" (= about the nature of things) which is
quite peculiar because it gives an exceptional view on the scientific knowledge at that
time. Some statements are original but many are taken from Epicurus, a Greek philoso-
pher (born in 341 B.C. in Samos, died in 270 B.C. in Athens) whose works survived only
partially to the present. Quotations (volume and lines) refer to the critical edition by C.
Bailey in 1968 [82]:

> *Principio hoc dico, quod dixi saepe quoque ante,*
> *in terra cuiusque modi rerum esse figuras;*
> *multa, cibo quae sunt vitalia, multaque, morbos*
> *incutere et mortem quae possint accelerare.* [VI: 769/772]

[Firstly I repeat once again what I reported often before,
in the earth there are the components of various substances;
many may be used as food and are useful to life
and many may be the cause of illness and anticipate death.]

Nonne vides etiam terra quoque sulphur in ipsa
gignier et taetro concrescere odore bitumen;
denique ubi argenti venas aurique sequuntur,
terrai penitus scrutantes abdita ferro,
qualis exspiret Scaptensula subter odores?
Quidve mali fit ut exhalent aurata metalla !
Quas hominum reddunt facie qualisque colores!
Nonne vides audisve perire in tempore parvo
quam soleant et quam vitai copia desit,
quos opere in tali cohibet vis magna necessis ?
Hos igitur tellus omnis exaestuat aestus
exspiratque foras in apertum promptaque caeli.   **[VI: 806/817]**

[Don't you see that sulphur is also produced by the earth itself
and bitumen thickens giving off a horrible smell;
at last, where the silver and gold veins are worked,
and the hidden depth of the earth is searched with tools,
what odours are released by the subsoil of Scaptensula   ?
What noxious emanations are released by the golden mines !
What look and what kind of complexion they give to miners !
Have you never seen or heard how they usually
die young and how poor is the life of those
who are compelled to such works by necessity ?
So the earth spreads all these emanations
and gives them off in the free and open air.]

Therefore, T.L. Carus' reference to a miner's disease is probably the oldest record of radiogenic lung cancer (which happens to be the oldest type of radiation-induced malignancy known) because he quotes a work written by Epicurus about 200 years before . It is difficult to know for certain if the disease was caused by silicosis or radon: nevereless it must be pointed out that in the vicinity of Mt. Pangaion (where the mines are quoted by T.L. Carus) the indoor radon concentration is above 100 Bq/m^3 and the outdoor gamma-ray dose is around 100 nGy/hour [83]. Therefore, these values (among the highest of Greece) support the hypothesis that, in such mines, radon concentrations were high enough to induce malignancies in local miners. On the other hand, the mining technology of that time would have hardly produced large amounts of dust to cause frequently silicosis.

8. The legacy to future mankind

Much effort has been devoted to understanding the behaviour of long-lived radionuclides (e.g.: ^{239}Pu, ^{99}Tc). If, from one point of view, such emphasis can to some extent be justified, it could equally be argued that the importance of long-lived radionuclides (particularly in the frame of waste disposal) seems to have been overestimated. In fact, it is rather difficult to extrapolate the impact of radioactivity into the far future from its use today.

If one of our ancestors (e.g. Neanderthal man) had forecasted the needs of our society based on his available knowledge, he should have identified the flint as the limiting factor for society's development and the degree of civilisation attainable. Of course, the results of such an analysis would have been completely wrong, notwithstanding a set of correct starting data, because other factors have substantially changed the situation in the meantime.

There is no reason to consider ourselves in a better position to assume that ionizing radiation will be a threat for populations in the far future: for example cancer will probably not be so important in the future as it is today.

Therefore, many problems attributed to the disposal of very long-lived radionuclides are simply false problems. In fact it cannot be morally justified to burden our society now with difficulties simply in order to avoid what are essentially hypothetical difficulties for the future. A passage of an Italian writer, Giuseppe Tomasi di Lampedusa, in his novel "The leopard" (chapter 1) describes quite correcly the right attitude which should be followed:

"For us a palliative which promises to last for one hundred years is equivalent to eternity. We may even be worried about our sons, perhaps about our grandchildren; but beyond those we may hope to caress with these hands, we have no obligations to fulfil."

Our knowledge of the behaviour of radionuclides in the environment has been extended considerably in recent years. For the description of almost every situation there exists a model; it is unfortunately not possible to list all these models in this publication.

The Chernobyl accident focused the attention on a problem whch seems to exist with any model applied. It appeared that the values, frequently used in these models, were very conservative, i.e. they guarantee a safe marge, much too pessimistic. Of course, it is convenient to use such conservative values in assessment models. However, for planning future radioecological countermeasures after an accident, data as close as possible to reality must be used, otherwise the countermeasures applied may imply consequences worse than those produced by the accident itself.

This criterion was clearly stated by ICRP [67; 69] but it was frequently neglected all over the world after the Chernobyl accident. Now, it is a rather hard task to convince public opinion that the limits adopted by political and administrative authorities were definitely too low and that the much higher limits, proposed by the experts in countermeasures, were chosen in the interest of the population and not for commercial or other interests.

It is worthwhile to realize that radioecological models are far better developed than models for other pollutants. Now is the time to yield the experience of many years and to investigate in greater depth the mechanisms which control the behaviour of other pollutants. The extent to which the chemical industry influences environmental quality is well recognized, whereas radionuclides have, in general, a negligible effect, the associated exposures being less than the fluctuations in natural background.

Research methods using radioactive tracers must not be regarded as a threat to mankind but as a tool which can play a major role in improving the quality of life.

9. Present situation of radioecological research.

During the "fallout period" a great number of laboratories were active all over the world in nuclear research centres, nuclear power plants and other institutions. After the moratorium of the atmospheric tests in 1963, only China and France continued. Their contributions, however, were limited respectively to the northern and the southern hemisphere on account of locations of the test sites.

Nevertheless, the steady decrease of fallout since the 1963 peak and thereafter the decrease below the detection limits for many radionuclides, together with a reduction of the discharges from the nuclear plants, led to lesser attention to the measurements of environmental radioactivity together with a decrease of competent laboratories.

The practical consequence is a widespread reduced capability to deal with radionuclide measurements in environmental samples. Obviously, the capacity for determining beta- and alpha-emitters suffered the largest reduction on account of the need for chemical treatment of the sample before measurement.

The determination of gamma-emitters can be carried out by high resolution spectrometry which is presently largely automatic and requires neither a special sample preparation nor much personnel. Notwithstanding such an advantage, this kind of determination is also endangered because the interpretation of data to be performed after any kind of measurement requires skilled people.

The normal turnover of experts in radioecology and, more generally, in radiation protection, in the very next future will be recruited from the organizations where they presently work. That will be those ones who have gained experience during the "golden age" of radioecology (i.e. up to the beginning of the 1980s). This occurrence will be rather deleterious for the capacity to face and solve the problems deriving by any radioactive environmental contamination that might occur in the future.

10. Conclusions

During the nuclear weapon fallout period the behaviour of the most important radionuclides (typically: ^{90}Sr , ^{137}Cs, $^{239,240}Pu$) in most critical pathways were studied; after the Chernobyl accident studies were extended to natural and semi-natural ecosystems and other less common radionuclides. A further step should be the study of the mechanisms regulating the behaviour of radionuclides in different compartments of the environment.

The interest to extend the knowledge in this field is not a purely academic one because such an extended knowledge could be essential to improve the level of radiation protection by actively enhancing some convenient environmental process and not simply by relying passively on the natural transfer pathways.

Semi-natural and natural ecosystems are of special interest in the context of environmental radioactivity. Such an interest occurred during the sixties when it appeared that ^{90}Sr along the foodchain lichen-reindeer-man, contaminated man. It showed at the same time the importance of artificial radionuclides as a tool for tracing and understanding nutrient turnover in ecosystems.

Semi-natural ecosystems may hide other critical pathways for radionuclides transfer to man and may also act as temporary sinks or long-term sources for radionuclides deposited from the atmosphere. Further research and modelling is therefore required [84].

But the main problem of the environmental aspects of radiation protection is found outside the domain of radiation protection itself. It has been clearly stated in an official 1988 paper by the Commission of the European Communities (obviously what is here referred to Europe can easily be extrapolated to other countries in the world): *"It should be emphasized, that radiation protection has now entered a particularly critical phase because the number of competent scientists has declined substantially and will continue to do so unless support for training and research can be provided. At a time when technologies involving radiation in general and nuclear power in particular meet increasing resistance by the public, a loss of Community expertise in this field could have economic and social consequences which could harm the European development to a considerable extent"* [85].

However, from the point of view of radiation protection, the development of the energy sources in the next future risks to be accompanied by a decrease of the present standards of safety if the crisis of the radiation protection laboratories is not solved in a short time by stopping the observed decline of the number of scientists involved and the general reduction of interest.

In conclusion, the main problem of radiation protection appears to be outside its own domain and depends heavily on the attitude of politicians and authorities. Unfortunately any wrong decision will not only affect the activity of the people involved in radiation protection, but, what is really the worst, will endanger the human society without any justification by reducing the high standards of protection which have been typical of the nuclear activities since the very beginning.

11. Acknowledgements

The author is particularly grateful to many colleagues who contributed to the collection of some important references and in particular to: A. Antonelli, A. Benco, L. Foulquier, G. Polikarpov, S. Rukhman.

12. References

1. Haeckel E., (1866) - *Allgemeine Entwicklungsgeschichte der Organismen*. Berlin.
2. Morgan K.Z., (1967) - *History of Damage and Protection from Ionizing Radiations*. In: Morgan K.Z. & Turner J.E. - *Principles of Radiation Protection*. John Wiley & Sons, Inc.: 14.
3. Grubbé E.H., (1933) - Priority in the therapeutic use of X-Rays. *Radiology,* **21**:156-162.
4. Capranica S., (1896) - Sulla azione biologica dei raggi di Röntgen. *Atti R. Acc. Lincei,* s. 5, Rend. Classe Sc. Fis. Mat. Nat., **5**(1): 416-417.
5. Capranica S., (1897) - Sull'azione biologica dei raggi X. *Atti R. Acc. Lincei,* s. 5, Rend. Classe Sc. Fis. Mat. Nat., **6**(1): 38-39.
6. Tarkhanov I., (1896) - [Experiments with the effects of Roentgen X-rays on living organisms] (in Russian). *Isv'st. CPB. Biolog. Laborat.* , **1**(3): 47-52.
7. Pacinotti & Porcelli, (1899) - *Azioni microbicide esercitate dai raggi Bequerel*. Firenze.
8. Curie P. & Bequerel H., (1901) - Action physiologique des rayons du radium. *Compt. Rend. Acad. Sc.*, Paris, **132**:1289-1291.
9. Sklodowska Curie M., (1904) - *Recherches sur les substances radioactives*. Gauthier-Villars, Paris.
10. Polikarpov G.G., (1993) - Radioecology. *Atti XXVIII Congr. Naz. A.I.R.P., Taormina, 13-16 Oct. 1993* (in press).
11. Gajewskaja N.V., (1923) - *Effect of roentgen rays on Artemia salina*. Verhandl. Internat. Vereinig. Limnologie Stuttgart, **1**: 359.
12. Vernadsky W., (1929) - *La Biosphère dans le cosmos*. Paris.
13. Vernadsky W., (1929) - [On the concentration of radium by living organisms] (in Russian). *Doklady Academii Nauk of USSR*, Ser. A, N° 2: 33-34.
14. Stoklasa J. & Penkava J., (1932) - *Biologie des Radiums und der Radioactiven Elemente. Erster Band, Die Biologie des Radiums und des Uraniums*. Verlag von Paul Parey, Berlin, XIV.
15. Deleage J.P., (1991) - *Histoire de l'écologie*. Ed. La Découverte, Paris.
16. Platt R.B. & Mohrbacher J.A., (1959) - Studies in Radiation Ecology: 1. The Program of Study. *Bull. Georgia Acad. Sciences,* **17**: 1-30.
17. Webb J.H., (1949) - The Fogging of Photographic Film by Radioactive Contaminants in Cardboard Packaging Materials. *Phys. Rev.,* **76**(1): 375-380.
18. Eisenbud M., (1990) - *An environmental odyssey: people, pollution and politics in life of a practical scientist*. Univ. Washington Press, Seattle & Washington.
19. Eisenbud M. & Harley J.H., (1953) - Radioactive Dust from Nuclear Detonations. *Science,* **117**: 141.
20. Garrigue H., (1949) - Sur la radioactivité de l'atmosphère. *Compt. Rend. Acad. Sc.*, Paris, **228**:1583-1584.
21. Garrigue H., (1950) - Sur la radioactivité naturelle de l'atmosphère. *Compt. Rend. Acad. Sc.*, Paris, **230**:1272-1274.
22. Garrigue H., (1950) - Création d'un avion laboratoire et perfectionnement des appareils pour l'étude des faibles activités de l'atmosphère. *Compt. Rend. Acad. Sc.*, Paris, **230**: 2279-2280.
23. Garrigue H., (1951) - Recrudescence d'anomalies dans la radioactivité de l'air à moyenne altitude. *Compt. Rend. Acad. Sc.*, Paris, **232**: 827-829.
24. Garrigue H., (1951) - L'invasion d'air radioactivif d'origine atomique et son influence sur les précipitations atmosphèriques. *Compt. Rend. Acad. Sc.*, Paris, **232**: 1003-1004.
25. Garrigue H., (1951) - Recherches sur la radioactivité de l'air libre. *Compt. Rend. Acad. Sc.*, Paris, **233**: 860-862.
26. Garrigue H., (1951) - Recherches de radioactivivité au sommet du Puy de Dôme. *Compt. Rend. Acad. Sc.*, Paris, **233**: 1447-1448.
27. Garrigue H., (1952) - Recherches sur les précipitations atmosphériques à l'échelle du globe. *Compt. Rend. Acad. Sc.*, Paris, **234**: 1571-1573.
28. Garrigue H., (1952) - Sur la radioactivité anormale de l'atmosphère. *Compt. Rend. Acad. Sc.*, Paris, **235**: 1498-1499.
29. Garrigue H., (1953) - Observations sur les impuretés contenues dans l'air libre. *Compt. Rend. Acad. Sc.*, Paris, **236**: 2309-2311.
30. Garrigue H., (1953) - Prospection de la radioactivité de l'air. *Compt. Rend. Acad. Sc.*, Paris, **237**: 802-803.
31. Garrigue H., (1953) - Sur la radioactivité de l'atmosphère d'origine atomique. *Compt. Rend. Acad. Sc.*, Paris, **237**: 1232-1233.

14

32. Garrigue H., (1954) - Recherches sur la radioactivité de l'atmosphère. *Compt. Rend. Acad. Sc.*, Paris, **238**: 2074-2075.

33. Garrigue H., (1954) - Appareil bétamètre et gammamètre adapté à l'étude des dépôts radioactifs recueillis en vol. *Compt. Rend. Acad. Sc.*, Paris, **239**: 1207-1208.

34. Garrigue H., (1955) - Sur la radioactivité de l'atmosphère d'origine atomique. *Compt. Rend. Acad. Sc.*, Paris, **240**: 178-180.

35. Garrigue H., (1955) - Nouvelle recrudescence d'activité d'origine atomique dans l'atmosphère. *Compt. Rend. Acad. Sc.*, Paris, **240**: 1453-1455.

36. Garrigue H., (1955) - Radioactivité de l'atmosphère. *Compt. Rend. Acad. Sc.*, Paris, **241**: 1971-1972.

37. Garrigue H., (1956) - Radioactivité de l'air et des précipitations. *Compt. Rend. Acad. Sc.*, Paris, **243**: 584-585.

38. Garrigue H., (1958) - Mesures sur la radioactivité d'origine atomique de l'air et des précipitations. *Compt. Rend. Acad. Sc.*, Paris, **246**: 3089-3091.

39. Garrigue H., (1959) - Sur la pollution radioactive d'origine artificielle de l'atmosphère et du sol. *Compt. Rend. Acad. Sc.*, Paris, **249**: 737-738.

40. Stefanizzi A., (1950) - Radioactivity of atmospheric precipitates. *J. Geophys. Res.*, **55**: 373-378.

41. Meinke W.W., (1951) - Observations on Radioactive Snows at Ann-Arbor, Michigan. *Science*, **113**: 545.

42. Cigna A., (1951) - Ripercussioni anche in Italia degli scoppi atomici di Las Vegas ? *Incontri*, Ist. Gonzaga, Milano, **7**(5): 84-87.

43. Cigna A.A., (1993) - Considerations on the Environmental Aspects of Radiation Protection. *Proc. NEA Workshop Paris 11-13 January 1993*, OECD: 41-52; also as: *Rapporto ENEA RT/AMB/93/22*.

44. Santomauro L. & Cigna A., (1953) - Alcune misure sulla radioattività delle precipitazioni atmosferiche. *Ann. Geofis.*, Milano, **6**(3): 381-387.

45. Abribat M., Pouradier J. & Venet A-M., (1952) - Sur les produits radioactifs artificiels existant dans l'atmosphère de la région parisienne. *Compt. Rend. Acad. Sc.*, Paris, **235**: 157-159.

46. Abribat M. & Pouradier J., (1953) - Évolution de la teneur en radioéléments artificiels existant de l'atmosphère de la région parisienne. *Compt. Rend. Acad. Sc.*, Paris, **237**: 1233-1235.

47. Abribat M., Pouradier J. & Venet A-M., (1955) - Évolution de la radioactivité de l'atmosphère de la région parisienne. *Compt. Rend. Acad. Sc.*, Paris, **240**: 2310-2311.

48. Lecolazet R. & Hée A., (1952) - Sur la mesure de la radioactivité de l'air atmosphèrique à l'aide du gammamètre. *Ann. Géophys. Fr.*, **8**(3): 320-322.

49. Tanaevsky O. & Vassy É., (1952) - Variations de la radioactivité naturelle et artificielle l'atmosphère. *Compt. Rend. Acad. Sc.*, Paris, **241**: 38-40.

50. Tanaevsky O. & Vassy É., (1955) - Natural and artificial radioactivity of the atmosphere. *Ann. Géophys.*, Paris, **11**(4): 486-490.

51. Jeanno C., Tanaevsky O., Labeyrie J. & Vassy É., (1952) - Radioactivité de l'air et des précipitations dans la région parisienne. *J. Mécan. Phys. Atmosphère*, **1**(2): 1-23.

52. Rose D.C. & Katzman J., (1952) - Radioactive Deposits Found at Ottawa after the Atomic Explosions of January and February 1951. *Canadian J. Phys.*, **30**: 11-116.

53. Haxel O. & Schumann G., (1953) - Über die radioaktive Verseuchung der Atmosphäre. *Naturwissenschaften*, **40**(17): 458-459.

54. Nishiwaki Y., (1955) - Effects of H-bomb tests in 1954. *Atomic Scientist J.*, **4**(5): 279-288.

55. Szalay A. & Berényi D., sen., (1955) - Unusual radioactivity observed in the atmospherical precipitations in Debrecen (Hungary) between Apr. 22-Dec. 31, 1951. *Acta Physica Acad. Sc. Hung.*, **5**(1):1-14.

56. Gedeonov L.I., (1957) - Radioactive contamination of the atmosphere. *Soviet J. Atomic Energy*, **2**: 313-325 (English translation of: *Atomnaya Energiya*, **2**: 261-271).

57. Ravera O. & Vido L., (1961) - Misura del Mn-54 in popolazioni di *Unio pictorum L.* (Molluschi, Lamellibranchi) del Lago Maggiore. *Mem. Ist. Ital. Idrobiol.,Pallanza*, **13**: 75-84.

58. Ravera O. & Gaglione P., (1962) - Concentrazione di Mn-54 da fall-out in relazione all'età e all'organo di *Unio mancus elongatulus* (Molluschi, Lamellibranchi). *Rend. Ist. Lomb. Sc. Lett.*, Milano, (B) **96**: 157-164; also as: Report EUR 254i.

59. Amavis R., Smeets J., Branca G., Breuer F., Cigna A., (1974) - Developpement et application du concept de la capacité radiologique en radioprotection. *Proc. Symp. "Environmental Behaviour of Radionuclides Released in the Nuclear Industry"*, Aix-en-Provence 1973, IAEA, Vienna: 583/592.

60. C.E.C., (1975) - Organization and operation of radioactivity surveillance and control in the vicinity of nuclear plants. *Radiological Protection 2, Report EUR 5176e*.

61. Agre A.L. & Korogodin V.I., (1960) - [The distribution of radioactive pollutions in a stagnant reservoir] (in Russian). *Medicinskaja Radiologija*, **1**: 67-72.

62. Schiager K. J., (1992) - Putting the health back into health physics. *Health Physics*, **63** (3): 257-258.

63. Polikarpov G.G. & Ivanov V.N., (1961) - (in Russian: Action of ^{90}Sr- ^{90}Y on developing *Engraulis* spawn). *Vopr. Iktiol.*, **1**(3): 20-26.

64. Polikarpov G.G. & Ivanov V.N., (1962) - (in Russian: Harmful effects of ^{90}Sr- ^{90}Y on the early development of *Mullus, Crenilabrus, Trachurus* and *Engraulis*). *Dokl. Akad. Nauk. SSSR*, **144**(1): 219-222.

65. Polikarpov G.G., (1977) - Personal communication.

66. Polikarpov G.G., (1977) - Effects of ionizing radiations upon aquatic organisms (Chronic irradiation). *Atti della giornata sul tema alcuni aspetti di radioecologia. XX Congr. Naz. A.I.R.P.*, Bologna: 25-46.

67. ICRP, (1978) - Recommendations of the ICRP. Publication 26, *Annals of the ICRP,* **1**(3), Pergamon Press.

68. ICRP, (1984) - Protection of the Public in the Event of Major Radiation Accidents: Principles for Planning. Publication. 40, *Annals of the ICRP,* **14**(2), Pergamon Press.

69. ICRP, (1987) - Lung Cancer Risk from indoor Exposures to Radon Daughters. Publication 50, *Annals of the ICRP,* **17**(1) Pergamon Press.

70. Rutherford, E., (1905), *Harper's Magazine,* February.

71. UNSCEAR, (1962) - *Report of the U.N. Scientific Committee on the Effects of Atomic Radiations.* New York.

72. UNSCEAR, (1972) - *Report of the U.N. Scientific Committee on the Effects of Atomic Radiations.* **1**: Levels; **2**: Effects. New York.

DEVELOPMENT OF RADIOECOLOGY IN EAST AND WEST

ASKER AARKROG[1] and GENNADY G. POLIKARPOV[2]
[1] *Risø National Laboratory*
P.O. Box 49, DK-4000 Roskilde, Denmark.
[2] *Institute of Biology of the Southern Seas*
Prospekt Nakhimova 2, CIS 335011 Sevastopol, Crimea, Ukraine

1. Introduction

The need for development of radioecology in the western world, which in the context of this publication covers western Europe, the American continents, south-east Asia and Australia, came up in the USA in the forties. The creation of large amounts of man-made radionuclides in connection with the development of nuclear weapons during the second world war made radioactive contamination of the environment a reality.

In 1955 Odum [1] introduced the term radiation ecology at the First International Conference on the Peaceful Uses of Atomic Energy in Geneva. In 1959 Odum [2] defined "Radiation Ecology as the discipline, which is concerned with radioactive substances, radiation and the environment. " He mentions two rather distinct phases of radioecology, which require different approaches: "On the one hand we are concerned with the effects of radiation on individuals, populations, communities, and ecosystems. The other important phase of radiation ecology concerns the rate of radioactive substances released into the environment and the manner by which the ecological communities and populations control the distribution of radioactivity". Today, in 1995, many will prefer to consider the last phase as radioecology, while the first phase may better be referred to as radiation ecology. But in the early days radiation ecology and radioecology was interchangeable. The need for a radioecological approach in radiation protection was realized, when it appeared that permissible concentrations applied for air and water did not necessarily ensure that doses to man were kept below the permissible limits. Chamberlain *et al* [3], at the above mentioned 1956 Geneva conference, concluded that "levels of ^{131}I in the atmosphere, which are lower by a factor of 1000 than those which create a breathing hazard to man, may be dangerous if the fission products are deposited on herbage, which is subsequently consumed by grazing animals."

2. Milestones in radioecology

The development of western radioecology may be characterized by a number of events, which are related to specific sources of environmental contamination.

F. F. Luykx and M. J. Frissel (eds.), Radioecology and the Restoration of
Radioactive-Contaminated Sites, 17–29.
© 1996 *Kluwer Academic Publishers.*

2.1 NUCLEAR WEAPONS FALLOUT, PROJECT SUNSHINE

Global radioactive contamination of our environment was initiated by the testing of the first nuclear weapon at the Operation TRINITY site in New Mexico U.S. on 16 July 1945. Already two years after the TRINITY-detonation the first radioecological field group was organized by Warren and Bellany from the University of California. The group represented interests in mammals, reptiles, birds, insects, vegetation, soil and radiation [4]. A few years later in 1952 the testing of the first thermonuclear device took place in the Pacific and radioactive fallout became worldwide distributed. In the summer 1953 at the suggestion of Libby a major effort was made in the United States to critically re-examine the potential hazards from radioactive fallout from testing of thermonuclear weapons or from a nuclear war [5]. At this time no one had measured ^{90}Sr levels in man or his environment at great distances from the test sites. By special developed low-level counting equipment the Lamont Geological Observatory at the Columbia University in the U.S. carried out these first measurements of ^{90}Sr in our environment.

The samples were collected in July 1953 and consisted of marine shells, surface soil, cheese and calf bones. The concentrations varied from zero to 0.4 Bq ^{90}Sr/gCa. (Table 1).

TABLE 1. First Measurements of ^{90}Sr in Human Environment (July 1953). (Lamont Geological Observatory, Columbia University)[6]

Sample	Bq ^{90}Sr/g Ca
Ancient marine shells (blank)	<0.007
Modern shells (Long Island) (living organism)	<0.0007
Lamont campus surface soil	0.40
Munster Cheese, Madison, Wisconsin	0.02
Calf Bones, Wisconsin (two year old)	0.01
Calf Bones, Montana (6 month old)	0.01

During the fifties and sixties the studies of ^{90}Sr continued. The term sunshine-unit (Strontium Unit) for pCi ^{90}Sr (g Ca)$^{-1}$ was introduced. Milk appeared to be one of the most important pathways of ^{90}Sr (and other radioactive substances) to man. Comar *et al* [7] applied the term observed ratio (OR) to relate the ratio of ^{90}Sr to Ca in a biological system to that in the source system. The Sr/Ca OR expresses the overall discrimination,

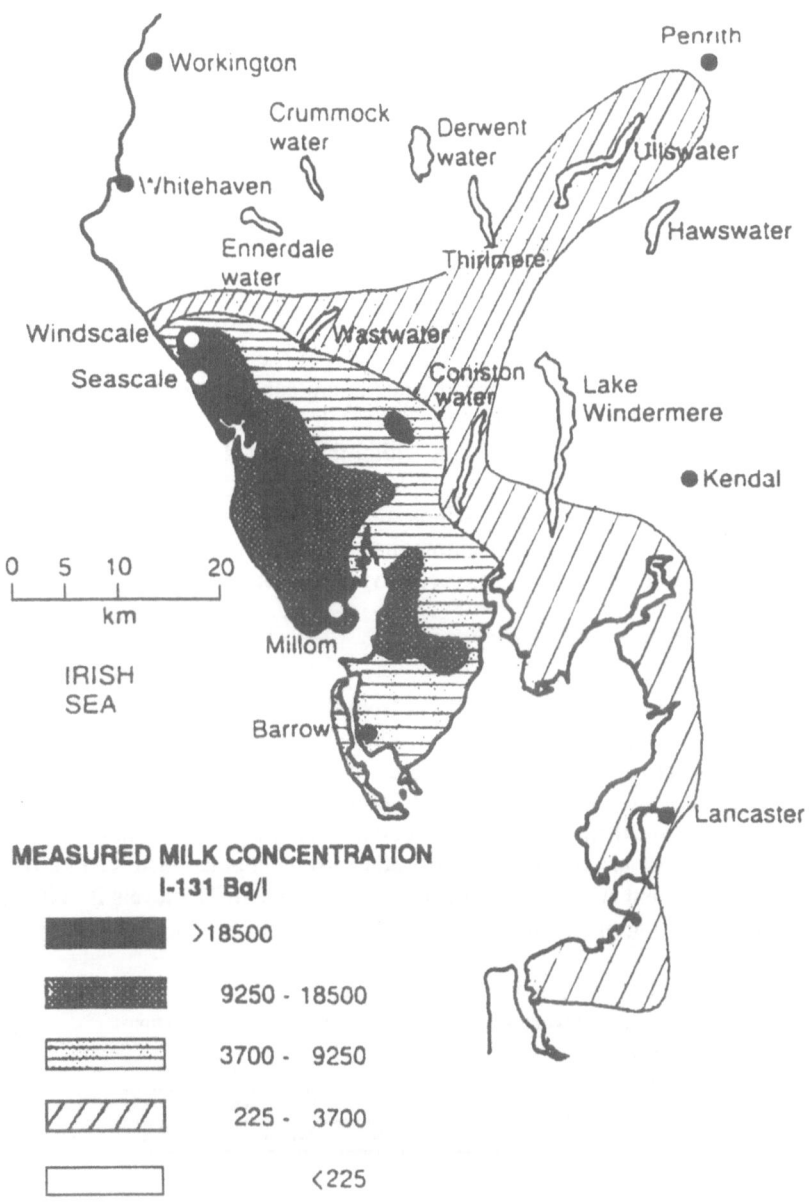

Figure 1. Maximum Activity Concentrations in Milk and Extent of
Restrictions instigated following the Windscale Accident, 1957. [9].

that is observed in the transfer of the two elements, for example from fodder of a cow to its milk:

$$OR_{milk/fodder} = (^{90}Sr/Ca)_{milk} / (^{90}Sr/Ca)_{fodder}$$

This OR is 0.1 for cows milk; i.e. during the transfer of Sr and Ca from fodder to milk, the cow discriminates against Sr by a factor of ten compared to Ca. It was further observed that $OR_{infant\ bone/milk}$ was 0.25 [8].

2.2 THE WINDSCALE ACCIDENT

The Windscale accident was the first large-scale accidental release of radioactive nuclides from a nuclear reactor. It occurred on 10-11 October 1957 and 0.6-1 PBq ^{131}I was released to the environment. Figure 1 shows the maximum concentrations of ^{131}I in milk after the accident. Milk containing more than 3700 Bql^{-1} of ^{131}I was banned from consumption. The ban lasted from 25 to 44 days within the affected area. This precaution reduced the maximum dose equivalent to a child's thyroid from 360 to 160 mSv [9]. From a radioecological point of view the Windscale accident demonstrated that compliance with the maximum permissible concentrations of ^{131}I in air did not ensure that dose limits for children thyroid had not been exceeded via the milk pathway [10].

2.3 LICHEN-REINDEER-MAN FOOD CHAIN

In the early sixties a number of reports [11, 12] from the Nordic countries showed that the Laplanders in northern Scandinavia contained high body burdens of ^{137}Cs. It became evident that this was because reindeer, which is an important diet constituent of Lapps, eat lichens during the winter. Lichens are slowly growing plants with a large surface to weight ratio, which makes them able to collect contaminants e.g. ^{137}Cs from global fallout from the atmosphere very efficiently. Also other Arctic populations breeding reindeer or caribou showed enhanced body burdens of ^{137}Cs [13-15]. The levels in reindeer are evidently higher during the winter, when the reindeer feed on lichen (see Table 2).

TABLE 2. Caesium-137 concentrations (Bq kg^{-1}) in reindeer and caribou from Arctic regions. [16]

REGION	1964 Winter	1964 Summer	1965 Winter	1965 Summer
Swedish Lapland	2,200	890	2,500	740
Finnish Lapland	1,850	-	2,700	-
Russia, Murmansk	-	-	4,100	-
Russia, Yakutia	-	-	630	-
Alaska	1,220	480	1,440	110
Canada	520	250	300	200
Greenland	610	-	290	50

The doses received by some of these Arctic people was the highest seen any where, as a result of nuclear weapons testing, apart from areas contaminated by local fallout.

The radioecological studies of the lichen-reindeer-man food chain resulted into a Nordic-American cooperation with regular symposia. The international symposium in Stockholm in 1966 on Radioecological Concentration Processes was a result of this cooperation [16].

2.4 THE CRITICAL GROUP CONCEPT

During 1964-1966 about 75 TBq ^{106}Ru were discharged from the nuclear reprocessing facility Windscale (later called Sellafield) to the Irish Sea. At that time 170 individuals living in South Wales had a daily median consumption rate of laverbread of 160 g. Laverbread is made of the red algae Porphyra, which concentrates ruthenium from seawater. It was calculated that the annual dose to a member of this group of people was 7.4 mSv. Had the dose exceeded 15 mSv, the British Authorities would have asked Windscale to reduce their discharge of ^{106}Ru. This was an example of a so called critical group (the laverbread eaters) and a critical radionuclide (^{106}Ru) [17], which are decisive for the discharge limits for a nuclear plant (Figure 2).

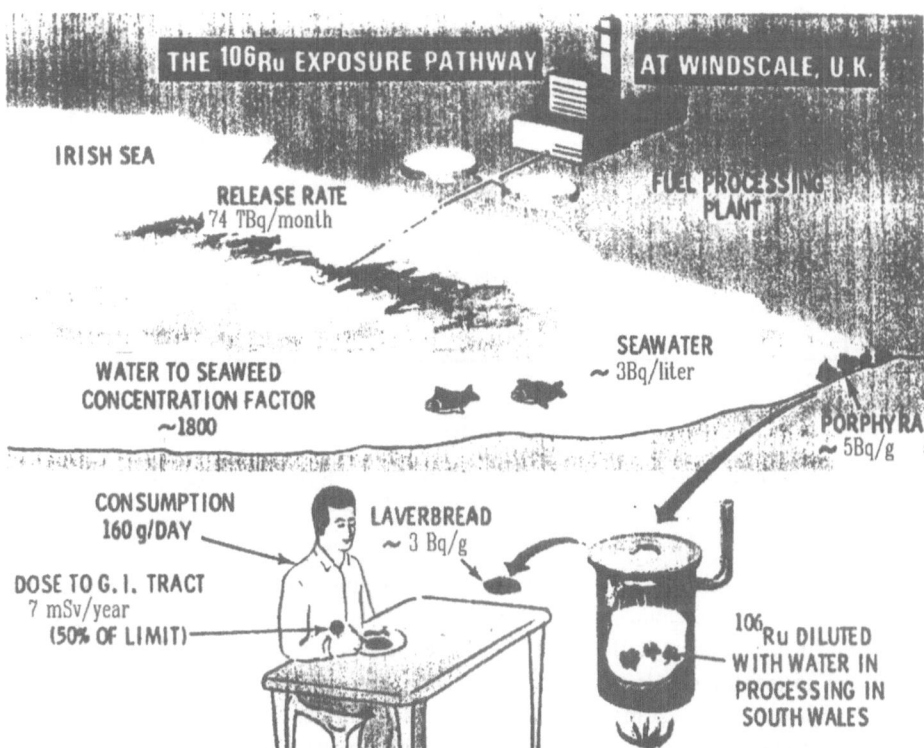

Figure 2. The critical exposure pathway associated with the U.K. atomic energy plant at Windscale is the release of ^{106}Ru to the Irish Sea, its accumulation by seaweed, and the consumption of the seaweed by man. [18].

2.5 PLUTONIUM RELEASES

During an in-flight refuelling of a B-52 bomber over Palomares in Spain in 1966 an accident happened and plutonium from two of the nuclear weapons in which the conventional explosives detonated was dispersed over an agricultural area. In order to reduce the possibility of resuspension, removal of vegetation and ploughing to a depth of 25 cm was carried out. Lung counting and urine analysis on a number of individuals have shown that the radiation risk for the local population can be considered as insignificant [19].

In January 1968, two years after Palomares another B-52 accident happened in Thule, Greenland. The aircraft crashed during an emergency landing on sea ice 11 km west of Thule Airbase. The conventional explosives in the four nuclear weapons exploded by the impact and 6 kg plutonium were dispersed. Most of the plutonium was recovered but about 0.5 kg (~1 TBq) went to the bottom of the sea, when the ice melted during the summer.

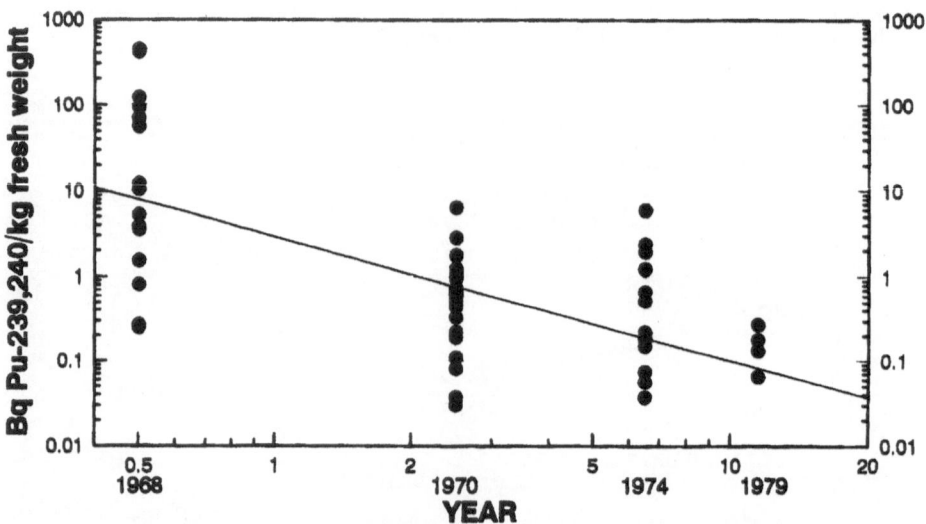

Figure 3. Plutonium in shrimp samples collected at Thule 1968-1979. Unit: Bq/kg fresh weight.
$Bq\ kg^{-1} = 2.9\ T^{-1.46}$ (T time in years since accident)

Sediments and benthic animals have since then contained enhanced plutonium levels. However, the effective environmental half life of the plutonium has shown to be short compared to the physical half life of ^{239}Pu (24000 year). Figure 3 illustrates how the plutonium concentrations in shrimps from Thule, living near to the bottom, have decreased with time. The estimated effective dose commitment to a hypothetical critical group which has been living of local marine products from Thule since the accident is 100-200 µSv and thus of no significance from a health point of view.

2.6 SELLAFIELD DISCHARGES

Figure 4 shows how the waterborne discharges from the Sellafield reprocessing plant in the UK increased to a maximum during the mid seventies.

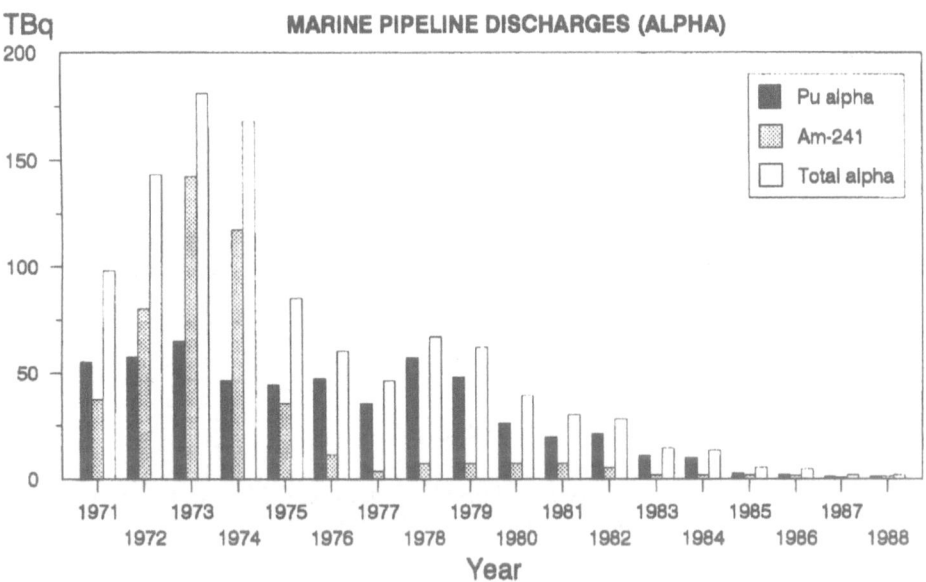

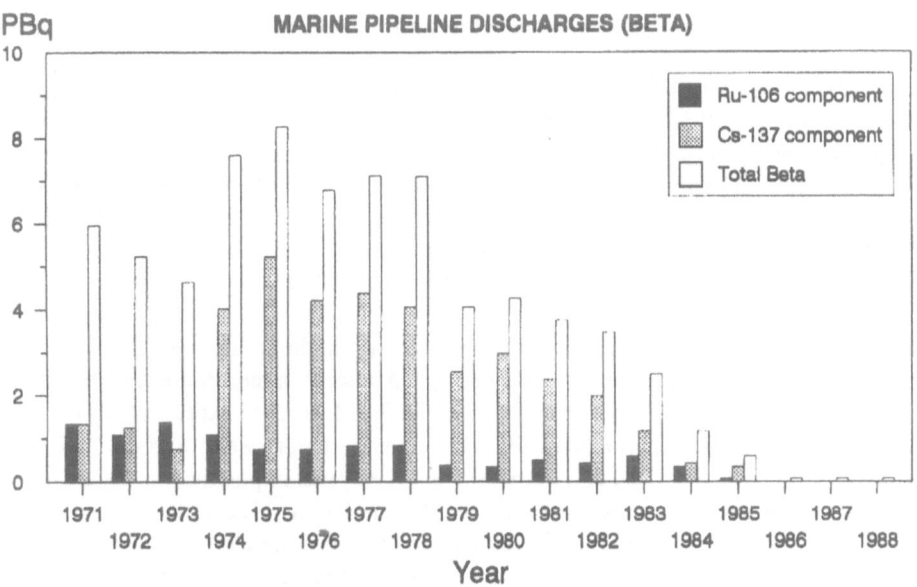

Figure 4. Waterborne discharges from Sellafield since 1971. [20]

In 1975 about 5 PBq ^{137}Cs was released. The study of the environmental behaviour of radiocaesium in the North Atlantic became a major radioecological activity in western Europe. A special project, code named MARINA [21], was initiated by the European Commission, in order to assess the radiological impact on the member states from these waterborne releases. The critical group for the ingestion pathway at Sellafield comprised consumers of locally-caught fish and shellfish. The maximum dose rate to this group was reached in 1981 by about 3.5 mSv/y. Most of this dose was due to ^{137}Cs. In 1986, the dose rate had become less than 0.5 mSv/y and now most of the dose was due to Am and Pu discharged to the Irish Sea [21].

Based on the liquid discharges from Sellafield and on measurements of water and seaweed collected in the NE Atlantic, models have been built used for dose assessments, in particular for discharges from nuclear installations in Western Europe. A box model MARIN 1 was thus developed under the CEC MARINA project [22]. This model predicts water concentrations in the various water bodies of the NE Atlantic Ocean. The model was tested on the ^{99}Tc data collected in the period 1980-1983. The model was able to predict the observed concentrations within a factor of 3 for any part of the NE Atlantic.

2.7 SPECIFIC DEVELOPMENTS IN THE EAST

During the first steps of prehistory of radioecology, at the end of the last century and the beginning of the current century, science was internationally connected. The situation changed because of the changes of the political structure of tsarist Russia into the USSR. Practically, the nineteen twenties were not yet a very closed period for Soviet science. Contacts got lost in the late thirties. That is why, for example, in the nineteen twenties, Prof V.I. Vernadsky from time to time read lectures on geochemistry at the Sorbonne in Paris and young talented Russians were sent to the USA and Germany.

In the 19th century the science of Russia was tight connected with the Western science. Therefore the first "radioecological" investigations dealt with animals (vertebrates: Amphibia and Cyclostomata; invertebrates: Diptera and Lepidoptera as well as plants: Vallisneria spiralis) and was held at Sankt-Petersburg, in the Physiological Laboratory of the Academy of Sciences, by Prof Ivan R. Tarkhanov in 1896, already one year after the discovery of Roentgen rays in 1895 in Germany.

To feel the real scientific atmosphere of that time, it is better to cite the introductory paragraph from an article by Tarkhanov on "Experiments on the impact of Roentgen X-rays on living organisms" [24]: "At one of the meetings in the spring of this year [1896], held at the S.-Petersburg Biological Laboratory and after an interesting lecture about X-rays by Prof P.P. Van-der-Vliet, it was proposed by Prof Lesgaft to express an opinion on possible and desirable experiments with X-rays, having a purpose to find out their significance in the processes of life". Vernadsky attached great importance to the phenomenon of radioactivity in nature. After the visit to the UK in 1908 in his capacity as a corresponding member of the British Association of Sciences, he organized in 1911 the first Russian radioecological expedition and wrote soon his first studies on biogeochemistry, paying much attention to the role of radioactivity in the biosphere. In

1922 Vernadsky published "A note on the organization of the State Radium Institute of the Russian Academy of Sciences" [26].

It is remarkable that in 1922 he wrote the following scientific prognoses: "In the processes of life, I gave, among others things, my following views: before to make any detailed investigations with them, it is necessary to determine, whether they affect and in which way, the main processes of life: the metabolism of living formations, the course of their development and the main vital function irritability of living formations. If it will be found that these vital processes relate not indifferently to X-rays, then one may await that these rays will be able to influence also the whole of life of living formations in either direction. The director of S.-Petersburg Biological Laboratory was in sympathy with the plan of biological experiments with X-rays, outlined shortly: by myself -on animals- and by Mr Polovtsev -on plants-, and promised his help in the purchase of all apparatus for the laboratory, which are necessary for experiments with Roentgen rays. Waiting for it and working at the Physiological Laboratory of the Academy of Sciences during almost the whole summer, I was occupied with some questions, as outlined above. Thank to the kindness of Academician Prince Golitsin, who granted me all equipment, which is necessary for getting X-rays".

In the second decade of the 20th century Tarkhanov studied effects of ionizing radiation on wildlife. In 1923 Prof Nina S. Gajewskaya published results on the influence of roentgen rays on a brine-shrimp Artemia salina. Her ideas and suggestions for the future investigations were radioecological ones. She wrote in her article "Influence of Roentgen rays on Artemia salina": "It is necessary to note, that salt from salt lakes near Sevastopol is radioactive, therefore Artemia salina in natural conditions is influenced by, although weak, however permanent impact of gamma-rays of radium" [27]. Pechenko described the beginning stage of depression and consequent normalization of a protozoa in the article "On influence of Raraysu pon Amoeba" [28]. Effective doses of ionizing radiation and latent periods were determined in 1930 and following years in irradiated invertebrates Hydra fusca (Coelenterate), Planaria polychroa (Plathilmintes) and Planorbis marginatus (Mollusca) by Prof S.A. Nikitin (1932, 1938) at the Odessa Roentgenology Oncology Institute. Radiation effects of irradiation by roentgen rays in different doses were studied by Dr G.V. Somokhvalova (1935, 1938) on fishes Lebistes reticutatus and Carassius carassius.

The other direction of research in radioecology in the USSR before World War II was connected with the name of Academician Vladimir I. Vernadsky and his scientific school. He created both the Radium Institute in Petrograd (in 1922) and the Biogeochemical Laboratory - later the Institute of Geochemistry and Analytical Chemistry- (in 1928) in Moscow. He was the first President of the Academy of Sciences of Ukraine (Kiev) and a Professor on the Moscow, Kiev and Taurian (Crimean) Universities. In his report on a visit to the Ilmen mountains in the Urals he reports "On the necessity of the investigation of radioactive minerals of the Russian Empire", S.-Petersburg, 1910.

In 1922 he wrote the prognoses "We are approaching a great overturn in life of mankind, which can not be compared with past experience. Time is not far, that man will receive atomic energy into his hands. Such source of power, will give him the possibility to build his life, as he will wish. It may happen in the nearest years, it may happen in hundreds years. However, it is clear, that it must be. Will he be able to profit from this power, to direct it towards good, but not to self-annihilation? Does he develop the ability to use the power, which will be inevitably given to him by science? Scientists must not shut their eyes to possible consequences of their scientific work. They have to feel their responsibility for the consequences of their discoveries. They must connect their work with a better organization of whole mankind" [29].

In 1929 V. Vernadsky published his work "On the concentration of radium by living organisms " [29], in which significant differences were noted in the accumulation ability of radium of different species of duckweed (Lemna) from different water bodies of the USSR. The ratio of concentrations of Ra-226 in living Lemna minor from the Petergof ponds (near Sankt-Petersburg) was 14-47 (on wet weight basis) and in Lemna gibba and Lemna trisulca from the Kiev ponds it was 200-477.

So, Vernadsky introduced a new notion into science - the ratio of concentrations of radioelements in organisms and in water. The name was translated into different languages: "koeffitsient nakopleniya", "concentration factor", "facteur de concentration", "Anreicherungsfactor [31-33]. Prof N. Timofeev-Resovsky, who proposed the Russian name ''koeffitsient nakopleniya", often told his students a scientific slogan: "Nakoplenie koeffitsientov nakopleniya" = "Concentrate concentration factors". At the same time, because of better reproducibility, Timofeev-Resovsky recommended to relate the concentration factors to "Dry Weight".

The political disintegration of the World, started in the thirties and transformed finally into the Cold War after World War II, had a severe influence on the destiny of science and scientists. In nuclear science the separation was a total one.

A few years after the USA, perhaps in 1949 or somewhat earlier, the releases of untreated liquid radioactive wastes began from the Chelyabinsk-60 enterprise, now called "Mayak", into the Techa River, connected with the Kara Sea via a chain of rivers. In 1956 liquid radioactive wastes began to be disposed into the Karachay Lake, which is now the most radioactive lake in the World.

Timofeev-Resovsky used since the end of the forties at "Chelyabinsk-40" the term radiation biogeocenologie. The term is widely used in 1950 and later years, and in particular since 1955 at one of the important radioecological institutions, the Institute of Biology of the Ural Branch of the USSR Academy of Sciences with its Biostation "Miassovo". Afterwards it was renamed into Institute of Ecology of Plants and Animals of the Urals Division of the Russian Academy of Sciences with its Biophysical Station at Zarechny. At this moment, "radiation biogeocenology" should be seen as one of the main scientific fields of radioecology.

2.8 THE CHERNOBYL ACCIDENT

The accident in the nuclear power plant at Chernobyl and the political "fission" of the Soviet Union was the beginning of the end of the separation between eastern and western radioecologists. It also triggered the disclosure of the unknown nuclear accidents in the Urals in 1957 and 1967, and the use of nuclear tactic weapons in the Orenburg Region. It became clear that there existed a severe negligence of radiation safety in areas around the nuclear test grounds in Kazakhstan and Altay, in the uranium mines and in the nuclear waste disposal sites in the CIS.

Nowadays the process of amelioration of scientific exchange between West and East, triggered by the International Union of Radioecology and now, by the NATO Advanced Study Institute, transforms the former political meanings "East and West" into pure geographic meanings.

The Chernobyl accident was, also from a scientific point of view, a last milestone. It was since 20 years the height of terrestrial radioecological research. Experience gained during the studies of fall-out from nuclear weapons testing was partly lost because many of those engaged in radioecological science had moved to other fields. Although the Chernobyl accident brought new results to radioecology, many observations were in fact not new. It was not surprising that ^{131}I and ^{137}Cs were the radiologically important radionuclides. It was also to be expected that certain food chains, e.g. lichen-reindeer-man were very sensitive to radioactive contamination. Nor was it a surprise that fish from oligotrophic lakes contained high radiocaesium levels. It was predictable that the time of the year of the contamination influenced the radioactivity levels in crops, particularly vegetables. It was well known that fruits would be expected to show enhanced radiocaesium levels due to translocation. Nevertheless, the Chernobyl accident taught some lessons which had not received much attention during the global fallout studies in the fifties and sixties.

One of the important lessons was that the composition of the debris from a reactor accident may change with distance. This is illustrated in Table 3.

TABLE 3. Estimated releases and deposition of long-lived radionuclides from the Chernobyl nuclear accident. All activities are decay-corrected to 26 April and are approximate. [34].

Radionuclide	Total released radioactivity (PBq)	Radioactivity deposited in European part of USSR (PBq)
^{137}Cs	100	30
^{134}Cs	50	15
^{90}Sr	8	7
^{106}Ru	35	25
^{144}Ce	90	75
^{110m}Ag	1.5	0.5
^{125}Sb	3	2
^{239,240}Pu	0.055	0.05
^{238}Pu	0.025	0.02
^{241}Pu	5	4
^{241}Am	0.006	0.005
^{242}Cm	0.6	0.55
^{243,244}Cm	0.006	0.005

28

It appeared that while about two thirds of the radiocaesium was deposited outside the USSR, only 10% of the transuranic nuclides and about 15% of the ^{90}Sr, left the European part of the USSR. Hence, the radioecological situation changed with the distance from Chernobyl, because the contamination levels altered and the composition of the contamination differed.

The Chernobyl accident also drew the attention to the importance of resuspension [35], the radioecology of natural and semi-natural ecosystems, in particular the high radiocaesium uptake by mushrooms [36], and to the special circumstances in urban environments, where most people in the western world are living [37].,

3. References

1. Odum, P.E. (1956) Consideration of the total environment in power reactor waste disposal. *Proc. Intern. Conf. on the Peaceful Uses of Atomic Energy*, **13**: 350-353, United Nations, Geneva 1955.

2. Odum, P.E. (1959), *Fundamentals of Ecology*, W.B. Saunders Company. Philadelphia,P.A.. 452-486.

3. Chamberlain, A.C., Loutit, J.F., Martin, R.P. and Scott Russell, R. (1955) The behaviour of I-131, Sr-89 and Sr-90 in certain agricultural food chains. Proc. Intern. Conf. on the Peaceful Uses of Atomic Energy., **13**, 360-363. United Nations, Geneva 1955.

4. Larson, K.H. (1961) Continental Close-in fallout: It's history measurement and characteristics, *Radioecology*,19-25 edt. Schultz V. & Klement Jr., A.W. Reinhold Publishing Corp., New York.

5. Rand Corporation (1953) Report R-251-AEC, Project Sunshine, Santa Monica, Calif.; Aug. 1953. Available from: Rand. Corp., Santa Monica, CA.

6. Kulp, J.L. and Schulert, A.R. (1962) Sr-90 in man and his environment. Vol. I Summary. pp 385. Geochemical Laboratory, Columbia University, Palisades, New York.

7. Comar, C.L.; Wasserman, R.H. and Nold, M.M. (1956) Strontium-Calcium discrimination factors in the rat, *Proc. Soc. Exp. Biol. Med*, **92**: 859.

8. Loutit, J.F.; Russell, R.S.; Bruce, R.S. and Bartlett, B.O. (1960) The ratios of ^{90}Sr in milk in the bones of infants. *Nature* **201**, 770.

9. Jackson, D. and Jones, S.R. (1990) Reappraisal of environmental countermeasures to protect members of the public following the Windscale nuclear reactor accident, 1957, *Comparative Assessment of the Environmental Impact of Radionuclides Released during Three Major Nuclear Accidents: Kyshtym, Windscale, Chernobyl*. Luxembourg 1-5 Oct. 1990. 1015-1039; CEC.

10. Wolff, A.H. (1959) Milk concentration in the Windscale incident, *Public Health Reports*, **74**, 41-43.

11. Lidén, K. (1961) Caesium-137 burdens in Swedish Laplanders and reindeer., *Acta Radiol.* **56**: 237.

12. Miettinen, J.K.; Jokelainen, A.; Roine, P.; Lidén, K.; Naversten, Y. (1963) 137Cs and potassium in people and diet - a study of Finnish Lapps. *Ann. Acad. Sci. Fennicae*, Ser. A, II, Chem. 120.

13. Hanson, W.C.; Palmer, H.E.; Griffin, B.I. (1964) Radioactivity in northern Alaskan Eskimos and their foods, summer 1962. *Health Physics*, **10**, 421-429.

14. Hanson, W.C. (1966) Radioecological concentrations processes characterizing Arctic ecosystems, B. Åberg; F.P. Hungate, eds. *Radioecological concentration processes*. Oxford: Pergamon Press; 183-191.

15. Ramzaev, P.V.; Shamov, V.P.; Troizkaj, M.N.; Lebedev, O.V.; Ibatulin, M.S. (1965) Indirect assessment of total body burden of 137Cs in people. *Medizinskaj Radiol.*, **6**, 22-28.

16. Åberg, B. & Hungate, F.P. (eds.) (1966) *Radioecological Concentrations Processes*. Oxford: Pergamon Press pp. 1040.

17. Dunster, H.J. (1955) The discharges of radioactive waste products into the Irish Sea. Part 2: The preliminary estimate of the safe daily discharge of radioactive effluent. *Proc. international conference on the peaceful use of atomic energy*, Geneva. New York: UN Publ. **9**. 712-715.

18. National Academy of Sciences (1971) *Radioactivity in the marine environment*. Washington D.C. pp 272.

19. NEA (1981) T*he environmental and biological behaviour of plutonium and some other transuranium elements*. OECD, Paris pp 116.

20. BNFL (1994) *Annual report on radioactive discharges and monitoring of the environment* 1993 Vol. 1. British Nuclear Fuels plc. Risley, Warrington, Cheshire, UK pp 89.

21. Camplin, W.C.; Aarkrog, A. (1990) Radioactivity in North European waters, *The radiological exposure of the population of the European Community from radioactivity in North European marine waters. Project MARINA*. Report EUR 12483 Commission of the European Communities, Luxembourg, 199-305.

22. CEC (1990) *The radiological exposure of the population of the European Community from radioactivity in North European marine waters Project "Marina"*. Report EUR 12483, Commission of the European Communities, Luxembourg, pp 571.

23.Charles, D.; Jones, M.; Cooper, J.R. (1990) The radiological impact on EC member states of routine discharges to north European waters. *The radiological exposure of the population of the European Community from radioactivity in North European marine waters. Project MARINA*. Report EUR 12483 Commission of the European Communities, Luxembourg, 414-560.

24. Tarkhanov, I. (1896) Experiments on the action of Roentgen's X-rays on the animal organism. *Izvestya*. Sankt-Peterburgskoy Biol.. Lab., 1(3), pp. 47-52. (in russian)

26. Vernadsky, V. (1922), *Iz. Ross. Akad. nauk*, 6 seria, 16.

27. Gajewskaya, N.S., (1923), *Verhandl. Intern. Vereinig. Limnologic*. Stuttgart, 1, 359).

28. Pechenko. V.F., (1922), *Vestnik of Roentgenol. and Radiol.*, 2, N 1

29. Vernadsky, V. (1922), *Essays and speeches*, II, Petrograd, issue 2, 124 p.

30. Vernadsky, V. (1929), *Doklady Academii Nauk SSSR*, Ser. A, N 2, pp. 33-34.

31. Timofeev-Resovsky, (1957) Use of radiations and radiation sources in experimental biogeocoenology, *Bot. zh.* 42(2), 161-194 (in russian)

31a. Spooner G.M. (1949) Observations of the absorption of radioactive strontium and yttrium by marine algae, *J Marine Biol. Assoc.* U.K. **28**(3), 587-625.

32. Timofeeva-Resovskaya Ye.A., Popova E.I., Polikarpov G.G. (1958) Concentration of chemical elements from aqueous solutions by fresh water organisms. 1. Concentration of radioisotopes of phosphorus, zinc, strontium, ruthenium. cesium and cerium by various species of fresh water mollusca, *Byull. Mosk. obshch. ispyt. prirody biol'*.,**63**(3), 68-78 (in russian).

32a. Timofeeva-Resovskaya Ye.A, (1963) Distribution of Radioisotopes in the main components of fresh water bodies (monograph) *Tr. inst. biol., Akad. Nauk SSR, Ural'skiy filial*, **30**, 1-78 (in russian) (JPRS 21, 816. (in english))

33. Whicker F.W. & Schultz V., (1982) *Radioecology: Nuclear Energy and the Environment*. Vol I. CRC Press, Inc. Boca Raton, Florida. 212p.

34. WHO (World Health Organization, (1989). Health hazards from radiocaesium following the Chernobyl nuclear accident. Report on a WHO working group, *J. Envir. Radioact.*, **10**, 257-295.

35. Garland J.A., Pattenden N.J,; Playford K. (1992) *Resuspension following Chernobyl*. IAEA-TECDOC-647. International Atomic Energy Agency, Vienna, 9-33.

36. Johanson K.J., Bergström R.V., Bothmer S, Karlen G. (1990) Radiocaesium in wildlife of a forest ecosystem in central Sweden. *Transfer of radionuclides in natural and semi-natural environments*. London: Elsevier Applied Science; 183-193.

37. Roed, J. (1987) Dry deposition in urban and rural areas in Denmark. *Radiat. Prot. Dos.* **21**,33-36.

INVENTORY OF NUCLEAR RELEASES IN THE WORLD

ASKER AARKROG
Risø National Laboratory
P.O. Box 49, DK-4000 Roskilde, Denmark

1. Introduction

Releases of radioactive substances to the environment have in principle always occurred. Erosion of land, run-off with rivers and volcanic eruption have thus always contaminated the geo-hydro-sphere with natural occurring radionuclides. In the present context we shall , however, deal with man-made (anthropogenic) radionuclides only.

The release of these artificial radionuclides begins in the early forties of this century, when man began to explore the fission process in nuclear reactors and nuclear weapons.

In the following we shall consider 3 types of radioactive releases to the environment.
-nuclear explosions
-routine releases from nuclear installations
-accidental releases

2. Nuclear explosions

The main input of artificial radioactivity on a global scale has come from nuclear weapons testing in the atmosphere.

2.1 HIROSHIMA AND NAGASAKI

Nuclear weapons have only been used two times in war. In August 1945 the USA deployed two fission bombs over the Japanese cities Hiroshima and Nagasaki. Although these weapons killed a large number of people and caused great material damages, they have left very little environmental contamination with radioactivity.
The maximum deposit of ^{137}Cs measured in Nagasaki was 5.3 kBq m^{-2} in 1986, and the maximum ^{239,240}Pu level was 1.8 kBq m^{-2} [1].

2.2 MILITARY TEST EXPLOSIONS

The two major test sites for atmospheric testing of thermonuclear weapons have been Novaya Zemlya in the Arctic part of the former USSR and the Bikini and Eniwetok

F. F. Luykx and M. J. Frissel (eds.), Radioecology and the Restoration of
Radioactive-Contaminated Sites, 31–43.
© 1996 *Kluwer Academic Publishers.*

32

Islands (US) in the Pacific Ocean. Furthermore China has carried out tests at LopNor in Western China, France at Muoroa in the Pacific Ocean and the UK at the Christmas Islands in the Pacific. USA, UK and USSR stopped atmospheric testing by 1962 and since 1980 there has been no atmospheric testing carried out by any country.

About 423 atmospheric explosions had occurred until 1980 (Table 1). The total explosive yield amounts to 545 Mt TNT, 217 Mt from fission and 238 Mt from fusion. Local fallout deposited close to the test sites, amounts to 12%, tropospheric fallout, which is deposited in a latitude band around the latitude of the test site 10%, and global fallout which mainly is deposited on the same hemisphere as the test site 78% [2-4].

TABLE 1. Number and yield of atmospheric nuclear explosions [2-4].

YEAR	NUMBER	Estimated yield (Mt)	
		Fission	Total
1945-1951	26	0.8	0.8
1952-1954	31	37	60
1955-1956	44	14	31
1957-1958	128	40	81
1959-1960	3	0.1	0.1
1961-1962	128	102	340
1963	0	0.0	0.0
1964-1969	22	10.6	15.5
1970-1974	34	10.0	12.2
1975	0	0.0	0.0
1976-1980	7	2.9	4.8
1981-1990	0	No further tests	
1945-1990	423	217.4	545.4

TABLE 2. Deposition density in the 40°-50°N latitude belt from atmospheric nuclear testing [2-4]

Radionuclide	Deposition density[a] (Bq m^{-2})
^{89}Sr	20,000
^{90}Sr	3,230
^{131}I	19,000
^{137}Cs	5,200
^{140}Ba	23,000
^{238}Pu	1.5
^{239}Pu	35
^{240}Pu	23
^{241}Pu	730
^{241}Am	25

[a]) total deposition until 1990 not corrected for decay.

Figure 1 shows the global distribution of global fallout illustrated by ⁹⁰Sr. In Table 2 the deposition of other important radionuclides in weapons fallout is given for the 40°-50° northern latitude band, which represents the global maximum belt of fallout from weapons testing.

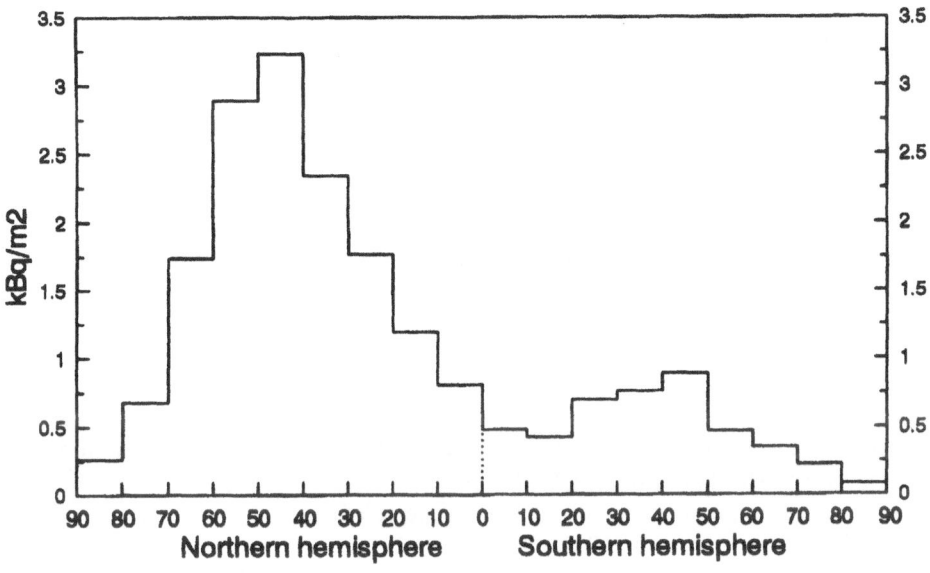

Figure 1. Integrated deposition density of Strontium-90, kBq m⁻² [2].
(Cumulative deposition is 0.45 x integrated in 1995)

Underground nuclear testing is still going on. Venting from such explosions have occasionally been detectable distant from the test sites [5]. The activity seen is usually dominated by ¹³¹I. The environmental impact is negligible due to the very low concentrations.

2.3 CIVILIAN NUCLEAR EXPLOSIONS

More than 100 nuclear explosions have been carried out for peaceful purposes in the USA and in particular in the former USSR. Although contamination has been measured far away from the explosions e.g. ¹⁸¹W from the US and ¹³¹I from the Soviet explosions, the contamination has not contributed significantly to the global effective dose [2-4].

3. Releases under normal operation

3.1 NUCLEAR REACTORS

Under normal operation of nuclear power plants or experimental reactors the discharge of radioactive substances to the environment are regulated by the authorities. The releases are kept low in order to protect man and environment and the doses received from normal operation of nuclear plants are thus in general negligible. Discharges are made both to the atmosphere and the aquatic environment. The reported discharges vary by several orders of magnitude, even for the same type of reactors. The by UNSCEAR [2-4] reported normalized releases, shown below, should therefore be considered with this information in mind.

3.1.1 Airborne discharges

The discharges from nuclear reactors to the atmosphere mostly consist of noble gases viz. ^{41}Ar, ^{85}Kr, ^{85m}Kr, ^{87}Kr, ^{88}Kr, ^{131m}Xe, ^{133}Xe, ^{133m}Xe, ^{135}Xe and ^{138}Xe. All these radionuclides are short lived except ^{85}Kr with a half life of 10.8a and they are not of radioecological interest, the main pathway to man being inhalation and external radiation from the clouds. More interesting are the releases of tritium and ^{14}C. The average normalized release from LWRs (PWR and BWR) of tritium is about 3TBq $(GW_ea)^{-1}$, from HWRs it is nearly 500 TBq$(GW_ea)^{-1}$ and for GCRs 9TBq$(GW_ea)^{-1}$. FBRs are reported to have a mean release of nearly 100 TBq tritium $(GW_ea)^{-1}$.

In case of ^{14}C the average normalized releases are 120 GBq$(GW_ea)^{-1}$ for PWRs, 450 for BWRs, 4800 for HWRs and 540 for GCRs.

Iodine-131 is the radioecologically most important fission product released with airborne effluent from nuclear reactors. The average normalized releases are for PWR's 0.93 GBq ^{131}I $(GW_ea)^{-1}$, for BWR's 1.8, for HWR's 0.19, for GCR's 1.4, and for LWGR's 14.

Particulates, mostly consisting of corrosion and fission products, are also discharged from nuclear reactors to the atmosphere. The normalized average releases vary between 0.2 and 12 GBq $(GW_ea)^{-1}$ for the different reactor types.

3.1.2 Waterborne discharges.

The dominating radionuclide in liquid effluent from nuclear reactors is tritium. The average normalized release of tritium from PWRs is 23 TBq $(GW_ea)^{-1}$, 0.62 from BWRs, 377 from HWRs, 120 from GCRs, 11 from LWGRs and 3 from FBRs. In Table 3 are shown the most important radionuclides, apart from tritium, found in waterborne discharges from reactors.

TABLE 3.Normalized average release from LWRs with liquid effluent , GBq$(GW_ea)^{-1}$,[2-4]

	^{51}Cr	^{54}Mn	^{55}Fe	^{59}Fe	^{58}Co	^{60}Co	^{65}Zn	^{110m}Ag	^{134}Cs	^{137}Cs
PWR	2.8	1.9	5.6	0.22	27	7.6	0.016	2.4	2.3	4.6
BWR	21	5	1.9	1.7	2.8	11	1.3	0.06	1.3	2.9

3.2 NUCLEAR FUEL REPROCESSING

Throughout the seventies and the early eighties, the most important source to radioactive environmental contamination from the nuclear fuel cycle was nuclear fuel reprocessing. Sellafield in the UK, which was the main source of contamination, has in the later years significantly reduced its liquid discharges, in particular ^{137}Cs, as it appears from Figure 2. [6].

3.2.1 Airborne discharges
The normalized releases per $GW_e a$ were in the last half of the eighties 12.3 PBq ^{85}Kr, 41 TBq ^{3}H, 2 TBq ^{14}C, 5.7 GBq ^{129}I and 1.5 GBq ^{137}Cs. Tritium, ^{14}C, ^{85}Kr and ^{129}I are dispersed globally and thus contribute to the global rather than the local and regional dose.

3.2.2 Waterborne discharges
The normalized releases per $GW_e a$ from liquid effluent from nuclear reprocessing were 643 TBq ^{3}H, 39 TBq ^{106}Ru, 36 TBq ^{90}Sr, 13 TBq ^{137}Cs, 0.54 TBq ^{14}C and 32 GBq ^{129}I. As only 4% of the nuclear reactor fuel (1985-1989 data) is reprocessed, activity releases from reprocessing are presently much less than they otherwise would have been. From a dose point of view, the liquid effluent from reprocessing are nearly four times more important than the airborne discharges [2-4].

3.3 DISPOSAL OF RADWASTE

The low level and intermediate level waste arising from the nuclear fuel cycle are mostly disposed by shallow burial although some of this waste until 1982 was also dumped in the ocean. The environmental contamination from radwaste is almost entirely due to ^{14}C and is difficult to quantify.

The future permanent disposal of highlevel waste may be a potential source to contamination of the ground water. However, if the disposal is carried out properly e.g. in deep salt mines or in deep sea sediments, it is unlikely that the activity should ever enter the biosphere.

In 1993 it was reported that radwaste throughout the years has been dumped in shallow waters in the Arctic close to Novaya Zemlya. This waste also comprises a number of naval nuclear reactors. The detailed composition of the waste has not been reported, but the total amounts are today (1995) in the order of 10 PBq [7-9].

In the early days of nuclear weapons development in the former USSR medium-level radwaste was disposed directly into the Techa river from the plutonium production at Chelyabinsk-40 in the Urals. Techa is a part of the upstream Ob-river-system. During 1949-1956 about 12 Bq ^{90}Sr and 12 Bq ^{137}Cs were discharged into the Techa. Some of this activity may have reached the Arctic Ocean, but most of it seems to have remained in the upper parts of the Techa river system, where large reservoirs have been constructed to retain the waste [10].

36

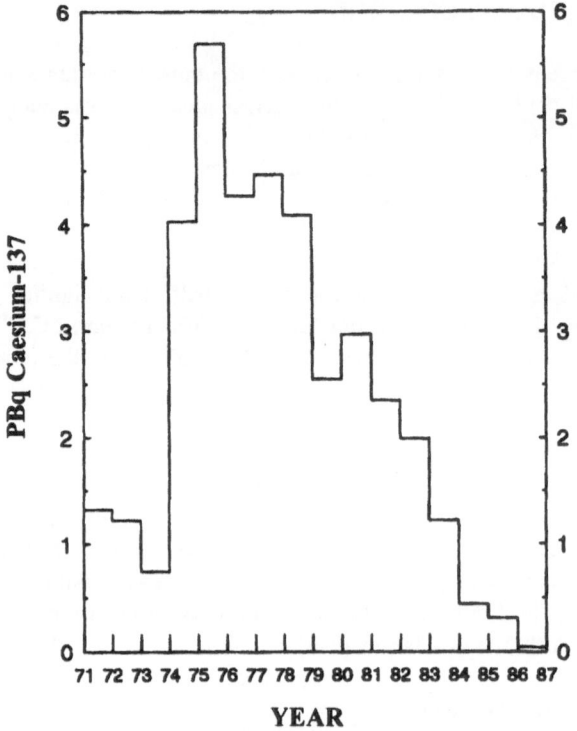

Figure 2 Annual liquid discharges from Sellafield 1971-1986 [6].

3.4 HOSPITALS, INDUSTRY AND RESEARCH

Hospitals are the main users of unsealed radionuclides outside the nuclear fuel cycle. Most of these radionuclides, e.g. 78h ^{67}Ga, 6h ^{99m}Tc, 13h ^{123}I and 74h ^{201}Tl are fairly short lived and thus of little environmental concern. Others, such as ^{3}H, ^{14}C, ^{32}P, ^{35}S and ^{131}I may be detected in the effluent from hospitals. Research and educational institutions use primarily ^{3}H, ^{14}C, ^{32}P and ^{35}S and in industry ^{147}Pm (ß:0.224 MeV) is applied in relatively large quantities. Beside this application of unsealed isotopes, there is a use of sealed sources e.g. ^{60}Co, ^{137}Cs, ^{226}Ra and ^{241}Am, which in case of accidents, may contaminate the environment.

UNSCEAR [2-4] has calculated the annual normalized production in the developed countries of radioisotopes used in medical, educational and industrial applications in GBq per 10^6 population. ^{3}H becomes 90, ^{14}C:30, ^{85}Kr:10, ^{125}I:30, ^{131}I:180 and ^{133}Xe:1600. The dominating contributors to the collective effective dose are ^{14}C (90% of the dose), ^{131}I (9%) and ^{3}H (1.5%).

4. Accidental releases

Throughout the years, accidents causing a release of radioactive material to the environment have occurred. Most of these accidents have only been of local or regional concern. However, the amounts of activity released by an accident may be substantial higher than those released during the whole lifetime of normal operation of the nuclear facility in question.

4.1 REACTOR ACCIDENTS

4.1.1 Chernobyl 1986.

The accident, which so far has released the greatest amounts of activity and which was of global concern, was the accident in the Chernobyl 4 reactor in Ukraine in April-May 1986.

The reactor was a graphite moderated, light-water-cooled system (LWGR) called RBMK-1000. The installed electrical generating capacity was 1GW. The accident followed some engineering tests of a generator. Basic operating safety rules were being violated. Most control rods were thus withdrawn from the core and crucial safety systems were switched off. A prompt critical excursion occurred April 26, 1986, and explosive energy was released, lifting the 1000 tonne cover plate of the reactor. The graphite started to burn and the fire lasted for 10 days before the release of activity from the reactor could be stopped. Contamination was spread over large areas and was measurable in most parts of the Northern Hemisphere. The approximate (± 30%) amounts of long lived radionuclides are reproduced from a WHO report [11] in Table 4.

TABLE 4. Estimated releases and deposition of long-lived
radionuclides from the Chernobyl accident.
All activities are decay-corrected to 26. April 1986 [11].

Radionuclide	Release in PBq	Activity in PBq deposited in the European part of the USSR
^{137}Cs	~ 100	~ 30
^{134}Cs	~ 50	~ 15
^{90}Sr	~ 8	~ 7
^{106}Ru	~ 35	~ 25
^{144}Ce	~ 90	~ 75
^{110m}Ag	~ 1.5	~ 0.5
^{125}Sb	~ 3	~ 2
^{239,240}Pu	~ 0.055	~ 0.05
^{238}Pu	~ 0.025	~ 0.02
^{241}Pu	~ 5	~ 4
^{241}Am	~ 0.006	~ 0.005
^{242}Cm	~ 0.6	~ 0.55
^{243,244}Cm	~ 0.006	~ 0.005

From a radioecological point of view the 630 PBq ^{131}I which were released during the first weeks after the accident caused most concern. Later the attention changed to radiocaesium, which will be the main contributor to the ingestion dose and the external dose from the Chernobyl accident [2-4]. From Table 4 it appears that the relative distribution of the various radionuclides changes with the distance from Chernobyl. E.g. ^{90}Sr and Pu are relatively more abundant close to the accident site compared to ^{137}Cs far away from the reactor [12].

4.1.2 Windscale 1957.

The other nuclear reactor accident of environmental concern was the Windscale accident in October 1957. A military air-cooled graphite-moderated natural-uranium reactor used for plutonium production caught fire during the liberation of Wigner energy in the graphite. Emissions from the Windscale lasted for 18 hours. Approximately 0.6-1 PBq ^{131}I, 22-94 TBq ^{137}Cs, 5 TBq ^{89}Sr, 0.22 TBq ^{90}Sr, 8.8 TBq ^{210}Po, 5 PBq ^{3}H and 1.6 GBq ^{239}Pu were released. The activity was detectable in most parts of Western Europe, but the majority of the activity was deposited in the UK [13].

4.2 WASTE DEPOSITORIES

4.2.1 Kyshtym.

Between Cheliabinsk and Ekaterinburg in Russia plutonium production for military use began in the forties. The production plant called MAYAK (Chelyabinsk-40) is located near Kyshtym and high level radwaste as nitrate-acetate from the plutonium production were stored here in large water-cooled tanks. A failure in the cooling system dried out the waste in one of the tanks and on 29 Sept. 1957 an explosion equivalent to 70-100 tonnes TNT occurred. The total deposition from the accident was 74 PBq. Most of the activity consisted of short lived fission products in particular ^{95}Zr and ^{144}Ce. However, 2 PBq of ^{90}Sr were also released by the accident. The waste was depleted in ^{137}Cs of which 30 TBq only were released. The deposition occurred in a cigar shared narrow band up to a distance of a few hundred km in NNE direction involving parts of the Chelyabinsk, Ekaterinburg and Tyumen provinces [14-16].

4.2.2 Lake Karachay.

In connection with MAYAK a number of reservoirs for disposal of high level radwaste has been established.
One of these is the 0.45 km^2 Lake Karachay which contains 0.74 EBq ^{90}Sr and 3.6 EBq ^{137}Cs. In the summer 1967 dust from the shore line of the lake containing about 20 TBq ^{90}Sr and ^{137}Cs in the ratio 1:3 was dispersed by wind over an area of 1800 km2 to a distance of 75 km. The huge amounts of radwaste in Lake Karachay makes this open waste repository a potential source of environmental contamination in the Urals [10, 17].

4.3 LOST SUBMARINES

In the world oceans a number of wrecked "nuclear" submarines resides on the ocean floor. The US has lost 2 and the former USSR 6 of which two may have been raised. Most losses have occurred in the North Atlantic. The submarines probably all carry nuclear weapons [18].

The last accident occurred in April 1989, 180 km SW of Bear Island in The Barents Sea, where a submarine Komsomolets from the USSR Navy sank to 1500 m depth. The core inventory in this submarine is 2.8 PBq ^{90}Sr and 3.1 PBq ^{137}Cs. The nuclear weapons contain according to a report by a joint Russian-Norwegian Expert Group about 16 TBq Pu [19].

In sofar only traces of radiocaesium, which may come from the submarine, have been detected close to the wreck. In connection with the accidents with nuclear submarines minor amounts of ^{131}I have occasionally been released to the atmosphere.

4.4 LOST NUCLEAR WEAPONS

The US Strategic Aircommand has at two occasions lost aircrafts carrying nuclear weapons which has resulted in environmental contamination with transuranic elements, viz. in 1966 at Palomares in Spain and in 1968 at Thule in Greenland [20].

In Spain 2.26 km^2 of uncultivated farm land and urban land were contaminated by ^{239,240}Pu. The present contamination level is less than 1.2 MBq ^{239,240}Pu m^{-2}. It is estimated that the total inventory of this contamination is of the order of 0.1 TBq ^{239,240}Pu.

In Thule the contamination is found in sea sediments in the Bylot Sound west of the Thule Airbase. The total inventories are 1 TBq ^{239,240}Pu, 0.02 TBq ^{238}Pu and 0.1 TBq ^{241}Am. The contamination covers a seabed area in the order of 1000 km^2 [21].

4.5 LOST RADIATION SOURCES

Various accidents involving sealed sources for medical or industrial purposes have been reported. In 1983 16.7 TBq ^{60}Co from a broken therapy source were mixed in metallic products in Mexico . In 1984 at Mohammedia, Morocco, a source of ^{192}Ir was lost. A man found the source and brought it home. Eight persons died after having received doses between 8 and 25 Sv [2-4]. In Sept. 1987 a 50.9 TBq ^{137}Cs therapy source was dismantled by junk dealers in Goiania in Brazil. Soil, vegetation, water, fish, sediments and people were contaminated [22, 23].

4.6 SATELLITES

In 1964 a US satellite (SNAP-9A) burned up in the stratosphere over the Indian Ocean. The accident dispersed 0.63 PBq ^{238}Pu in the atmosphere. Most of the activity was deposited in the Southern Hemisphere. This accident has been the main reason for the enhanced ^{238}Pu/^{239,240}Pu in global fallout. Atmospheric nuclear test explosions have injected only 0.33 PBq ^{238}Pu [24].

A Soviet satellite, Cosmos 954, powered by a nuclear reactor reentered the atmosphere over Canadian NW Territories in January 1978. Radioactive debris were dispersed over a 1000 km path. The core contained approximately 3 TBq ^{90}Sr, 0.2 PBq ^{131}I and 3 TBq ^{137}Cs. About 75% of the material remained in the stratosphere and have thus been globally distributed [25].

5 Conclusion

UNSCEAR's 1993 report [4] has summarized the inventory of nuclear releases in the world and provides calculated collective effective doses from these releases (Table 5). It is evident that atmospheric nuclear weapons testing is the main contributor to the dose with ^{14}C being the most important radionuclide. The Chernobyl accident comes next; in this case it is ^{137}Cs which delivers most of the dose.

The total releases of the two radioecologically important, long lived fissions products ^{90}Sr and ^{137}Cs have been 616 and 1020 PBq respectively.

TABLE 5. Estimates of radionuclide releases and collective effective dose from man-made environmental sources of radiation [4].

Source	Release (PBq)						Collective effective doses[a] (man Sv)	
	^{3}H	^{14}C	Noble gases	^{90}Sr	^{131}I	^{137}Cs	Local and regional	Global
Atmospheric nuclear testing								
Global	240,000	220		604	650,000	910		22,300,000
Local								
Semipalatinsk							4,600	
Nevada							500[b]	
Australia							700	
Pacific test site							160[b]	
Underground nuclear testing			50		15		200	
Nuclear weapons fabrication								
Early practice							8,000[c]	
Hanford							15,000[d]	
Chelyabinsk							1,000	10,000
Later practice							30,000[e]	
Nuclear power production								
Milling and mining							2,700	
Reactor operation	140	1.1	3,200		0.4		3,700	
Fuel reprocessing	57	0.3	1,200	6.9	0.004	40	4,600	
Fuel cycle							300,000[e]	100,000
Radioisotope production and use	2.6	1.0	52		6.0		2,000	80,000
Accidents								
Three Mile Island			370		0.0006		40	
Chernobyl					630	70		600,000
Kyshtym				5.4		0.04	2,500	
Windscale			1.2		0.7	0.02	2,000	
Palomares							3	
Thule							0	
SNAP 9A								2,100
Cosmos 954				0.003	0.2	0.003		20
Ciudad Juarez							150	
Mohammedia							80	
Goiania						0.05	60	
Total							380,000	231,000,000
Total collective effective dose (man Sv)								**23,500,000**

[a] Truncated at 10,000 years.
[b] External dose only.
[c] From release of ^{131}I to the atmosphere.
[d] From releases of radionuclides into the Techa river.
[e] Long-term collective dose from release of ^{222}Rn from tailings.

42

6. References

1 Kudo, A., Mahara, Y., Santry, D.C., Miyahara, S. & Garrec, J-P.(1991) Geographical distribution of actionated local fall-out from the Nagasaki A-bomb. *Journ. of Environmental Radioactivity* **14**, 305-316 .

2. UNSCEAR (1982), Reports on ionizing radiation to General Assembly. United Nations, New York.

3. UNSCEAR (1988), Reports on ionizing radiation to General Assembly. United Nations, New York.

4. UNSCEAR (1993), Reports on ionizing radiation to General Assembly. United Nations, New York.

5. Bjurman, B., DeGeer, L-E., Vintersved, I, Rudjord, A.L., Ugletveit, F., Aaltonen, H., Sinkko, K., Rantavaara, A., Nielsen, S.P., Aarkrog, A. and Kolb, W.(1990) The detection of radioactive material from a venting underground nuclear explosion. *J. Environ Radioactivity* **II**, 1-14.

6. BNFL(1978-1991) Annual Reports on Radioactive discharges and monitoring of the environment. British Nuclear Fuels. PLC, Risley, Warrington, Cheshire, UK.

7. Sivnitsey, Y. (1993) Study of Nuclides Composition and Characteristics of Fuel in dumped Submarine Reactors and Atomic Icebreaker, *Lenin* Part 1. - Russian Federation, Moscow, December.

8. Yefimov, E.I., Gromov, B.F., Pankratov, D.V.(1994) Radionuclides composition of activity in NPP with liquid metal coolant dumped in the Kara Sea and some assessment of radiological consequences of this activity release. Institute of Physics and Power Engineering, Obninsk, Russia. *Presentation at the IAEA Advisory Group Meeting on Assessment of the Consequences of the Dumping Radioactive Waste into the Arctic Seas.*, Vienna, 16-19 May, 1994.

9. Petrov, O.J.(1993) Facts and Problems Related to Radioactive Waste Disposal in Seas Adjacent to the Territory of the Russian Federation, *Materials for a Report by the Government Commission on Matters Related to Radioactive Waste Disposal at Sea, Created by Decree No. 613 of the Russian Federation President, October 24, 1992*, Small World Publishers, Moscow.

10. Nikipelov, B.V., Nikiforov, A.S., Kedrovsky, O.L., Strakkov, M.V., Drozhko, E.G.; (1990) Practical rehabilitation of territories contaminated as a result of implementation of nuclear material production defence programmes, *draft chapter for book to be published by VNIPI prom-technologii, Moscow)* .

11. WHO (1989) Health hazards from radiocesium following the Chernobyl nuclear accident. *J. Environ. Radioactivity* **10**, 257-295 .

12. Aarkrog, A. (1988) The Radiological impact of the Chernobyl debris compared with that from nuclear weapons fall-out, **6**, 151-162 .

13. Jackson, D. and Jones S.R.(1990) Reappraisal of Environmental Countermeasures to Protect Members of the Public Following the Windscale Nuclear Reactor Accident, 1957 EUR 13574, 1015-1055, CEC, Luxembourg.

14. Nikipelov, B.V. (1989) Experience in managing the radiological and radioecological consequences of accidental release of radioactivity which occurred in Southern Urals in 1957. IAEA SM 316/55.

15. Trabalka, J.R. and Auerbach, S.I (1990) A Western perspective of the 1957 Soviet nuclear accident. Paper presented at CEC-IUR seminar on comparative assessment of the environmental impact of radionuclides released during three major nuclear accidents: Kyshtym, Windscale and Chernobyl, Luxembourg 1-5 Oct., 1990.

16. Ternovskij, I.A., Romanov, G.N., Federov, E.A., Teverovskij, E.N. and Kolina, Yu.B(1989) Radioactive cloud trace formation dynamics after the radiation accident in the South Urals in 1957: migration processes, IAEA SM 316/55.3.

17. Cochran, T.B. and Norris, R.S.(1990) Soviet nuclear warhead production. Natural Resources Defence Council, Washington, DC 20005, NWD 90-3.

18. Ølgaard, P.L.(1993) Nuclear ship accidents, description and analysis. pp 29. Technical University of Denmark, Department of Electrophysics.

19. Joint Russian-Norwegian Expert Group. Radioactive contamination at dumping sites from nuclear waste in the Kara Sea.(1994) Results from the 1993 expedition). NRPA, Østerås, Norway.

20. Wrenn, M.E.;(1974) Environmental levels of plutonium and transplutonium elements, WASH-1359, 89 .

21. Aarkrog, A., Dahlgaard, H., Nilsson, K. and Holm, E. (1984) Further studies of plutonium and americium at Thule, Greenland, *Health Physics* **46**, 29-44 .

22. Amaral, E.C.S., Vienna, M.E.C., Godoy, J.M., Rochedo, E.R.R., Campos, M.J., Pires do Riso, M.A., Oliveira, J.P., Pereira, J.C.A. and Reis, W.G.(1991) Distribution of Cs-137 in soils due to the Goiania accident and decisions for remedial action during recovery phase, *Health Physics* **60**, 91-98.

23. Godoy, J.M., Guimarâes, J.R.D., Pereira, J.C.A. and Pires do Rio, M.A. (1987) Cesium-137 in the Goiania waterways during and after the radiological accident, *Health Physics* 60, 99-103. IAEA, Fusion Reactor Safety. IAEA TEC DOC-440, 1987.
24. Krey, P.W. Atmospheric burnup of a plutonium-238 generator.(1967) *Science* 158, 769-771 .
25. Tracy, B.L., Prantl, F.A. and Quinn, J.M.(1984) Health impact of radioactive debris from the satellite Cosmos 954, *Health Physics* 47, 225-233.

ECOLOGICAL CONSEQUENCES OF THE ACTIVITIES AT THE "MAYAK" PLANT

G.N. ROMANOV AND Ye.G. DROZHKO
"Mayak" Experimental Research Station
CIS-454060 Chelyabinsk-60, Russia

1. Introduction

The first nuclear complex for the manufacture of military plutonium in the former Soviet Union, now operating under the name "Mayak", was commissioned in 1948. The enterprise is located in the north of the Chelyabinsk region near the town of Kyshtym in the upper reaches of the river Techa, which has a catchment area covering the vast Kasli-Irtyash system of mountain lakes (Figure 1). Over more than 40 years of operation of "Mayak" have created a number of radioecological problems arising from
   1) the former radioactive contamination of a large area around its site;
   2) the accumulation of radioactive waste on the surface of the plant site, which presents a danger of further radioactive contamination to the region.

The immediate causes of these ecological problems were
   - the imperfection of the plant's nuclear technology in the early stages of operation;
   - the absence of an effective solution for handling radioactive waste produced by the nuclear industry;
   - a lack of knowledge of the behaviour of radioactive substances in the environment and of the health consequences of radioactive contamination in the 1940s-1950s.

2. Radioactive contamination of the environment

Radioactive contamination of the environment around "Mayak" took place mainly in the first half of the period under consideration. The main sources of contamination were
   - the release of liquid radioactive effluents into the river Techa from 1949 to 1956;
   - the release of radioactive substances into the atmosphere as a result of an accident in 1957;
   - routine releases of radioactive substances into the atmosphere.
The characteristics of the main sources of radioactive contamination are given in Table 1.

F. F. Luykx and M. J. Frissel (eds.), Radioecology and the Restoration of
Radioactive-Contaminated Sites, 45–55.
© *1996 Kluwer Academic Publishers.*

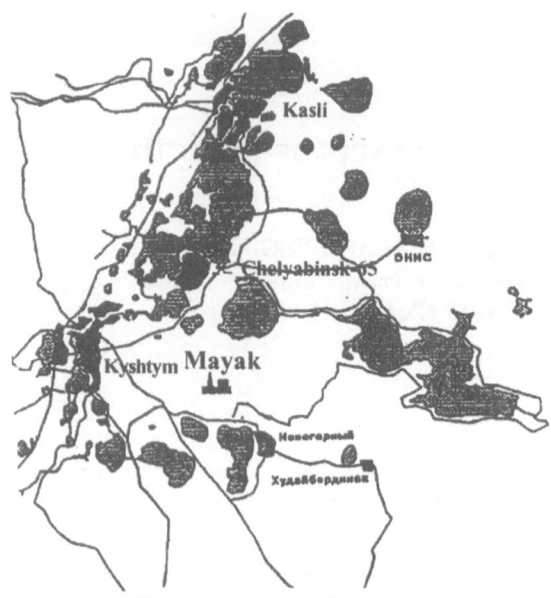

Figure 1. Location of "Mayak"

The specific feature of the radiological situation around "Mayak" today is the presence of long-lived ^{90}Sr and ^{137}Cs, which accounts for the long-term character of radioactive contamination of the environment and human food chains.

2.1 EFFLUENTS INTO THE RIVER TECHA

In the period from 1949 to 1956, the plant released about 76 million m^3 of liquid waste into the upper reaches of the river Techa. This waste, predominantly medium-active, had a total activity amounting to 100 PBq (2.75 MCi) (Table 1), of which 6 PBq ^{90}Sr (160,000 Ci) and 12 PBq ^{137}Cs (335,000 Ci) . The main part of the effluents was released in 1950-1951. Releases into the river were stopped with the construction of the first stage of a system of storage ponds for low-active liquid waste in the upper reaches of the river (1956).

The releases caused long-term radioactive contamination of the river system Techa-Isset-Tobol-Irtysh-Ob along its entire length, with radioactive effluents reaching the Arctic Ocean (the stretch of the effluence has not yet been determined). The highest levels of radioactive contamination were observed in the upper and middle reaches of the river Techa. About 28,000 people were evacuated from settlements located in this part of the area. Restrictions on the use of the river and its contaminated floodplain cover the entire length of the river Techa over an area of 80 km^2.

At present, the concentration of ^{90}Sr in the water of the river Techa exceeds the admissible concentrations, as set by the Russian Standards of Radiological Safety, in the first 40 km only; however, the concentration of ^{90}Sr and ^{137}Cs in the floodplain soils is so high that it precludes uncontrolled use of the land and does not ensure the safety of the population. The ^{90}Sr and ^{137}Cs contamination amounts to several TBq/m^2 in the bogged upper reaches of the river and up to 1 TBq/km^2 near the estuary. The bogged upper reaches of the river are estimated to contain about 40 TBq of ^{90}Sr and 200 TBq of ^{137}Cs. This part of the river is a permanent secondary source of radionuclides (primarily ^{90}Sr), which is washed off by runoff, particularly during floods.

The levels of contamination in the river Isset are about one order of magnitude lower than in the lower reaches of Techa, and another order of magnitude lower in the river Tobol.

2.2 THE 1957 ACCIDENT

The explosion of a tank with radioactive waste on 29 September 1957 resulted in radioactive contamination of territories in the Chelyabinsk, Sverdlovsk and Tyumen regions ("The Kyshtym Accident"). As a result, around 70 PBq (Table 1) were deposited in the area of the so called East-Ural Radioactive Trail (Figure 2), in which ^{90}Sr accounted for 2 PBq . The area of the trail within the boundaries set by the lower limit of detection of 3.7 GBq ^{90}Sr/km^2 amounted to about 20,000 km^2; the area within the boundaries set by the level of contamination of 74 GBq ^{90}Sr/km^2, adopted as the admissible safety limit for the population, is about 1,000 km^2.

The distribution of the levels of radioactive contamination over the territory of the East-Ural Radioactive Trail is satisfactorily approximated by the common models for a single atmospheric release from a terrestrial point source.

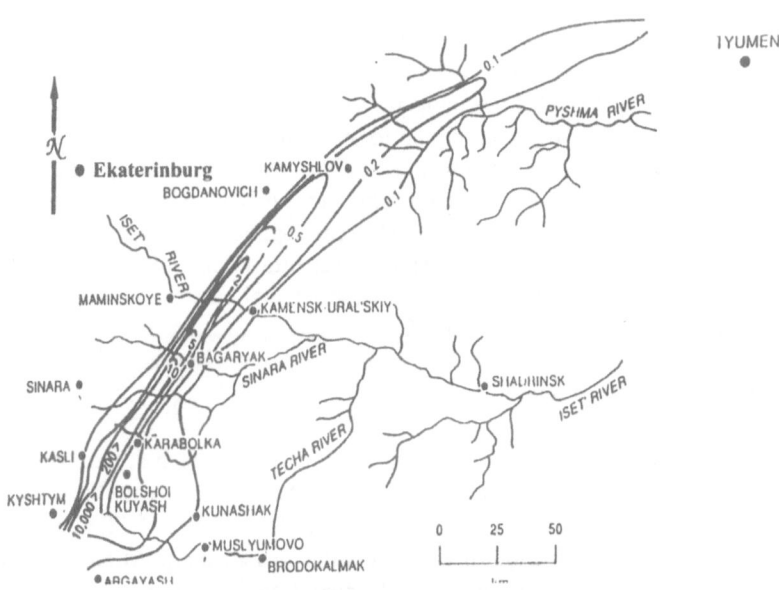

Figure 2. East-Ural radioactive trail
Decimal figures: levels of contamination by ^{90}Sr, Ci/km^2 (37 GBq/km^2)

TABLE 1. Sources and Consequences of Radioactive Contamination

Parameter	Discharges into river Techa (1949-1956)	Release due to the 1957 accident	Resuspension (1967)
Total Beta, PBq	2.75×10^6	2×10^6	600
Composition, % of active mixture			
^{89}Sr	8.8	trace amounts	-
^{90}Sr	5.8	2.7	17
^{95}Zr	6.8	12.5	-
^{106}Ru	13	1.8	-
^{137}Cs	12.2	0.036	48
^{144}Ce	-	33	9
^{33}P	26.8	-	-
Pu	trace amounts	trace amounts	
Surface of Contaminated area (S=area, km²; L= length, km)	Flood-plain L=1,000 S=2,000 Restricted zone 80 km²	> 74GBq /km2 ^{90}Sr: S = 1,000; >3.7GBq /km² ^{90}Sr: S = 20,000 Restricted zone 1,000 km²	> 74GBq /km² ^{90}Sr: S = 30; >3.7GBq/km² ^{90}Sr:S = 1,800
Exposed population	124,000	300,000	40,000

As a result, the trail has a clear axial maximum of contamination levels which decreases linearly with the from the sourcedistance and rapidly in the lateral directions. The maximum levels along the central axis amounted to 70 to 150 TBq ^{90}Sr per km².

The presence of long-lived ^{90}Sr accounts for the persistence of the trail and its permanent presence in the environment and human food chains. Restrictions on economic use of the contaminated areas above the level of 74 GBq ^{90}Sr/km², imposed after the accident, were partly lifted in 1961. By 1982, farming and forest engineering activities had been resumed in 82 % of the restricted area. This corresponds to maximum levels of initial contamination of 3.7 TBq ^{90}Sr/km². The central (most contaminated) part of the trail, which covers an area of 19 km², is now an experimental base for studying the long-term environmental behaviour of ^{90}Sr and ^{137}Cs.

2.3 RESUSPENSION IN 1967

In the spring of 1967 a draught exposed the bed of Lake Karachai, into which medium-active liquid wastes had been released since 1951. Within two weeks the wind dispersed about 22 TBq over the surrounding territory (Table 1), of which 4 TBq ^{90}Sr and 11 TBq ^{137}Cs. The territorial distribution of contamination was a superposition of several single fall-outs, which resulted in a complicated general configuration. This contamination covered the areas of the East-Ural Radioactive Trail and territories south-east and south of the plant.

The area of contamination within the boundaries delineated by the level of contamination of 3.7 GBq ^{90}Sr/km^2 amounted to about 1,800 km^2; however, the levels of contamination were relatively low as compared to the 1957 accident. The total area within the boundaries of 74 GBq ^{90}Sr/km^2 was 15-17 km long and 2-3 km wide.

2.4 ATMOSPHERIC RELEASES

Routine stack releases of radioactive substances into the atmosphere were maximal during the first 10-15 years of operation when the plant was increasing its production while the gas-cleaning facilities were still insufficient. The contribution of these releases to the radioactive contamination of the territory was barely detectable compared to that from the accidents. Subsequent estimates showed that the integral contribution of atmospheric releases to the amounts of ^{90}Sr and ^{137}Cs in the monitoring zone of the plant did not exceed several percent of the total amounts from all other sources, mainly accidents.

2.5 CURRENT RADIOLOGICAL SITUATION

The current radiological situation in the plant's monitoring zone is characterised by the indicators given in Table 2. Apart from the East-Ural Radioactive Trail with the initial contamination of 2-4 TBq ^{90}Sr km^2 and the flood-plain of the river Techa, all the characteristics of the radiological situation meet the requirements of radiological safety for the population.

TABLE 2.Current Radiological Situation Around the"Mayak" Plant
(R=70 km)

Indicator	Limit values
Concentration in air, μBq/m^3	
Total Beta	330-2600
^{90}Sr	11-260
Total alpha	0.3-2.6
Intensity of fallout ,	
GBq /km^2 year	0.7-1.8
Dose rate from gamma-radiation, μR/h.	10-18
Contamination density	
(except rivers Techa and EURT),	
GBq/km	
^{90}Sr	4-19
^{137}Cs	4-74
Concentration in farm produce,	
Bq/kg	
- milk: ^{90}Sr	0.6-7.8
^{137}Cs	2.8-18.5
- potatoes: ^{90}Sr	0.2-3.7
^{137}Cs	0.3-5.2
Effective dose to the population,	
μSv/year	120 – 240

3. Radioactive effluents from the "Mayak" plant

In the course of its operation "Mayak" has accumulated, and stored under various conditions, liquid and solid long-lived radioactive wastes of various levels of activity. The total activity of the wastes amounts to 28,000 PBq (Table 3), of which 226,000 PBq are liquid wastes. Strictly controlled technological storage of the high-active and part of the medium-active wastes in special repositories does not create any ecological problem; however, the use of the system of natural and artificial surface reservoirs (Figure 3) as long-term storage for medium- and low-active waste presents an important ecological problem.

The characteristics of the storage ponds for liquid wastes are presented in Table 4. As can be seen from this table, the 'famous' lake Karachai stores 4,400 PBq of long-lived activity, while the cascade of ponds No. 3, 4, 10, 11, storing 6.5 PBq of activity, contain 311 million m³ of contaminated water. All of these surface reservoirs present a clear potential danger of radioactive contamination to the environment.

3.1. LAKE KARACHAI

Initially Lake Karachai presented a natural water body with a surface area ,depending on moistening conditions, of 26-51 ha and an average depth of 1.5 m. Since 1951 it has served as a place of discharge for medium-active high-salt liquid wastes. At present it contains a total activity of 4,500 PBq (40 % ^{90}Sr + ^{90}Y; 60 % ^{137}Cs) of which 7 % are contained in the water, 52 % in the loose bottom sediments and 41 % in the loamy bedrock.

The concentration of ^{90}Sr in water amounts to 44 MBq /l , 6.3- 126 GBq /kg in the bottom sediments and 15- 520 MBq /kg in the bedrock. The respective values for ^{137}Cs are 0.5 GBq /kg, 26- 150 GBq /kg; 0.4- 4 GBq/kg,

^{239}Pu + ^{241}Am concentrations amount to respectively: 120 kBq/kg, 10-63 MBq/kg , 7-300 kBq/kg.

During the use of this lake about 3.5 million m³ of contaminated water have filtered down through the bedrock, which has led to the formation of a lens of contaminated water covering an area of about 10 km². Observations show that the front of the lens is moving horizontally at a rate of up to 80 m/year towards the south, which creates a danger that the contaminated water could penetrate into the underground water supply reservoirs and break through into the surface water bodies.

3.2 CASCADE OF RESERVOIRS

A cascade of reservoirs, No. 3, 4, 10, 11, was created in the upper reaches of the river Techa, which earlier had been flowing from Lake Kyzyltash (reservoir No. 2). The dams and dikes of the diversion channels (constructed to drain water from the Kasli-Itryash lakes and from the lake Ulagach employed to remove ash from the local fossil-fuel power station) are mainly from rock, which does not exclude the penetration of contaminated water into the river Techa at the tail race of closing dam No. 11. Lately the annual

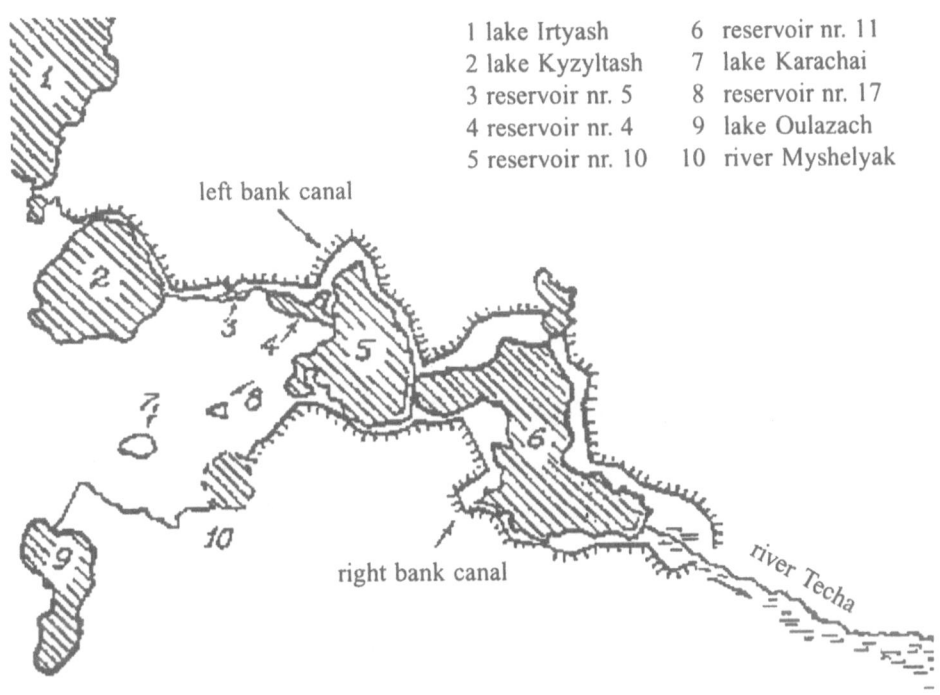

1 lake Irtyash   6 reservoir nr. 11
2 lake Kyzyltash   7 lake Karachai
3 reservoir nr. 5   8 reservoir nr. 17
4 reservoir nr. 4   9 lake Oulazach
5 reservoir nr. 10   10 river Myshelyak

left bank canal

right bank canal

river Techa

Figure 3. Scheme of hydrotechnical structures at "Mayak"

inflow of ^{90}Sr into the river Techa through the hydrotechnical structures has been around 0.56-1.5 TBq, of which up to 50-80 % by infiltration of water through the body of dam No. 11 (up to 10 mln m^3 per year). Such infiltration losses are determined by the hydrostatic pressure in reservoirs No. 10 and 11.

TABLE 3. Radioactive Wastes at "Mayak" (1994)

Class of waste	Volume (liquid wastes) Mass (solid wastes)	Total activity, PBq	Conditions of storage
	Liquid wastes,m^3		
High-active	29×10^3	18,100	Technological storage
			Technological storage
Medium-active	0.6×10^3	0.2	Reservoirs
	3.3×10^6	4,400	Reservoirs
Low-active	410×10^6	7.4	
	Solid wastes, t		
High-active:			
- vitrified	0.47×10^3 (m^3)	5,200	Technological storage
- untreated	25×10^3	480	Underground repositories
Medium- and low-active untreated	400×10^3	1.5	Underground repositories

TABLE 4. Storage Reservoirs for Medium- and Low-Active Liquid Wastes of "Mayak"

Reservoir	Area, km^2	Volume of water, $10^6\ m^3$	Total activity,
		Medium-active	PBq
Nr 9 (Lake Karachai)	0.15 *	0.3	4400
Nr 17	0.17	0.3	74
		Low-active	TBq
Nr 2 (Lake Kyzyltash)	119	83	810
Nr 3	0.5	0.75	630
Nr 4	1.3	4.1	250
Nr 10	19	76	4,100
Nr 11	44	230	1,600
Nr 6 (Lake Tatysh)	3.6	17.5	11

* Current surface area reduced from 0.45 to 0.15 km^2 as a result of liquidation work

4. Ecological problems

The ecological problems due to the operations in the "Mayak" plant are treated in two ways.

The first covers problems related to the rehabilitation of the environment which has suffered from radioactive contamination.

The second deals with the reduction of the risk of nuclear accidents associated with the radioactive wastes in surface storage which can lead to extensive radioactive contamination of the environment.

4.1 RADIOLOGICAL REHABILITATION OF TERRITORIES

A long-term programme has been developed and adopted for the radiological rehabilitation of the contaminated territories. It is aimed at the rehabilitation of the river Techa and its flood plain, the most contaminated areas of the plant's site and the underground repositories containing solid radioactive wastes as well as some areas within the East-Ural Radioactive Trail.

The rehabilitation programme for the river Techa provides for:
1. A considerable reduction (and termination the future) of the discharge of water through the diversion channel from the Kasli-Irtyash system of lakes with the purpose of drying the most contaminated upper reaches of the river and preventing radionuclides from penetrating into the river.
2. Reduction in the penetration of radionuclides with infiltrating water through the hydrotechnical structures by lowering the level of water in reservoirs Nrs 10 and 11.
3. Development of radioecological safety criteria for living and economic land use in the flood plain of the river Techa.
4. Development of methods and technologies for the decontamination and ecological rehabilitation of the river.
5. Practical work on radiological rehabilitation of the river and its flood plain in the upper and middle parts, primarily in the populated areas.

Radiological rehabilitation of the plant's site includes a set of measures aimed at its decontamination and ecological remediation.

The territory of the East-Ural Radioactive Trail, in principle, does not need radiological rehabilitation (if it is to be understood as a set of decontamination measures) since the current levels of activity satisfy the standards of radiological safety for the population. The programme, however, provides for further regulation of farming and forest use in the contaminated areas, primarily through optimisation of land use, which will permit the exposure of the population to be reduced further.

4.2. REDUCING THE RISK OF NUCLEAR ACCIDENTS

The main sources of the risk for nuclear accidents associated with the concentration of radioactive wastes at "Mayak" are Lake Karachai, the lens of contaminated water under it and the surface storage ponds for low-active wastes. Nuclear accidents can occur with a certain probability by a tornado; its consequences could be particularly unfavourable if it should pass over Lake Karachai. Another possible cause is a discharge of contaminated water from the cascade of industrial reservoirs into the river Techa due to spilling or destruction of the dams and dikes by an earthquake, failure of the structures, sabotage or hostilities.

The research and engineering programme for the reduction of the risk of nuclear accidents at "Mayak" pursues the following objectives.:
1. To reduce the existing and currently produced amounts of radioactive wastes of all types and levels of activity. This is to be achieved by developing and intro-

54

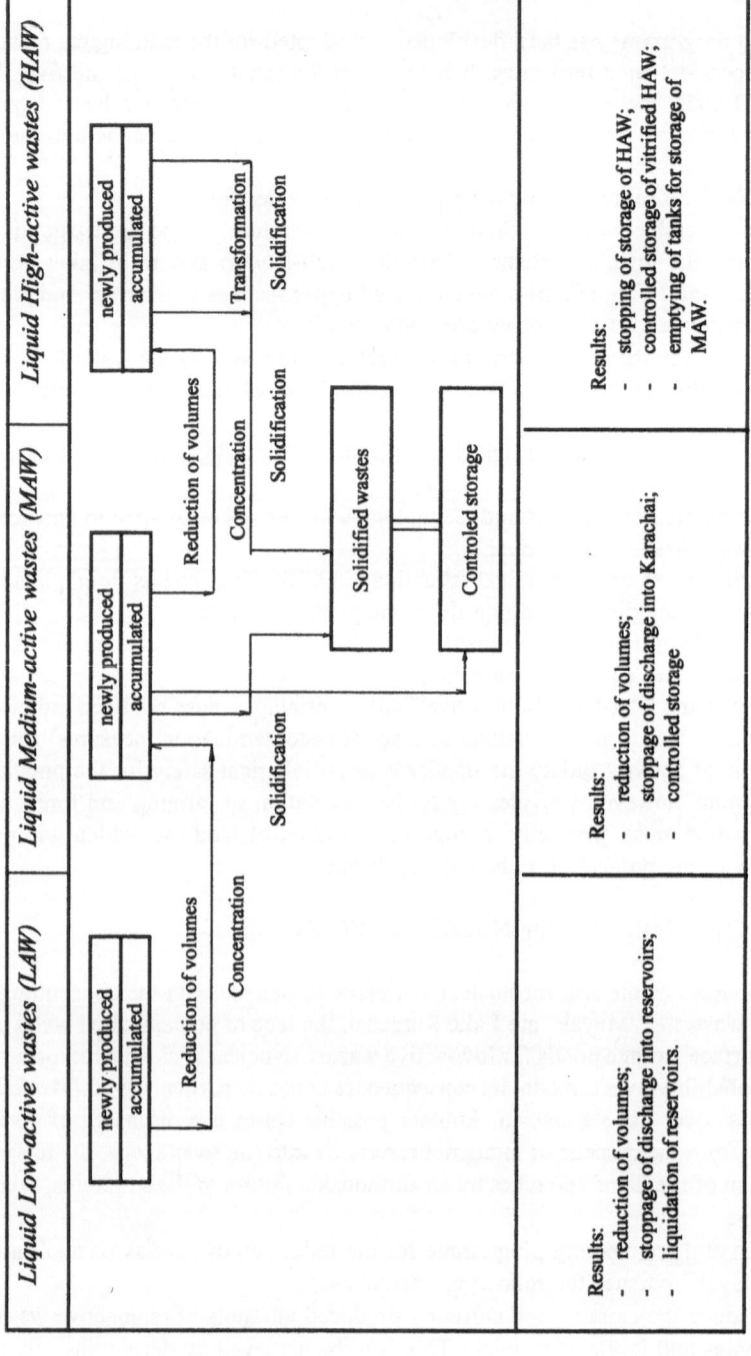

Figure 4. Closed-circuit scheme for liquid radioactive waste

ducing new low-waste treatment technologies into the basic production as well as new methods for radioactive waste disposal. The global objective of the latter is to concentrate liquid wastes with subsequent vitrification for safe storage (Figure 4). By mid-1994 "Mayak" had vitrified 30 % of the earlier accumulated waste and all the newly produced high-active liquid wastes, having obtained as a result of this 1506 t of glass, which is stored in 3148 containers.

2. To reduce the discharge of low-active liquid wastes into the cascade of surface reservoirs and to stop the discharge of medium-active wastes into Lake Karachai (by implementing the measures under item 1).

3. To liquidate Lake Karachai. Liquidation work has been carried on since 1986. As a result, the area of the lake has decreased from 45 ha to 15 ha. The area of the lake is filled with rock after sinking hollow concrete blocks on the bottom in order to fix the loose bottom sediments. Liquidation will be finished after stopping the discharge of medium-active wastes.

4. To stop the lens of contaminated underground waters under Lake Karachai by means of hydrogeological structures.

5. To stabilise and then lower the level of water in the cascade of industrial reservoirs by reducing the volume of discharge of low-active wastes and changing the hydrological regimes of the reservoirs. This objective could be reached faster by using reservoir No. 10 as a cooling reservoir of the Southern-Ural Nuclear Power Station planned for construction.

6. To improve the technical reliability of the hydrotechnical structures.

The above objectives, of course, do not exhaust all possible solutions to reduce the risk of nuclear accidents. An ideal variant would be liquidation of the artificial reservoirs and decontamination of the natural waste storage reservoirs, but this solution belongs to the far future.

CURRENT RADIOECOLOGICAL CHARACTERISTICS OF THE SITUATION IN THE AREA OF THE EAST-URAL RADIOACTIVE TRAIL

G.N. ROMANOV
"Mayak" Experimental Research Station
CIS-454060 Chelyabinsk-60, Russia

1. Introduction

The East-Ural Radioactive Trail (EURT) was formed as a result of a nuclear accident at the "Mayak" plant on 29 September 1957 when malfunctions in the cooling system of a tank with medium-active waste caused a chemical explosion of a mixture of nitrate-acetate salts and dispersal of radioactive matter in the atmosphere. An area of about 20,000 km² in the Chelyabinsk, Sverdlovsk and Tyumen regions was contaminated with 74 PBq of activity including 2 PBq of ^{90}Sr (2.7 % of the activity of the radionuclide mixture), the predominant part being represented by ^{144}Ce, ^{95}Zr and ^{106}Ru. The explosion fallout products contained also ^{137}Cs, the activity of which was only 1.3% of that of ^{90}Sr.

The geographical distribution of the contamination levels is sufficiently well approximated by a model with a single fallout from a ground point source; the trail has a clearly marked axis with monotonically exponentially decreasing contamination with distance and rapidly falling contamination crosswise. The maximum initial levels of contamination along the axis of the trail reached 150 TBq ^{90}Sr/km² ; the minimum level of contamination in the axial and lateral peripheral parts, equivalent to the detection limit for ^{90}Sr in the 1950s, was 3.7 GBq ^{90}Sr/km².

In this report, the term "East-Ural Radioactive Trail" (EURT) means the contaminated area delineated by an initial contamination level of 74 GBq (2 Ci) ^{90}Sr/km², which was considered safe for the population. The territory under consideration covers an area of about 1,000 km², measuring 105 km x 8 to 9 km; its main part is located within the Chelyabinsk region.

The initial spatial macro-distribution of contamination has not changed with time, because the natural lateral migration of the radionuclides was not very intensive.

Today, 38 years after the accident, the distribution, behaviour and migration of radionuclides in the natural environment and biological systems are determined by steady-state processes. All the short-lived radionuclides have decayed and the migration of ^{90}Sr and ^{137}Cs has reached an equilibrium; the same applies to the biogeochemical processes.

F. F. Luykx and M. J. Frissel (eds.), Radioecology and the Restoration of
Radioactive-Contaminated Sites, 57–67.
© 1996 *Kluwer Academic Publishers.*

2. Physiographic characteristics of the region

EURT is located in the Transural plain of Russia. The terrain is smooth, slightly hilly. The hydrographic system is represented by several small rivers; the runoff terminates in small lakes and bogs located in depressions.

The climate of the region is continental. The average monthly temperatures are -19°C for January and +18°C for July. The average annual amount of precipitation is 350-400 mm. The region is located in a zone of insufficient humidification characterized by draughts. This results in a relatively high evapotranspiration and absence of contact between atmospheric precipitation and ground waters.

The bedrock is represented by clays and loams. The typical soils are leached chernozem and grey forest soil with a sufficiently large sorption capacity. The exchangeable Ca content of the soils is 15-25 meq/100 g of soil.

The basic part of the region is located in a forest steppe zone with typical vegetation and fauna. Forests constitute 30-40 % of the total area, the typical species being birch with occasional pine. Agroecosystems cover 40-50 % of the total area.

3. Behaviour of ^{90}Sr and ^{137}Cs in soil

The long-term behaviour of radionuclides in the top soil is determined by the following basic factors:
1. the chemical characteristics of the radionuclides and their physical and chemical form in the soil,
2. the agrochemical characteristics of the soil, primarily those which determine the intensity of sorption-desorption processes,
3. the soil moistening conditions,
4. the physical processes of vertical (within the soil) and lateral (on the surface) transfer,
5. the biological processes involving plant and animal organisms,
6. anthropogenic activities.

Acting jointly, these factors reduce the radionuclide content of the top soil (if it is not disturbed, for instance, by ploughing), thus reducing resuspension and runoff and modifying the intensity of uptake of radionuclides by biological systems.

The ratio of ^{90}Sr and ^{137}Cs forms in the soil has been stable over the last 20-30 years (Table 1).

The exchangeable forms of ^{90}Sr make up a large fraction (about 80 %), which accounts for the high bio-geochemical mobility of ^{90}Sr. ^{137}Cs is less mobile, the fraction of its mobile forms being about 15 %.

TABLE 1.Forms of ^{90}Sr and ^{137}Cs in EURT soils (%)

Forms	^{90}Sr		^{137}Cs	
	Leached chernozem	Grey forest soil	Leached chernozem	Grey forest soil
Water soluble	1.4	2.2	0.12	0.14
Exchangeable	82	82	12	19
Non-exchangeable	10	16	88	79

TABLE 2. Rate of natural decontamination of top soil(0-20 cm) with intact structure

Radionuclide	$T_{1/2}$(year)	λ(year^{-1})	vertical migration rate in 0-20 cm layer, cm/year
^{90}Sr	5.5-13	0.053-0.126	0.3-0.5
^{137}Cs	15-32	0.022-0.046	0.15-0.20

Figure 1 demonstrates the change of the distribution of ^{90}Sr in the profile of various soils with time. The differences in the velocity of vertical migration between various soils are determined primarily by differences in the sorption capacity: the sod-podzolic soils, which contain less exchangeable Ca than the chernozems, demonstrate deeper penetration.

We are convinced that the maximum estimated penetration depth for ^{90}Sr depends on the average depth of the penetration of atmospheric precipitation into the soil, which is the main mechanism of vertical convective migration.

Estimates of the self-cleaning characteristics (natural decontamination) of the top soil from ^{90}Sr through vertical migration are given in Table 2. The effective decontamination half-life is estimated at about 8 years.

4. Lateral transfer of ^{90}Sr and ^{137}Cs in the top soil surface

The degree to which radionuclides are involved in horizontal migration along the soil surface (surface and subsurface runoff, resuspension) depends on the intensity of natural physical processes and the radionuclide contents of the top soil. These processes involve soil particles in case of resuspension and soluble forms of radionuclides and soil particles in case of runoff. Thus, the rate of natural decontamination of the top soil is one of the main factors varying the rate of horizontal migration. Let us consider some concrete values.

Table 3 shows that the transfer of radionuclides with surface runoff over the period under consideration is reduced by a factor of 5 to 150. Resuspension (Table 4) is also reduced by 2 to 3 orders of magnitude. As can be seen from this table, the stabilisation of horizontal migration at a rate considerably lower than the initial rate is in agreement with the decrease in contamination of the top soil.

60

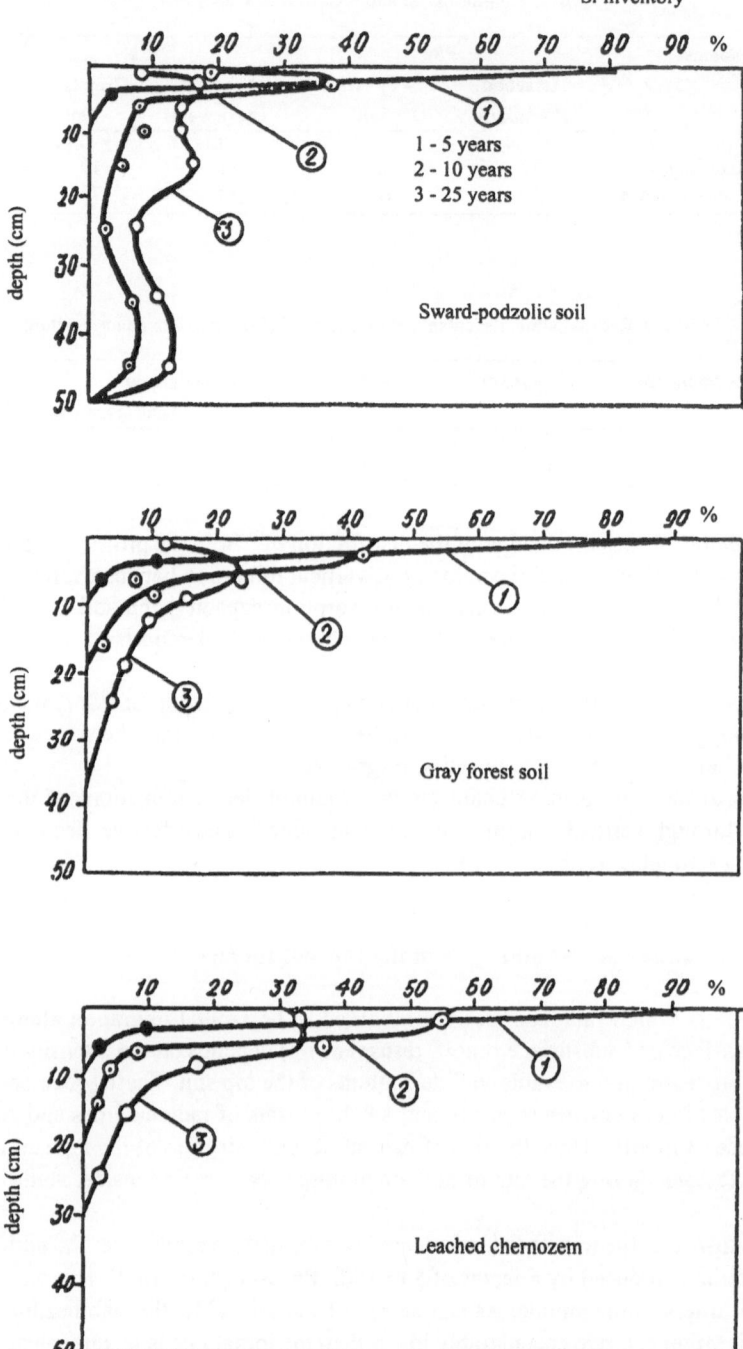

Figure 1. Evolution of ^{90}Sr distribution in soil

TABLE 3.Annual surface runoff*
(% of initial amount in catchment area)

Period	^{90}Sr	^{137}Cs
First 5-10 years	0.2-0.3	1.10^{-4}
Last 20 years	0.002-0.05	6.10^{-5}

*) Fraction of solid runoff: ^{90}Sr : 65 %, ^{137}Cs : > 90 %

TABLE 4. Resuspension of radionuclides

Period	resuspension factor $\frac{Bq/m^3}{Bq/m^2}$	resuspension rate $\frac{Bq/m^2 s}{Bq/m^2}$	annual removal of radioactivity % of initial amount in area
First 5 years	$10^{-7} - 10^{-6}$	$10^{-9} - 10^{-7}$	10^{-1}
Last 20 years	10^{-9}	$10^{-11} - 10^{-9}$	$10^{-3} - 10^{-2}$

In conclusion, resuspension has extended the initial macro-distribution of radioactive contamination towards the eastern peripheral part of the trail. It went on for the first two years and led to the expansion of the trail (4 GBq ^{90}Sr/km^2) by about 50 km^2.
Subsequently, neither of the processes of lateral migration led to any noticeable decrease of the amount of radioactive substances within the trail, nor did they change the established macro-distribution of contamination.

5. Relation between radionuclide content of soil and vegetation

Since this report deals with current steady-state processes, let us consider the relation between the radionuclide contents of vegetation and soil via the root pathway. (Incidentally, in the first two years after the accident, the non-root pathway accounted for 50-90 % of the resultant concentration of radionuclides in the vegetation. Its role became secondary or negligibly small after the first 5-10 years.)
The uptake of ^{90}Sr and ^{137}Cs by wild plants through the root system depends on a number of factors.
The first and most important factor is that the uptake of radionuclides by a specific plant species is directly proportional to the level of contamination, which provides evidence that the development and productivity of plants are not suppressed at contamination levels up to 40-80-120 TBq ^{90}Sr/km^2.
The second factor shows that the root uptake of Ca depends on the plants needs, and not on the amount of Ca in the soil. Sr behaves as Ca and is taken up in proportion to the amount of Ca. The main Ca source in the soil is the exchangeable Ca pool. The larger this pool, the lower the ratio Sr/Ca and thus the lower the uptake of Sr. In general, the uptake of Sr is inversely related to the amount of exchangeable Ca in the soil. The same reasoning applies to K and Cs.

For the same level of radioactive contamination, the uptake of Sr by beans is relatively high; this can be explained by the high Ca content of beans. The uptake is also high on sod-podzolic soils, because they show a low sorption capacity (i.e., low exchangeable Ca content).

The root uptake of ^{137}Cs is maximal for sod-podzolic sandy soils; an increase in the proportion of clay in the soil leads to a reduction in accessible Cs due to the inclusion of Cs atoms into the crystalline lattice of the clay minerals.

Table 5 presents average values of the accumulation coefficients (plant-to-soil concentration ratio) of ^{90}Sr and ^{137}Cs by wild plants typical for the soil conditions and plant species in the EURT. Soil is the main depot of radionuclides in any ecosystem because the deposition of radionuclides on biological systems, including the existing biomass, prevents any substantial decrease in the amount of radionuclides in the top soil.

Table 6 shows that in terrestrial natural ecosystems, about 90-99 % of the total ^{90}Sr or ^{137}Cs is contained in soil. The annual rate of inclusion of ^{90}Sr into the biological turnover does not exceed 0.1-0.5 % of its total content; for ^{137}Cs, this rate does not exceed 0.02-0.03 %.

TABLE 5. ^{90}Sr and ^{137}Cs accumulation factors in wild plants, m^2/kg

Type of vegetation	^{90}Sr	^{137}Cs
Herbs	0.2-2	0.03-0.5
Shrubs		
-leaves	0.4-0.5	0.1-0.2
-wood	0.04-0.05	0.03-0.04
Trees		
-leaves	1.5-2.5	0.01-0.06
-needles	1-1.5	0.005-0.01
-wood	0.2-0.4	0.001-0.005
Mushrooms	0.002-0.2	0.002-0.7

TABLE 6. Distribution of ^{90}Sr and ^{137}Cs in terrestrial natural ecosystems of EURT, in % of amount in ecosystem

Links of ecosystem	^{90}Sr		^{137}Cs	
	meadow	forest	meadow	forest
vegetation	0.1-0.3	6.7	0.05	0.5
sod, litter	0.3	6	0.05	0.4
mineral part of soil	99.4	87.3	99.9	99.1
annual involvement in biological turnover	0.1-0.3	0.5	0.02	0.03

Although the behaviour of radionuclides in fresh-water (lake) ecosystems is not considered in this report, we should nevertheless mention that, in these ecosystems, the main depot of radionuclides occurs in the bottom sediments (Table 7). Permanent chemical interaction between water and bottom sediments (the indicator being the distribution coefficient K_d), natural decontamination of the upper layer of the sediment due to downward migration of radionuclides as well as annual growth of the layer of bottom sediments with a decreasing concentration of radionuclides result in a continuous decrease in concentration of ^{90}Sr in lake waters. Our estimates show that the effective half-life of ^{90}Sr in lake waters and, hence, in fish and other biological components is about 8 years.

TABLE 7. Distribution of ^{90}Sr and ^{137}Cs in the
lake ecosystems of EURT, %

Links of ecosystem	^{90}Sr	^{137}Cs
Water	0.2	0.09
Vegetation	0.03	0.04
Plankton+benthos+fish	0.03	0.05
Bottom sediment	99.8	99.9

6. Uptake of radionuclides by agricultural plants

To estimate the root uptake of ^{90}Sr in agricultural plants, taking into account the needs of plants or their organs for Ca and the exchangeable Ca content of soil, Academician V.M. Klechkovsky and co-workers proposed the following formula:

CPI = the concentration of ^{90}Sr with respect to Ca in a plant:
    (level of contamination by ^{90}Sr relative to the exchangeable calcium content of soil)

$$CPI = \frac{(Bq\ ^{90}Sr/kg\ plant)\ /\ (g\ Ca/kg\ plant)}{(GBq\ ^{90}Sr/km^2)/(meq\ Ca/100\ g\ soil)}$$

Table 8 presents experimental CPI values for the first 10 years after the accident.

TABLE 8. Klechkovsky's Complex index
CPI, 10^{-9} km^2.meq Ca/ g Ca. 100 g soil
(first 10 years)

Produce	CPI
grain (cereals beans)	9
potatoes and vegetables	15
hay, silo (plough land)	14
hay of natural herbs	9

The attempt to derive an easily computable formula for the accumulation of ^{137}Cs in crops in different types of soil failed due to the effect of other interfering factors, primarily the absence of a clear dependence of root uptake of ^{137}Cs on the potassium content of plants and the absence of a clear relationship between the physicomechanical composition of soil and its exchangeable K content. Therefore, the root uptake of ^{137}Cs by crops grown on different soils can be characterized only by empirical data. Table 9, which contains data on the uptake of ^{90}Sr and ^{137}Cs by different plants on different soils, shows considerable differences in the accumulation of radionuclides in different produce.

TABLE 9. Proportionality coefficients of root uptake of radionuclides for crops
growing on soils of various types
(first ten years of radioactive contamination of the area),
Bq/kg/GBq/km2

| | Soil types | | |
Products	Sandy turf-podzol soil Ca=0.9 meq./100 g	Loamy-grey forest soil Ca=18 meq./100 g	Loamy heavy black-earth Ca=50 meq./100 g
	^{90}Sr		
Hay of wild herbs	170-670	8.5-33	3.0-12
Vegetative mass of field and vegetable plants	5.7-14	0.28-0.70	0.10-0.25
Grains (cereals)	3.1	0.16	0.056
Root-crops	2.0-4.0	0.10-0.20	0.04-0.07
Potatoes	2.2	0.11	0.04
Berries	0.5-1.5	0.03-0.08	0.009-0.03
	^{137}Cs		
Hay of wild herbs	13	1.9	0.68
Grain (cereals)	0.81	0.12	0.042
Leaf vegetable	1.2	0.16	0.07
Root-crops	0.3	0.04	0.01
Potato	0.3	0.05	0.02

Disregarding the effect of the agrochemical characteristics of soil, the highest uptake of ^{90}Sr and ^{137}Cs is observed in natural herbs and in the vegetative part of field and vegetable cultures. The most favourable soils, characterized by minimal levels of accumulation of radionuclides in produce, are those with a high sorption capacity (rich chernozem with an exchangeable Ca content of up to 50 meq/100 g). Sod-podzolic soils with a low sorption capacity show a maximum accumulation of radionuclides in crop. For the uptake of ^{137}Cs, the most unfavourable option is sod-podzolic soil, which contains a large amount of accessible exchangeable ^{137}Cs due to the relatively small proportion of clay minerals.

In the 1980s, the established numerical values for root uptake of ^{90}Sr by plant produce (Tables 8,9) were found to decrease with time. The latest estimates show that this phenomenon has been characteristic for the last 20 years and this has been proven for both pilot plots and farmland at the EURT periphery, where no limitations were imposed on agricultural production.

The effective half-life of ^{90}Sr amounts to 8-10 years for field crops and 15-20 years for natural herbs. An effective half-life of 8-10 years has also been found for ^{90}Sr in cow milk and in the diet of the local population.

As regards arable land, the decrease in ^{90}Sr uptake by crops at a rate of about 10 % per year is due for an equal extent to radioactive decay of ^{90}Sr, systematic removal of the crop from the field and biogeochemical processes. The first two factors can be confirmed by calculation; the last one can only be explained hypothetically, due to the absence of corresponding experimental investigations.

We put forward the following hypothetical sources for ^{90}Sr root uptake by crop, under the notion "biogeochemical processes".

1. The discrepancy between the probably time-dependent composition of ^{90}Sr exchangeable forms in soil solution and those replaced by ammonium acetate in laboratory conditions, which are permanent in time (the "ageing" of exchangeable forms in soil).
2. A reduction in biological accessibility of ^{90}Sr for root uptake from soil solution, within the root system zone of plants, due to
   a) reciprocal transformation of the cationic and anionic forms of ^{90}Sr;
   b) biochemical transformation of the dissolved forms of ^{90}Sr under the influence of soil microflora, resulting in an increase in the contribution of compounds of a greater molecular mass.
3. Vertical downward migration of ^{90}Sr beyond the boundaries of the main system of a plant's root system.

7. Uptake of radionuclides into trophic chains

Field studies have confirmed the continuous and interrelated movement of radionuclides along terrestrial and fresh-water trophic chains. However, in contrast to stationary field studies with plants, several migration indicators observed in real trophic chains are characterized by considerable uncertainties and approximate estimates, primarily for mammals and birds with extensive habitats.

In case of considerable contamination gradients over a distance of one to two hundred metres from the source, the only valid method for estimating the ratio of radionuclide concentrations in animals and top soil is to compare the statistical distribution curves of the frequency of corresponding indices for animals and soil; this is, however, problematic due to the difficulties in determining animal habitat areas and to insufficient statistics. The data presented in Table 10, therefore, should be regarded as approximate. They show, however, that both terrestrial and aquatic trophic chains may show differences in intensity of radionuclide transport. The presence of marked conventional concentrators is likely caused by the availability of more readily accessible, say, biogenic forms of ^{90}Sr and ^{137}Cs in the preceding food link. Regarded as concentrators may be amphibia, small rodents, herbivorous and omnivorous fish (^{90}Sr), carnivorous fish (^{137}Cs) and herbivorous aquatic birds.

TABLE 10.Radionuclide transport in trophic chains
(accumulation coefficients)

Trophic Chain Links	^{90}Sr	^{137}Cs
	Ground Chains	
	(with respect to soils)	
Soil saprophages	0.21	0.07
Insects (phytophages)	0.05	0.03
Amphibia	0.43	0.15
Reptile	0.19	0.11
Mammals	0.21	-
Small Rodents:		
- Hoofed	0.35	0.04
- Ungulate	0.23	0.02
- Predators	0.23	0.03
	Freshwater Chains	
	(with respect to water)	
Fish		
- vegetation-eaters	450	130
- mixed-food eaters	450	130
- predators	350	540
Near-to-water Birds		
- vegetation-eaters	3,700	-
- fish-eaters	150	100

8. Conclusion

The data on the behaviour of radionuclides in the eco- and agrosystems of the East-Ural Radioactive Trail throw some light on the mechanisms and rates of physicochemical and biogeochemical transformation of radionuclides, and on their migration in the environment and the biological systems. At the same time, they show that it is necessary to continue and, in a number of ways, expand the study of these processes.

The following problems should be regarded as being of primary importance in this respect:

1. The identification of the mechanisms responsible for physicochemical and biogeochemical transformation of ^{90}Sr and ^{137}Cs in top soil and bottom deposits as main vectors which determine further uptake of radionuclides into biological systems.

2. The development of dynamic models for the long-term migration of radionuclides in the main terrestrial biological and agricultural systems over long periods, say 50 to 100 years and experimental estimation of numeric values for the parameters of these models.

3. The development of statistic models for radionuclide transport models
   a) for conditions of living organisms having large habitats characterized by high contamination gradients;
   b) for conditions of transfer of radionuclides by living organisms outside the existing contaminated territory (transboundary transfer).

4. The development of dynamic models for the long-term behaviour of radionuclides in closed water bodies allowing for the role of hydromorphic, hydrochemical and hydrobiological characteristics of the water bodies.

All these problems are related to the rehabilitation of contaminated territories and to the handling of radioactive waste deposited on the surface. Solving these problems will contribute to the development of nuclear technology and nuclear power production.

POST-ACCIDENT DEVELOPMENT OF EXPOSURE CONDITIONS IN THE FORESTS OF THE 30-KM ZONE AROUND CHERNOBYL

P.I. YUSHKOV
Biophysical station of the Institute of Plant and Animal Ecology
Urals Division, Russian Academy of Sciences
ulitsa 8 Marta, 202
CIS-620219 Yekatarinburg, Russia

1. Introduction

The Chernobyl accident occurred on 26th April, 1986. The first post-accident examination of the surroundings revealed injuries to pine tree plantations caused by ionizing radiation. In later observations, other deleterious impacts on wood plants were found. The degree of radiation injury is determined by taking into account the radiation resistance of the plant species, the radiation level and some other characteristics. Therefore, one has to know the development of the radiation conditions during the first weeks after the reactor explosion as well as in the later periods to get a correct estimation and understanding of the radioactive release effects on the forests in the Chernobyl vicinity.

Radioactive releases from the destroyed reactor occurred in steps; the most intensive emissions were from 26th April to 6th May, 1986 [1]. Later in May, the releases from the reactor sharply decreased. The most injuring materials for biological objects were fuel particles and volatile radionuclides released from the reactor. A fraction of the volatile radionuclides condensed on soil and dust particles, forming condensed hot particles of specific activity lower than that of the fuel particles. The condensed particles accounted for 50% of the total amount of the hot particles [2]. It was found that 98% of the hot particles fell in a 25-km zone, around the reactor [3]. Some authors reported that from 6 to 8 tons of fuel particles were released into the atmosphere.

Even within the 30-km accident zone the polluted area had a very complicated mosaic of patches as regards dose rate and level and composition of radionuclide contamination. The most severely contaminated localities were to the west, north-west and north of the plant [4]. This was due to the force of the explosion, the speed and direction of the wind at the moment of the explosion and during the most intensive releases of radioactive substances from the reactor, special features of the relief, forestation of the locality as well as the state of aggregation of the radioactive material released.

The released materials contained large quantities of short- and long-lived radionuclides comprising those with substantial radiation effects on forest: ^{131}I, ^{106}Ru, ^{144}Ce-144, inert gases, ^{134}Cs, ^{137}Cs, ^{90}Sr [5, 6]. The total activity released from the reactor to the environment, was estimated at 1,800 PBq, including 400 PBq of ^{131}I and 800 PBq of ^{134}Cs and ^{134}Cs [7]. Irradiation of the forest in the 30-km zone was mainly caused by radioactive substances deposited on the plants after the accident.

F. F. Luykx and M. J. Frissel (eds.), Radioecology and the Restoration of
Radioactive-Contaminated Sites, 69–74.
© 1996 *Kluwer Academic Publishers.*

2. Exposure of forests

The contamination density of ^{144}Ce, ^{106}Ru, ^{90}Sr-90, ^{137}Cs and ^{134}Cs in the 10-km zone was 50-100 TBq /km² [8]. There were, however, areas still reaching values of 140-400 TBq /km² in October 1986 [9].

The maximum gamma dose rates at different times after the accident in 4 of the 16 areas examined in the 30-km zone are shown in Figure 1.

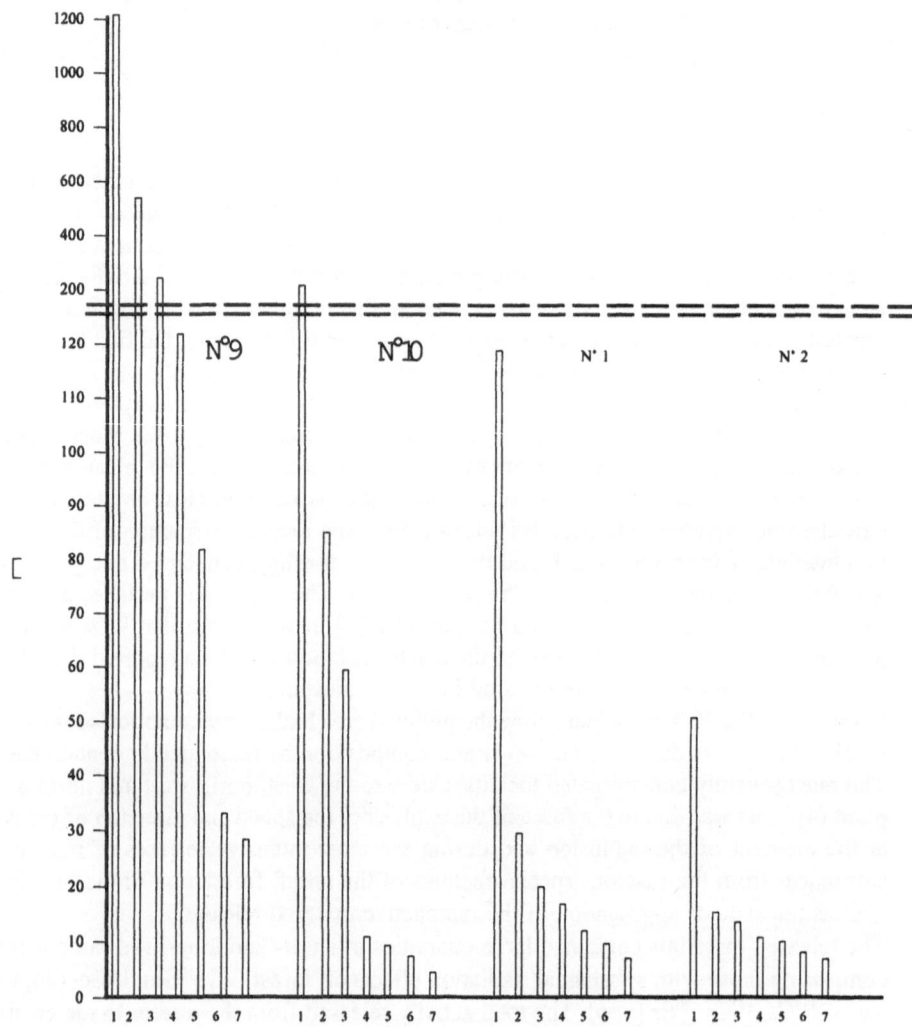

Figure 1. Gamma dose rates in pine-birch stands N°. 1, 2, 9, 10 in the 30-km zone
at different times after the accident;
measuring dates: 1: 01/10/1986, 2: 01/10/1987, 3: 01/10/1988,
4: 01/10/1989, 5: 01/10/1990, 6: 01/10/1991, 7: 01/10/1992

Pine tree plot No. 9 with 40- to 50- year-old trees, located 1.5 to 2 km to the west of the plant, was exposed to a maximum gamma dose rate of 1,200 mrem/h on 1 October 1986. This dose rate is respectively 2.6, 4.8, 10 and 24 times higher than that of plots Nos 10a, 1 and 2. The calculated absorbed dose up to 1 May 1987 was 100-180 Gy.

Plot No. 10a in a pine tree plantation of 40- to 60- year-old trees, located to the west of the destroyed unit, received by the end of 1986 an absorbed dose of 20-25 Gy and, in some, places 50 Gy.

Plot No. 1 is a pine tree forest located 5 km to the west of the plant. On 1 May 1987, the absorbed dose was 10-12 Gy and, in some places, it reached 50-60 Gy.

Plot No. 2 is a pine tree plantation, 25-35 years old in 1986, mixed with natural rebirth. It is located 4.5 km to the west of the plant and 2.5 km to the south of plot No. 1. At the beginning of May 1987, the absorbed dose was 3.5-5.0 Gy in the centre of the plantation and from 8-10 to 40-50 Gy on its edges.

3. Exposure periods

Some researchers identify several periods in the development of the exposure conditions in the forest plantations of the 30-km zone according to their radionuclide content, the quantity of radioactive material absorbed by the plantation, the exposure dose level and the decrease in dose rate. During the first weeks after the reactor accident, short-lived radionuclides were prevailing in the radioactive cloud over the forest plantation and the exposure of the trees was caused by direct radiation from the radioactive material pass-ing by and from radionuclides plant uptake. The latter comprised a great number of long-lived radionuclides (^{137}Cs, ^{90}Sr, ^{240}Pu, ^{238}Pu and ^{239}Pu) and the prevailing short-lived nuclides. Later, the contribution of long-lived radionuclides to the dose burden increased and became dominating with the decay of significant groups of short-lived nuclides. Therefore, the radiation impact on the forest in the 30-km zone can reasonably be con-sidered as due to chronic exposure with a varying dose rate. Hence, one cannot classify the first 15-20 days after the accident as a period of powerful acute irradiation exposure [8] within the approach mentioned above. The first period (26 April until 10-15 May 1986) in the development of the exposure conditions in the accident area plantation is characterized by a massive irradiation impact on forest ecosystems at dose rates that are maximum for that proper plantation. By the end of the period, the dose rate was reduced due to the decay of the short-lived radionuclides (^{131}I-, ^{65}Zr, ^{65}Nb) and the disappearance of radioactive inert gases (Table 1).

TABLE 1. Periods of development of the radiation conditions
in forests in the 30-km zone around Chernobyl

periods	date
first	26 April 1986 to 10-15 May 1986
second	mid-May to mid-Sept. 1986
third	mid-Sept 1986 to end of March 1987
fourth	April 1987 to 1995-96
fifth	after 1996-97

In the second period (from mid-May to the end of the leaf fall in 1986), the plantation was exposed to lower dose rates than in the first period, due to the decay of the short-lived nuclides. There was also an intensive self-cleaning effect of the top foliage due to the fall of leaves and part of the needles.

In the third period (September 1986 to April 1987), the dose rate further decreased slowly with a significant increase of the dose contribution from long-lived nuclides (^{137}Cs, ^{90}Sr).

At the beginning of the fourth period (April 1987 to 1995-1996), decontamination activities on the destroyed reactor site caused releases of radioactive substances and a secondary airborne contamination of the forest occurred. It should, however, be noted that radioactive releases from the Sarcophagus, built around the damaged reactor unit did not exceed 11 GBq/y [10], according to the measurements of 1988-1990, and therefore did not substantially influence the radiation situation even on the industrial site of the power plant.

At that moment, the fast self-cleaning of the top parts of the trees was complete and root uptake developed. It progressed more intensively in birch trees, slower in fir trees and much slower in pine trees [8, 9, 11]. One can assume that an equilibrium between the upstream and downstream of radionuclides in the trees was established during this period.

The dose burden to woody plants in this period was mainly caused by external exposure.

A relatively high content of radionuclides was found in the pine needles in the autumn of 1987 (Table 2). The highest concentration of 5 out of 6 radionuclides occurred in the needles of dead trees in plot No. 9.

TABLE 2. Radionuclide concentration in pine needles in the 30 km zone , kBq/kg fresh weight[8]

Plot Nr	^{95}Zr	^{95}Nb	^{106}Ru	^{134}Cs	^{137}Cs	^{144}Ce
9	814	1,554	4,070	1.6	4,670	13,357
1	10.2	20.8	96.9	19.5	69.6	190.0
2	7.7	15.5	59.2	17.1	721.5	150.2
3	3.0	4.9	19.4	5.0	21.5	4.6

The maximum concentration was that of ^{144}Ce, i.e., more than 13 MBq/kg. The minimum concentration was that of ^{134}Cs, i.e., 1.6 kBq/kg. In the plots exposed to lower contamination levels, the radionuclide concentration in the needles was also substantially lower. Thus, in plot No. 1, 5 km away from the Chernobyl plant, the concentration of radionuclides in pine needles was 42-80 times lower than in plot No. 9. The lowest concentration of all six radionuclides was found in plot No. 3.

The largest contribution to the total gamma dose absorbed by the needles in all forest areas examined was due to ^{144}Ce (46.7-52.5%), ^{106}Ru- (16.0-23.8%) and ^{137}Cs (16.0-22.0 %). The fraction of ^{134}Cs in the total activity absorbed by the needles was only 4.8-6.1%.

According to the data reported by E. Ostapenko in 1988, the uptake by the needles of more mobile radionuclides (mainly caesium) from soil increased [8]. In this investigation, the pine needles formed in 1987 and 1988 had the highest content of ^{137}Cs, the quantity of ^{134}Cs was lower and the lowest content was that of ^{144}Ce.

The increase of radionuclide content in the top organs of trees due to the increased uptake from soil during the first post-accident years was found in birch trees too [11, 12]. Thus, the concentration of ^{134}Cs and ^{137}Cs in seeds and leaves of trees in plot No. 1 near the plant increased in 1989 compared to 1988; there was an increase of 5 to 10 times, corresponding to 4.5 and 4.0 MBq/kg of air-dried weight. A great difference in ^{134}Cs and ^{137}Cs content from tree to tree was found in the structure elements of bark and leaves (a variation in the range of 34 to 90%).

In those years, a strong uptake of ^{90}Sr from soil into the ground organs of birch was found. In 1988, the concentration of this radionuclide in seeds and leaves of birch in plot No. 1 was respectively 150 and 140 kBq/kg of air-dried weight and it increased to 300 and 400 kBq/kg in 1989. Differences from tree to tree in the same plantation were very high (variation ranging from 26 to 98%).

By the end of the fourth period, the radiation conditions are expected to have become stable and determined by the long-lived radionuclides. The dropped seeds and juvenile plants will have received the highest dose.

4. Conclusion

The radiation conditions in the 30-km zone of the Chernobyl accident were characterized by

- different periods of exposure of forest areas, located in different directions from the power plant;
- non-uniformity in the degree of radiation impact, radionuclide content and aggregative state of the radioactive contamination of the territory;
- a variable radionuclide distribution in each plantation: it was highest at the side of the forest located opposite the plant, and levels of radioactive contamination of different parts of any given tree were different.

These factors and the resistance of species and individuals of woody plants against radiation determined the peculiarities of the radiation impact and the post-irradiation recovery of woody plants.

74

5. References

1. Belyaev S.T., Borovoy A.A. (1993) *On Radioactive Discharges ...*
2. Loshchilov N.A. et al. (1990) Ratios of fuel to condensate components in radioactive discharges resulted from the Chernobyl accident, *Geochemical Paths of artificial radionuclide migration in biosphere*. Gomel, 65.
3. Galyshkin B.A. et al. (1990) *Geochemical Paths of Artificial Radionuclide Migration in Biosphere*, Gomel.
4. Information presented by the Russian Federation to IAEA,(1988). The Chernobyl NPP accident and its consequences, *Atomnaya Energia*, V. 61, 301-320.
5. Israel Yu. A. et al. (1987) Radioactive contamination of nature environment in the Chernobyl NPP accident zone, *Meteorology and Hydrology*, N 2,5-18.
6. Israel Yu. A. et al. (1988) Ecological consequences of radioactive contamination of nature environment and the Chernobyl NPP accident area, *Atomnaya Energia*, **64**, 28-40.
7. International Chernobyl Project (1991) An assessment of radioecological consequences and protective measures, *Report of the International Consulting Committee*. Moscow, 95.
8. Kozubov G.M., Taskaev A. (1994) Radiobiological and radioecological investigations of wood plants, S- Petersburg, *Nauka*, 256.
9. Tikhomirov F.A., Shcheglov A.I., Sidorov V.P. (1993) Forest and forestry; radiation protection measures with special reference to the Chernobyl accident zone, *Science of Total Environment*, **137**. 289-305.
10. Borovoy A.A., Savintsev Yu.V. (1972) *Five Hard Years*, 102-158
11. Yushkov P.I. et. al. (1993) Cs-134+134 and Sr-90 content in birch tree bark during the first weeks after the Chernobyl accident, *Ecology*, **1**, 86-91.
12. Yushkov P.I., Chueva T.A., Kulikov N.V. (1990) Content of Cs-134, Cs-137 and Sr-90 in birch crowns in woods within 30-km of the Chernobyl power plant during the initial years following the accident. *Comparative Assessment of the Environmental Impact of Radionuclides Released during Three Major Nuclear Accidents: Kyshtym, Windscale, Chernobyl.* Luxemburg, EUR 13574, 1991, I, 333-339.

ASSESSMENT OF RADIONUCLIDE TRANSFER FROM SOILS TO PLANTS AND FROM PLANTS TO SOILS

P.J. COUGHTREY
LG Mouchel & Partners Ltd.
West Hall, Parvis Road
GB-KT14 6EZ West Byfleet, Syrrey
United Kingdom

1. Introduction

Plant surfaces may receive inputs of radionuclides by wet and dry deposition from the atmosphere and by direct irrigation. A proportion of this input can be lost via processes of resuspension and it is often difficult to separate those losses which are attributable to leaching during rainfall from processes of abscission, mortality and resuspension [1]. Seasonal factors have a major influence on these processes, affecting the surface area and physical characteristics of plant parts for particle capture, the degree of root uptake and translocation, and the magnitude of losses via abscission and mortality. All processes of uptake and loss from plants can be expected to be related not only to the maturity of the plant but also to the temperature, surface wetting, humidity, chemical form of the substance concerned and the physiological state of the plant.

Processes involved in the uptake of radionuclides by plants can be summarised as occurring in three steps [2], i.e. Solutes penetrate the plant cuticle and then the cell wall by limited or free diffusion. Adsorption occurs at the plasma membrane. Uptake into the cytoplasm takes place.

The absorption of trace elements by plant roots from solution has been suggested to show the following characteristics [3]:
It operates effectively at low concentrations in solution.
At low concentration the rate of absorption varies almost linearly with concentration.
It is active being sensitive to temperature and probably also aeration.
It reacts strongly to H^+ and other ions.
Its intensity varies with plant species.

These comments refer to absorption from solutions but the real interactions between plants and soils are much more complex as evidenced by Nye and Tinker's review of 1977 [4].

F. F. Luykx and M. J. Frissel (eds.), Radioecology and the Restoration of
Radioactive-Contaminated Sites, 75–84.
© *1996 Kluwer Academic Publishers.*

Assessment of the transfer of radionuclides from atmosphere to plant, from plant to soil and from soil to plant relies on the use of models which attempt to take account of known physical, chemical and physiological phenomena. Figure 1 summarises many of these phenomena in the form of a generalised soil-plant model. The present paper summarises the approaches that have been taken in radioecology to assess the potential transfer of radionuclides between the various components of the system that can be recognised.

2. The soil plant system

2.1 SOILS

Soil can be considered in terms of four components: an organic component, a mineral component; a parent material component; and solution phases. Within any one component there may be several subcomponents, for example, there may be surface, dissolved and suspended organic subcomponents within the organic component. However, in many radioecological studies, soil is often considered simply as a series of layers of increasing depths. This takes no account of differences in physical and chemical properties of different horizons of a soil profile or between the organic and mineral phases. Furthermore, plants are often considered in isolation from soil with insufficient regard for their root component, not all of which is equally involved in the process of radionuclide uptake. A further confounding influence is the effect of microbial activity in the vicinity of the roots (the rhizosphere) which, for mycorrhizal plants, has a critical influence on nutrient and hence radionuclide absorption.

Figure 1 separates three main solid phase components, all of which can interact with the soil solution. Soil solution is included as an intermediary between any of the solid components and the plant root component. Besides acting as a vehicle for transport of radionuclides from solution to the plant, roots also act as a vehicle for return of radionuclides to the soil as a result of mortality or, in agricultural soils, ploughing. Within the plant, radionuclides may either be translocated from root to shoot (as in many cereal crops) or from shoot to root for storage (as may occur with root vegetable or tuber crops).

The behaviour of radionuclides in soils is complex and is not dealt with here. In the last decade there has been a rapid development in the models used to describe radionuclide transport in soils but this development has not been accompanied by a similar expansion in data for testing and validation [5].

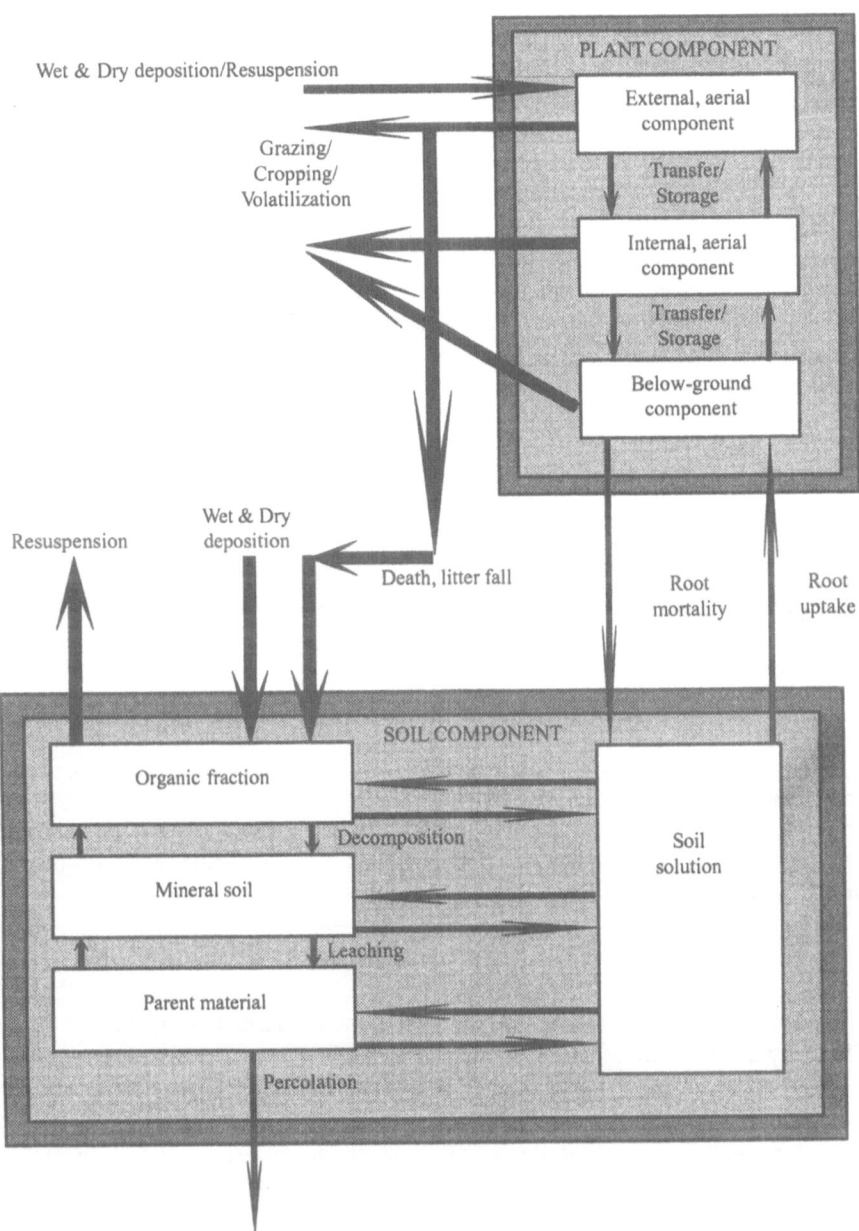

Figure 1. Conceptual model for radionuclide transport in the atmosphere-soil-plant system.

2.2 PLANTS

The above-ground parts of plants can be considered as at least two main components (Figure I); the external component which receives an aerial input via deposition of particulate matter from atmosphere and translocated material from the rest of the plant; and an internal component which mediates in absorption, translocation and storage processes.

2.2.1 Dry deposition to external plant parts

Dry deposition to plants from atmosphere is often estimated by means of a deposition velocity (Vg) which is simply the ratio of rate of deposition to ground per unit area divided by concentration per unit volume of air; it therefore takes the units of velocity (e.g. Bq/m2/s divided by Bq/m3). The concept of deposition velocity was introduced by Chamberlain and co-workers [6] and the parameter includes particle capture via the following processes.

Brownian motion
Sedimentation
Inertial impaction
Interception

The extent of particle capture is determined by the physical characteristics of the particles, their concentration in air, the ambient meteorological conditions and the physical and biological characteristics of the plant surface.

For particles of <0.3 µm diameter, the dominant transfer mechanism for particles close to the surface is Brownian motion which can be described mathematically using a molecular diffusivity approach.

For particles <0.1 µm diameter, Vg generally increases according to the following empirical relationship:

$$Vg = const \ (Sc)^{-n}$$

where *Sc is* Schmidt's number (the ratio of k, the kinetic constant of air, to D, the Brownian diffusivity of the particle) and n is a constant.

For particles >0.3 µm, the processes of sedimentation, impaction and interception become dominant. Turbulence in the boundary layer and surface roughness also influence the transfer process.

As a consequence, Vg is highly dependent on particle diameter, increasing by one to two orders of magnitude either side of the observed minimum at 0.3 μm. Typical values for different materials are summarised in Table 1.

TABLE 1. Typical deposition velocities for different materials.

Material	Vg range (m/s)
Reactive gases	10^{-2}
Small particles	10^{-3} to 10^{-2}
Unreactive gases	10^{-4} to 10^{-3}

2.2.2 Wet deposition to vegetation

The discussion of wet deposition processes for transfer of contaminants to vegetation is at least as complex as that for dry deposition. Nevertheless, wet deposition can be of considerable significance. Chamberlain proposed the following relationships:

$$Vgr = \Lambda h, \text{ and } Vgr = 2.3 \times 10^{-3} \text{ IW}$$

where Vgr is the wet deposition velocity, Λ is the washout coefficient, h is the equivalent length of the aerosol cloud, I is the rainfall rate and W is the washout ratio.

However there are very few studies that relate retention of particulates on plant surfaces to wet deposition flux from the atmosphere. Field studies are complicated since the same rainfall that deposits particles will also remove them from vegetation surfaces. The observation that cloud water sometimes contains large amounts of insoluble particles indicates another potential 'hidden' deposition pathway, the significance of which should not be underestimated [7].

2.2.3 Interception by vegetation

Because, in practice, it is generally difficult to separate deposition to vegetation from overall deposition to ground, it is common to introduce the concept of an interception fraction, i.e. that fraction of the total deposit which is initially intercepted by vegetation surfaces. Both initial interception and the subsequent retention of contaminants by vegetation are affected by humidity, precipitation and the form of the contaminant. In general, the proportion of a depositing contaminant intercepted by vegetation (r) decreases in the following order:

wet-deposited radionuclides > wet-deposited particles > particles dry-deposited on grass wet with rain > particles dry-deposited on grass superficially wet > particles dry-deposited on dry grass, with: small particles > large particles [8].

In a classic series of experiments, Chamberlain demonstrated a relationship between r and the dry weight density of vegetation (B in kg/m²) as follows: $r = 1-\exp(-\mu B)$

where μ was termed the absorption coefficient (with units of m²/kg). Typical values for μ for grass lie in the range 2.3 to 3 .3 m²/kg [1]. Table 2 provides some typical values for pastures in good growing conditions.

TABLE 2. Typical absorption coefficients for pastures in good growing conditions.

Pasture type	Absorption coefficient (m2/kg)	Pasture age (years)
Ryegrass	3.4	1
Ryegrass	1.3	1
Ryegrass/clover	2.4	2
Ryegrass	3.8	1
Ryegrass/clover	2.6	4
Timothy, fescue, clover	2.3	3
Timothy, fescue, clover	4.9	4
Agrostis/ryegrass	6.2	Permanent

An alternative approach developed by Chamberlain is to make use of a normalised specific activity *(NSA)* which is the ratio of concentration per kg in foliage to the rate of deposition of the contaminant per day per m2 and which therefore has the units of m2/d/kg. Values of *NSA* for radionuclides in the range 27 to 60 have been reported for grass in good growing conditions (Table 3) with much higher values appropriate to vegetation in poor growing conditions [1]. Values for leaves of trees and for non-radioactive contaminants are intermediate [9].

TABLE 3. Typical *NSA* values for good growing conditions.

Depositing material	NSA	Vegetation type
^{90}Sr	19-48	Herbage
^{90}Sr	61	Grass
^{90}Sr	31-66	Herbage
^{137}Cs	33	Grass and alfalfa
^{137}Cs	10-49	Alfalfa
^{210}Pb	50	
Lead	50	Grass
Fluorine	27-59	Herbage

Interception of wet deposited radionuclides is dependent on the rate of rainfall, the state of growth of the crop and the radionuclide [10].

2.2.4 Input to vegetation via resuspension

The effects of resuspension of radionuclides from soil to vegetation surfaces has been given little attention in the literature. Assuming that the effects of resuspension have been taken into account when calculating radionuclide concentration in air no specific calculation should be required. However, in practice, plants are often contaminated directly as a result of soil splash during rainfall. Dust loading can have a substantial effect on calculated soil-plant transfer ratios for both readily absorbed radionuclides such as ^{137}Cs and for unabsorbed radionuclides such as isotopes of plutonium [11] .

The extent of vegetation contamination by soil can be assessed in field conditions by measurement of the concentrations of elements such as titanium which are generally not well absorbed by plant roots. Whereas such measurements are valuable for edible products at harvest, they are not as useful for plant products consumed by animals unless available on a regular basis.

2.2.5 Foliar absorption of deposited radionuclides

Foliar absorption and the resulting translocation of radionuclides within plants should be clearly distinguished from external contamination of plant surfaces. There is relatively little information on the rate and extent of absorption of radionuclides in different chemical and physical forms by plant surfaces. Some authors have made a distinction between foliar, floral and plant-base absorption. The latter is poorly defined but may be the result of a combination of uptake by both above- and below-ground organs. It is, however, particularly important for some perennial species and for natural pastures where particles may become entrapped in the basal parts of plants or the superficial root matt.

2.2.6 Root absorption and the soil-plant ratio

There is a very substantial literature on the absorption of different elements by plant roots. Ion uptake can occur by passive and active processes and the interactions between influx and efflux are very complex. Many attempts have been made to model the uptake of radionuclides by plant roots with varying degrees of success. It is often difficult to apply such models to field conditions because of the lack of appropriate information on radionuclide concentrations in soil solution and of the distribution and effectiveness of absorption of plant roots in field conditions.

To overcome such difficulties, a simple approach has often been adopted based on the ratio of concentration of a radionuclide in vegetation or a given product to the concentration of that radionuclide in soil. This ratio does not, and should not be considered to, represent root uptake by plants. It simply represents the sum total of root uptake, translocation, foliar absorption, foliar translocation, foliar loss, root excretion etc. over the whole life of the plant to the time at which the concentration is measured. It is therefore very sensitive to a wide range of soil and plant parameters and the experimental or field conditions from which it is derived. It is also implicitly assumed that the derived ratio represents conditions at equilibrium - an assumption that is often difficult to justify in nature.

To overcome the difficulty of experimental artefacts and misinterpretation, the International Union of Radioecologists (IUR) developed a database of soil-plant transfer factors derived from standard experimental conditions. The contents of this database are summarised for important crops and radionuclides in Table VI of [10]. There have been several attempts to relate the observed transfer factor to soil physical and chemical factors but, in general, this has had little effect on reducing the overall variability that is generally observed.

More recently, IUR has supported an international initiative to develop a database of parameters describing the time-dependent behaviour of radionuclides in soils and plants. It is hoped that this will help with the development and testing of a new suite of models describing the soil to plant transfer of radionuclides for application to time-varying discharges and accidental events.

For the reasons given above, it is most important that, when site-specific data for soil-plant transfer factors exist, these should be used in assessments of radionuclide transfer. It is also important to determine whether the available data represent transfer from soil to plant in the absence of any direct deposition to plant from atmosphere or contamination by resuspension or soil splash.

2.2.7 Translocation

As for root absorption, the subject of translocation and mobility of radionuclides within plants is not well defined in many assessment tools. For root uptake, the effects of translocation are incorporated into the soil-plant transfer factor. For foliar-absorbed

The microorganisms influence radionuclides in various ways. Microorganisms need for their growth an oxidizing compound, usually nitrate. If no nitrate is available, ferric ions and other compounds are reduced resulting in an anaerobic system with a low redox potential. Some radionuclides as Pu, Am and Np exist as specimen which will be reduced. Also so called "hot particles" will solve better at low redox potentials. The major impact of microbial growth is the change of the specific surface of the Al-Fe colloids in the soil. Their adsorption capacity is substantially reduced. It is one of the reasons that for sediments and soils not the same Kd's can be used. Another interesting phenomenon is the chain: Organic-N, ammonium, nitrate, free N_2.

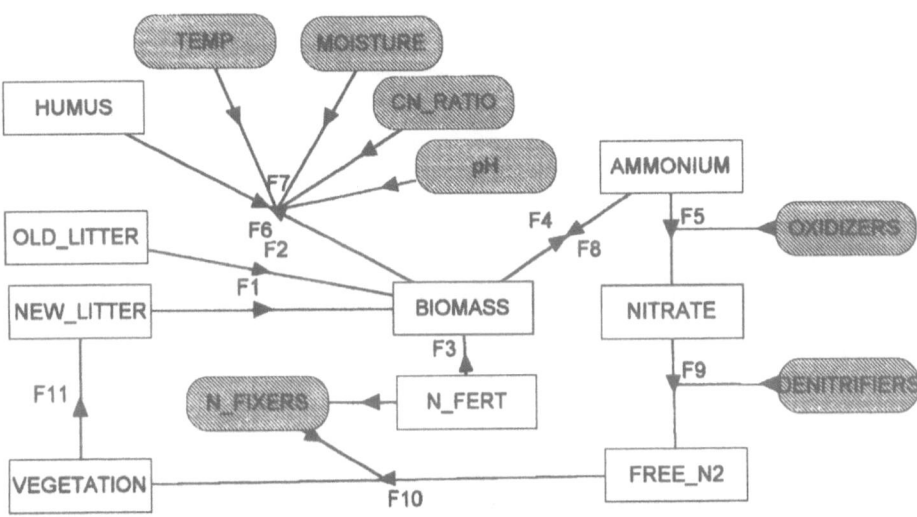

Figure 2. Simplified scheme of microbiological processes in soil. The boxes represent N and C quantities. Arrows between boxes represent N fluxes. Not all of them could be indicated. Eg the uptake of ammonium and nitrogen by vegetation is not indicated. The arrows between shaded symbols and fluxes represent influences which control the rates of fluxes. The temperature, moisture content and pH control the rate of all microbiological processes. They are only indicated for the production rate of humus. Also the C/N ratio in soil is a rate controlling variable. The oxidizers of ammonium, the denitrifiers of nitrates and the symbiontic N fixing microorganisms contain also nitrogen. In this scheme they are, however, only considered as rate controllers and not as part of the N-cycle. The symbols F1 - F11 represent the N fluxes between the indicated compartments. C fluxes are not shown.

In most soils in the temperate climate ammonium is rather fast oxidized to nitrate. In water logged soils at low temperatures, the oxidation rate of ammonium is very slow. The ammonium which forms after decomposition is not oxidized to nitrate and remains available in the organic layer. It competes with Cs ions for exchangeable sites. It is one of the reasons that Cs is better available in upland and Nordic organic, water logged, soils. The model shows also the impact of nitrogen fertilizers on the symbiotic nitrogen fixation. Nitrogen fertilizers inhibit nitrogen fixation and limit therefore the competition plant organs, soil solution or labile fractions, and the soil solid phase. The acquisition of data allowing the separation and quantification of transfers between these subcomponents requires novel approaches to the interpretation of past data and to the development of future experimental and field investigations. In particular, it is now clear that the behaviour of wet-deposited radionuclides or of particulates and gases deposited in damp conditions or to wet vegetation is very different to that of dry-deposited radionuclides. Further research is required to quantify such differences and to reduce the uncertainties associated with atmosphere-plant and soil-plant transfers.

4. References

1. Coughtrey, P.J. Thorne, M.C. (1983) *Radionuclide Distribution and Transport in Terrestrial and Aquatic Ecosystems. A Critical Review of Data. Volume 1.* A Balkema, Rotterdam.

2 Nishita, H., Wallace, A. and E.M. Romney (1978) *Radionuclide uptake by plants.* NTIS PB-287-43 1.

3 Loneragan, J.F. (1975) The availability and absorption of trace elements in soil plant systems and their relation to movement and concentrations of trace elements in plants. In: DJD Nicholson and AR Egan (Eds.) *Trace Elements in Soil-Plant-Animal Systems, pp. 109-134.* Academic Press, New York.

4 Nye, P.H. and Tinker, P.B. (1977) *Solute Movement in the Soil-Root System Studies in Ecology, Volume 4.* Blackwell, Oxford.

5 Coughtrey, P.J. (1988) Models for radionuclide transport in soils. *Soil Use and Management* 4:84-90.

6 Chamberlain, A.C. (1960) Aspects of the deposition of radioactive and other gases and particulates. *International Journal of Air Pollution* 3:6-88.

7 Unsworth, M.H. and Crossley, A. (1987) The capture of wind driven cloud by vegetation. In: P.J. Coughtrey, M.H. Martin and M.H. Unsworth (Eds.) *Pollutant Transport and Fate in Ecosystems, pp. 125-138.* Blackwell Scientific Publications, Oxford.

8 Eriksson, A. (1977) *Direct uptake by vegetation of deposited materials. I. Retention of nuclides and simulated fallout particles in pasture grass.* SLU-IRB-42, Uppsala.

9 Martin, M.H. and Coughtrey, P.J. (1982) *Biological Monitoring of Heavy Metal Pollution.* Applied Science Publishers, London.

10 IAEA (1994) *Handbook of Parameter Values for the Prediction of Radionuclide Transfer in Temperate Environments.* Technical Report Series No. 364. LAEA, Vienna.

11 Cawse, P.A. (1983) The accumulation of Cs-137 and plutonium-239+240 in soils of Great Britain and transfer to vegetation. In: P.J. Coughtrey, J.N.B. Bell and T.M. Roberts (Eds) *Ecological Aspects of Radionuclide Release, pp. 47-62.* Blackwell Scientific Publications, Oxford.

12 Aarkrog, A. (1983) Translocation of radionuclides in cereal crops. In: P.J. Coughtrey, J.N.B. Bell and T.M. Roberts (Eds) *Ecological Aspects of Radionuclide Release, pp. 81-90.* Blackwell Scientific Publications, Oxford.

SOILS ROLE IN THE RESTORATION OF TERRESTRIAL SITES CONTAMINATED WITH RADIOACTIVITY

M.J. FRISSEL
Senior Officer, IUR
Torenlaan 3, 6866-BS Heelsum, Netherlands

1. Soils

1.1 INTRODUCTION

For part of the population soil is only a material which supports roads and buildings or which can be used to construct dikes. Nomenclature is simple, for the first purpose one needs sand, for the second one clay. Soils scientists, however, have many ways to describe soils. Many popular soil names are based on a mixture of the appearance of the soil and its use in historical times. Some well known names as podsol (sandy soil which leached top layer), chernozem (black soil) and rendzina (calcareous soil) are from Russian or Polish origin and are relatively well defined by their name. Other names as peat bog soil define a soil less well. A very sophisticated system is based on a recommendation of the International Society of Soil Science, and propagated by FAO and UNESCO. Some old names are maintained -but having now a strict definition-, others have been added. Additions as "eutric" for fertile soils, "dystric" for infertile soils, "gleyic" -again Russian- for waterlogged soils complete the names. The peat bog upland soils are in this system called "dystric gleysols". It might well be that part of the high uncertainties which are so common in radioecological soil related parameters would be less if the proper names could be applied. For the time being there is a lack of well documented data and experience among radioecologists. Soil scientists, however, have more possibilities for classification, eg one based on texture. The most important fractions in this system are: clay (< 2 μm), silt (2 - 50 μm) and sand (50μm - 2 mm). The popular term "loam" is missing, loam is a mixture of fractions with a dominating medium texture fraction.

Yet, soil scientists see soils in different ways, not because of disagreement, but because of different interests. Some examples and their relation with radioecology are reported.

1.2 SOIL AS A TRANSPORT MEDIUM

Soils contain pores which allow vertical transport of water from the soil surface to the phreatic level. The quantity which moves downward depends on rainfall and evapotranspiration. Western Europe is closer to the desert than one might think; in large

F. F. Luykx and M. J. Frissel (eds.), Radioecology and the Restoration of
Radioactive-Contaminated Sites, 85–102.
© 1996 *Kluwer Academic Publishers.*

parts of Europe the annual precipitation surplus is close to 20 cm. -There exist excellent maps of precipitation surpluses and deficits [1]-. With a moisture filled pore space of 30 percent, a precipitation surplus of 20 cm means that the water front moves downward with a speed of 60 cm per year.

Almost all radionuclides adsorb on soil and migrate therefore much slower than water. A measure for the adsorption is the Kd, the distribution coefficient between adsorbed and liquid phase. Most Kds for Sr range from 200 to 500 L/kg , for Cs from 1000 to 5000 L/kg, for peat soils they may be much lower. The migration rate for radionuclides follows from:

$$V_{nuclide} = V_{water}/(1 + \rho.Kd/\theta) \qquad (1)$$

where V is the migration velocity, ρ the bulk soil density and θ the water filled pore space. With Kd = 300 l/kg, V_{water} = 60 cm/a , ρ = 1.4 kg/dm^3 and θ = 0.3, $V_{nuclide}$ is 0.04 cm/a. In the field migration rates of about 0.3 cm/a are observed. This is partly caused by mechanical transport of soil particles through cracks, by the trampling feet of cattle, etc. The Kd of Pu is much higher than for Sr and Cs, the mechanical transport is of course equal. In the upper soil layers the migration of Cs and Pu is therefore almost the same, Sr moves somewhat faster, indicating that transport via the water phase is significant. An equation which is able to take all these effects into account is:

$$\delta s/\delta t + \delta c/\delta t = D.\delta^2 c/\delta x^2 - v.\delta c/\delta x + \phi \qquad (2)$$

where s is adsorbed nuclide, c is nuclide in solution, v is interstitial velocity of water phase, x is vertical distance, D is apparent diffusion constant and ϕ is a source term which accounts for additional source terms and mechanical transport. Eqn (2) can also be applied to river systems and has been solved numerically by many authors, among them Konoplev [2] and Frissel et al. [3]. A disadvantage of eqn 2 is that many data, eg on adsorption and desorption, are required. An advantage is that it is easy to take into account all "disturbing" phenomena which results in a good agreement between ob-served and calculated data.

A much simpler model is a residence time model. It does not take into account any process, simply to each layer -which may differ in thickness- a residence time is attrib-uted. Residence time models are particularly suitable for steady state systems, for which the residence time can be defined as the "hold up" in a compartment divided by the "throughput". An example by Frissel et al. [3], using a uniform residence time for all layers up to a depth of 2.5 a/cm, is shown in Figure 1. The soil is a Mollic Gleysoil with a rather uniform clay profile, used as a permanent pasture. For a grazed pasture, the result is fair, and sufficient for all practical purposes. An example of a non-uniform distribution is shown by Strebl et al [4]. She describes a forest soil with a litter layer and a rather small humus rich layer up to 5 cm depth, there below sandy fractions dominate. Residence times vary from 6 a/cm in the top layer to 0.8 and 0.16 a/cm in the 5-10 and 10-20 cm layer resp. Tikhomirov et al [5] report migration data for a sandy forest soil. Figure 2 shows Tikhomirov's data in combination with calculated data by the author.

The residence time in the litter layer is set to 6 a/cm; there below residence times of 2 - 2.5 a/cm are applied. An advantage of residence times is that it is easy to compare data from different sites which each other. A related model is the half live model. The latter one is popular among radioecologists, they are used on half lives. The relation is $T_{1/2}$ - τ . 0.69 where τ is the residence time. An advantage is that the physical decay constant λ_{phys} can easily be combined with an ecological decay constant λ_{ecol}.
It will be clear that in general radionuclides can not be removed by leaching a soil. It takes to much time before radionuclides are migrated below the rooting zone.

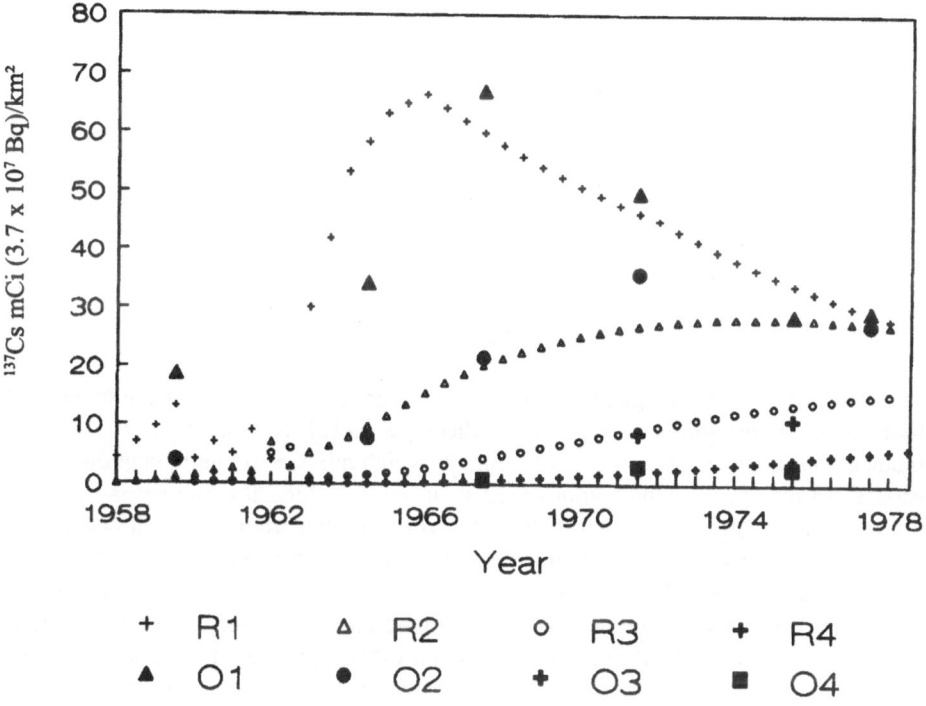

Figure 1. Migration of fallout ^{137}Cs in a clay soil, permanent pasture, near Alkmaar, Netherlands. Uniform residence time of 2.5 a/cm. R1-R4 calculated data. O1-O4 observed data for the layers 0-5, 5-10, 10-15 and 15-20 cm, respectively [3].

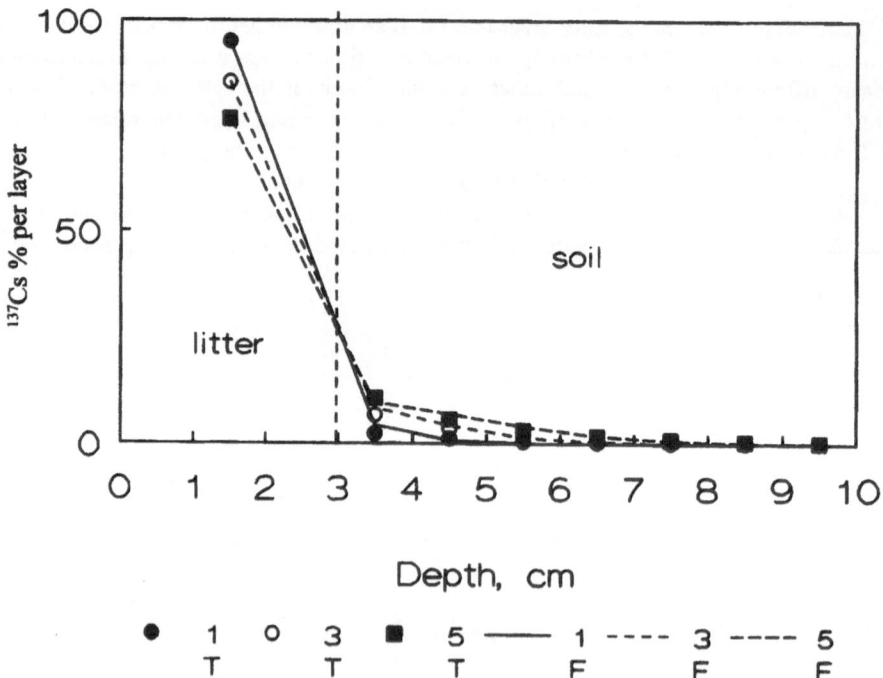

Figure 2. Comparison of observed migration data (T) by Tikhomirov [5]
and calculated data (F) by Frissel.
Sandy forest soil. Residence times :
Litter layer 6 a/cm, soil layer 3-10 cm 2.5 a/cm, soil layer 10-15 cm 2 a/cm.

1.3   SOIL AS A BIOLOGICAL SYSTEM

A soil is a powerful biological system. On a normal fertile arable soil the amount of
biomass within the soil is as large as 100 sheep per ha [6]. The major part is microbial
biomass, but there are numerous other animals which mix the soil more than one might
expect. An overview by the author of the main microbiological processes is shown in
figure 3. Soil organic matter is divided into different compartments which decompose
with different rates. There exist many alternative models with various degree of sophis-
tication and purpose. The core of many models is the growth and decomposition of
biomass [7]. For the growth a carbon compound and a sufficient nitrogen source is
essential. N is taken up by the growing biomass if C/N is above 20, if there is no nitrogen
source the growth of biomass stops. If C/N is lower than 20 nitrogen becomes available
as ammonium. Part of the decomposing biomass is considered as humus. This is a serious simplification.

The microorganisms influence radionuclides in various ways. Microorganisms need for their growth an oxidizing compound, usually nitrate. If no nitrate is available, ferric ions and other compound are reduced resulting in an anaerobic system with a low redox potential. Some radionuclides as Pu, Am and Np exist as specimen which will be reduced. Also so called "hot particles" will solve better at low redox potentials. The major impact of microbial growth is the change of the specific surface of the Al-Fe colloids in the soil. Their adsorption capacity is substantially reduced. It is one of the reasons that for sediments and soils not the same Kd's can be used. Another interesting phenomenon is the chain: Organic-N, ammonium, nitrate, free N_2.

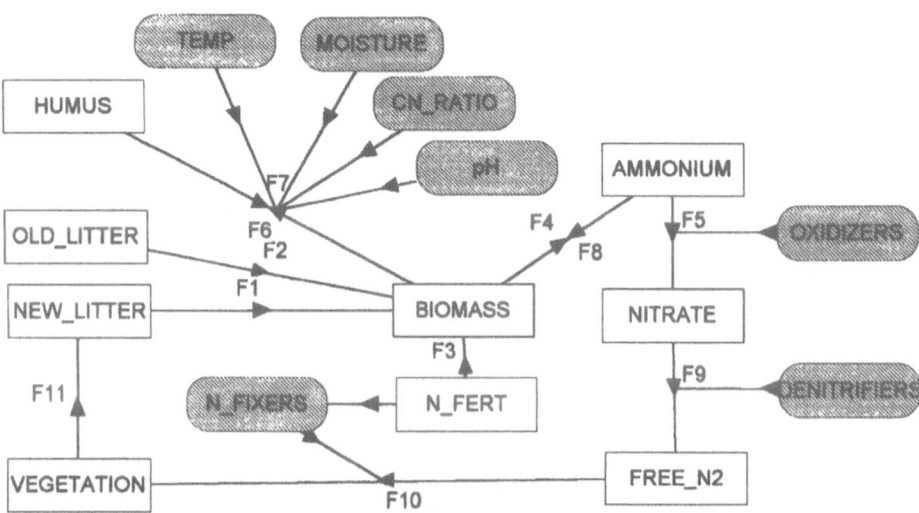

Figure 3. Simplified scheme of microbiological processes in soil. The boxes represent N and C quantities. Arrows between boxes represent N fluxes. Not all of them could be indicated. Eg the uptake of ammonium and nitrogen by vegetation is not indicated. The arrows between shaded symbols and fluxes represent influences which control the rates of fluxes. The temperature, moisture content and pH control the rate of all microbiological processes. They are only indicated for the production rate of humus. Also the C/N ratio in soil is a rate controlling variable. The oxidizers of ammonium, the denitrifiers of nitrates and the symbiontic N fixing microorganisms contain also nitrogen. In this scheme they are, however, only considered as rate controllers and not as part of the N-cycle. The symbols F1 - F11 represent the N fluxes between the indicated compartments. C fluxes are not shown.

In most soils in the temperate climate ammonium is rather fast oxidized to nitrate. In water logged soils at low temperatures, the oxidation rate of ammonium is very slow.

The ammonium which forms after decomposition is not oxidized to nitrate and remains available in the organic layer. It competes with Cs ions for exchangeable sites. It is one of the reasons that Cs is better available in upland and Nordic organic, water logged, soils. The models shows also the impact of nitrogen fertilizers on the symbiotic nitrogen fixation. Nitrogen fertilizers inhibit nitrogen fixation and limit therefore the competition between nitrogen fixing species and non-nitrogen fixing species. On pastures which receive annual applications of 100 to 200 kg N per ha nitrogen fixing species disappear. The root uptake of Cs by the latter species is considerably higher than the uptake by grass species [8[. Nitrogen fertilizers reduce therefore the Cs uptake on pastures which were not be fertilized before.

90

between nitrogen fixing species and non-nitrogen fixing species. On pastures which receive annual applications of 100 to 200 kg N per ha nitrogen fixing species disappear. The root uptake of Cs by the latter species is considerably higher than the uptake by grass species [8[. Nitrogen fertilizers reduce therefore the Cs uptake on pastures which were not be fertilized before.

1.4 SOIL AS A PHYSICAL CHEMICAL SYSTEM

Although one might assume that physical chemistry is far away from agriculture, this is not true. The physical chemical properties of soils are very important for agriculture and therefore also for radioecology. An important aspect of the difference between clay, silt and sand is the water holding capacity. Clay minerals swell in water thus storing important quantities of water in the soil. The vegetation can use this water during dry periods. A clay soil stores about three times as much water as a sandy soil; it is one of the reasons for the -potentially- high yields of clay soils.

A more important aspect, certainly for radioecologists, is the high exchange capacity of clay minerals. Nutrients, but also radionuclides, can be reversible adsorbed on the exchangeable sites and are, therefore, easily available for uptake by the plant root.

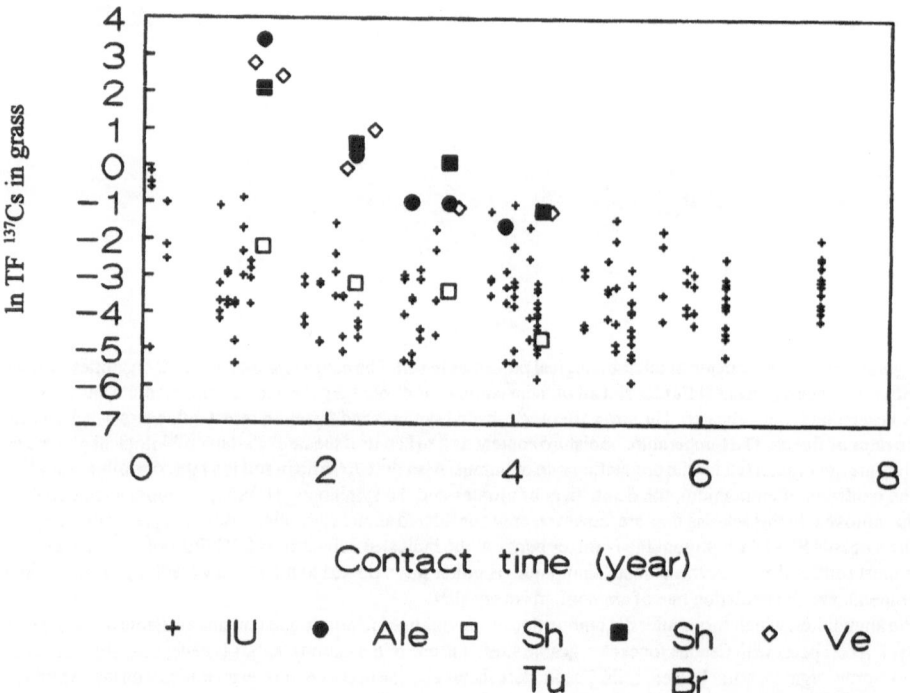

Figure 4. The uptake of Cs by grass as a function of the time that the Cs is present in soil.
The IUR values were mainly determined in north western Europe [10].
The other values were determined in the CIS countries and collected by Aleksakhin (Ale) [11], Shutov in the Tula region, black soils (Sh Tu) [12], Shutov in the Brinask region, sandy soils (Sh Br) and Vetrov (Ve) [13].

The type of clay mineral is also important. The surface areas of three representative clay minerals: Montmorillonite, illite and kaolinite are 400-900, 50-300 and 4-80 m^2 per g, respectively [9]. Adsorption capacities vary in a similar way. A complication is formed by illite. Illite platelets adsorb Cs ions between two platelets in such a way that the platelets are closely packed to each other. Cs desorbs therefore very difficult from illite. K and NH_4 ions, which have similar dimensions as Cs ions, promote desorption. Illite is a clay mineral which is very abundant, almost all soils contain traces of illite. Therefore almost all soils fix Cs. For field situations it is difficult to describe Cs fixation quantitatively. The uptake by plants as function of the time that Cs is present in soil provides a pragmatic and practical example of the fixation rate, an example is shown in figure 4. The Cs data in figure 4 suggest that Chernobyl Cs (the CIS data) is better available than nuclear weapons fallout Cs and laboratory prepared Cs (the IUR data). After some time the difference disappears and the fixation rate is for some years about 10 to 15 % per year. A first idea is that both Cs forms must be different. An other explanation is that pastures in the CIS are not fertilized with K, while in north western Europe (IUR data) fertilization is customary. The vegetation in the CIS therefore is K deficient and will take up all available K and Cs; in north western Europe this is not the case. Another explanation is that in the CIS the Cs was present in the top cm layer, thus causing more soil adhesion to the grass. The top layer of soil contains also more roots, thus promoting uptake. Yet another reason is that the vegetation on fertilized and non-fertilized pastures differs (see section 1.3). For Sr, all effects are less pronounced, the fixation rate is only a few percent per year. Because of the large annual variation in radionuclide uptake by crops this effect is hardly noticeable.

At this stage, an unavoidable key question is if it is possible to determine the availability of radionuclides in soils via a chemical method, eg with an extractant that simulates the plant root. It is a very old question. For many years farmers try to get an insight in the fertility status of soils and they have tried many types of extractants. Despite hundreds of experiments they failed. They solved the problem in another way. They applied regression analysis between extracted amounts, fertilizer applications and yields. This appeared to be very successful. So farmers apply various extractants, but do not assume that the extracted amounts simulate the available amounts. For radioecologists this approach is not possible; they cannot carry out hundreds of experiments and correlate uptakes, applications of amendments and extraction data. They have to try to analyse the soil chemically indeed. A very general characterisation scheme is shown in the left hand side of Table 1, a more refined scheme on the right hand side.

Sequential extraction techniques are mainly used to characterize "hot particles" in a soil. Hot particles are aerosol particles with a high specific activity. In a critical organ they may produce a dose of up to 10,000 mSv. When hot particles are deposited on a soil the physical chemical properties pH and redox potential determine eventually the chemical form of the radionuclides. Both properties can, however, not predict how much time is required before the equilibrium situation will be reached; therefore sequential extraction techniques are applied. The time required to solve hot particles in a soil lasts several years. Petryaev [15] showed how the fraction of available Sr near Chernobyl increased during three years; Sr uptake data by Aleksakhin [11] support this view. For Cs the fixation rate, about 30 % per year in the first years after contamination, hides the fact that Cs becomes available due to leaching of hot particles.

TABLE 1. Some extractants applied in sequential soil extraction.

General method		Method by Salbu et al [14]	
Extractant	Species	Extractant	Species
H_2O	Mobile compounds	H_2O	Mobile compounds
NH_4 acetate	Exchangeable cations	NH_4 acetate	Exchangeable cations
HCl	Acid soluble, non-exchangeable cations	Na acetate, pH=5	Carbonates
		$NH_2OH.HCl$ in acetic acid	Easily reducible Fe-Mn-oxides
		H_2O_2	Oxidizable org. matter
HNO_3 or $HClO_4$ + HF	Fixed compounds	HNO_3 7 molair	Acid digestible compounds

A new approach to determine the bioavailable radionuclides in a soil is to collect the soil solution. In earlier times this had to be done by centrifugation. A new method is to replace the soil solution by a liquid which does not mix with water [16]. It is a promising technique, although it remains difficult to simulate the plant root. The root is able to take up more radionuclides from the soil than the soil solution does.

1.5 SOIL TYPES

Perhaps soil types should have been the first paragraph of the section soils. But all mentioned topics as texture, hydrology, biology and physical chemistry of a soil determine the type of soil. As mentioned in the introduction there exist many systems for the nomenclature of soils. One important aspect of soil has not yet been mentioned, ie its pH. In many cases the sophisticated nomenclature describes also the pH. A striking example is rendzina, which indicates a calcareous soil. The pH will always be high; acid calcareous soils are non-existent.

On the contrary, a definition as sandy soils does not say anything on the pH. Sandy means only that the particle size of the soil particles is between 50 and 2000 µm. According to this definition even calcareous material may be called sand. Beaches of atoll islands exist of calcareous material indeed. The majority of sandy material exists, however, of quartz -mainly SiO_2-. This material has an extremely low exchange capacity; such soils are very infertile and radionuclides may easily be taken up. The statement that on "sandy soils " the root uptake is relatively high refers to rather acid quartz sandy soils of north western Europe and the northern CIS countries. For clay similar limitations exist. Most clay soil data refer to marine depositions of clays. These deposits contain shells and have therefore a pH not too far from 7. There exist, however, also clay deposits which are very acid. Examples are river deposits as Thionic fluvisols. They are only suitable as grassland. After liming -but in that case they are no longer very acid- they can be used for other crops too. Many data sets list loamy soils. As explained, this is a texture based definition. In practice almost all data for loam refer to the aeolian deposited soils known as loess or löss soils. Loess soils cover a wide band of Europe and Asia. The name peat soil refers to the composition; a major part of the soil consists of organic material.

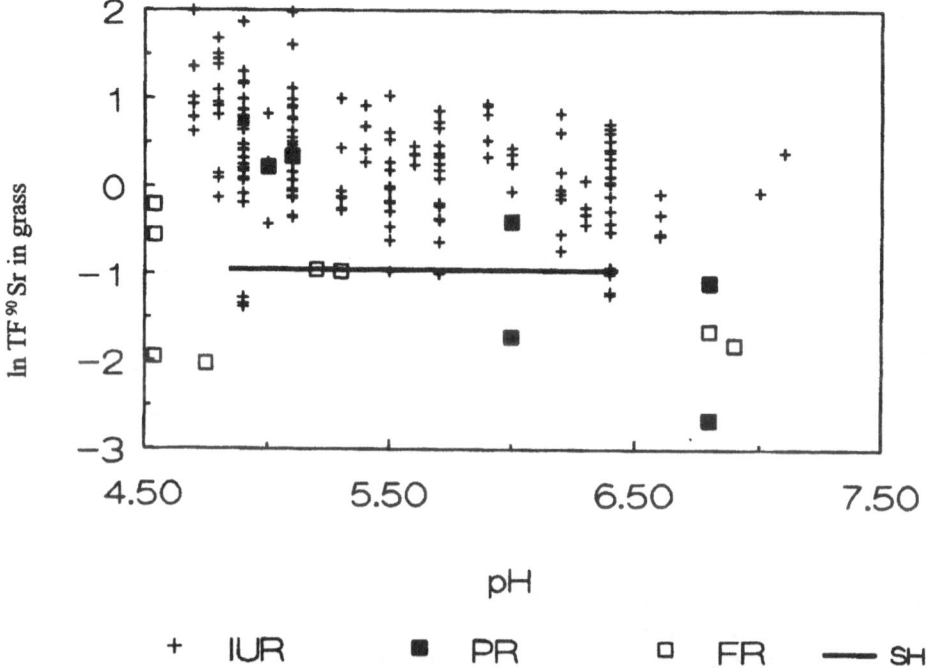

Figure 5. Root uptake of grass as a function of pH. Data sources: IUR [10], Prister [PR, 8], Shutov [SH, 12]. FR refers to recalculated data from 1965 by Frissel.

Both very fertile and very poor organic soils exist. Wet organic upland and Nordic soils may have extremely low pH values and redox potentials. Their productivity is almost zero. The appearance of such soils does not look too bad, but that is because also the decomposition rate of the organic material is low. When such a peat soil is transported to a warmer climate and is drained, the productivity becomes very high. It is never a good idea to move a peat soil to another location to use it for experiments. Temperature and moisture content determine the growth, not the chemical composition of the peat soil. Type of soil and pH are correlated. It is therefore a question of preference if the pH or the type of soil has to be used describing a system. Whatever the choice might be, additional remarks on region or pH are required. The IAEA-IUR "Handbook on transfer parameters in temperate climates" provides eg soil-to-plant transfer values for different soil types, but in addition a standardized pH is indicated [17]. In most data sets such an indication is missing. An example of the impact of pH on the root uptake is shown in figure 5. Another example of the application of the pH to classify soils is reported by Prister [18]. He uses pH groups of 4.5-5.5, 5.5-6.5 and 6.5-7.5 to classify remedial actions.

2. Countermeasures

2.1 OVERVIEW OF POSSIBILITIES

In general there exist four ways to reduce the uptake of radionuclides from soils:
1) Removal of the activity from the soil
2) Measures to store the radioactivity in the subsoil outside the reach of roots
3) Supply of materials which reduce the uptake
4) Change of crops

2.2 TIME SCALE

Countermeasures require intensive planning. Many measures are very time consuming and costly; others have an impact on the productivity of a soil, both in physical and economical sense. A division can be made according to a possible time scale [19]:

2.2.1 Short term actions
1) The main short term actions are not oriented to soils. They are limited to actions as prompt harvesting to save the crop and the removal of the contaminated crop to protect underlying soils.
2) The only effective action on a soil is to flatten it as much as possible before contamination occurs. It will facilitate the removal of the radioactivity afterwards. In most cases this is only an academic remark.
Note: Sometimes it is recommended to cover the soil with plastic sheets to protect the soil. It is overlooked that any vegetation will die off under the cover and that any rain or even dew will cause unsolvable practical problems.

2.2.2 Medium term actions
1) Ploughing
2) Fixing radionuclides by application of polyurethane foam and removal of foam afterwards
3) Fixing the upper soil layer by sowing grass and removal of the upper layer of soil as grass sod afterwards

2.2.3 Medium to long term actions
1) Soil treatment by adapted fertilization, special chemicals, etc
2) Special ploughing
3) Adapted irrigation

2.2.4 Long term actions
1) Removal of soils
2) Decontamination of soils
3) Change of land use
4) Abandonment of land

2.3 PLOUGHING

Ploughing seems a simple means to reduce the uptake of radionuclides. The only complication is that ploughing must guarantee that the radioactivity is brought below the rooting zone. And that is not simple. The real reason that soils are ploughed is to bring down surviving weeds and to cover seeds of weeds so that they do not compete with the crop. A few years ago, within the European community several so called no-till projects have been carried out. They failed, not because agricultural productivity decreased, but because weeds could not be kept under control. Normal ploughs are not designed to invert soil layers. It looks more as if they turn a soil by slightly more than 90 degrees. But even if the soil upper 20 cm layer is inverted it will not reduce the uptake. In fact it is hardly important were the radioactivity is situated in the rooting zone. As long as it is in the rooting zone, it will be taken up. And roots of many crops go down very deep, eg to a depth of 40 to 80 cm. Consequently special ploughs have to be designed which bring the contaminated top layer to a depth below 80 cm. Vovk [20] reports of a tractor drawn trench plough which was used to completely invert a thick (40-60) cm layer of soil placing the contaminated top 10-15 cm to the bottom and moving the deeper uncontaminated layers onto the top. Even with deep ploughing it might be useful to cover the contaminated deep lying layer with chemicals which prevent penetration of roots in that zone. Na_2CO_3, Na_2SO_4 and isopropylphenyl carbonate have been tried. Environmental aspects of such chemicals are questionable. If they are not persistent they will not work; if they are persistent they may by transport via the soil hydrologic system emerge far from their point of application and cause harmful effects.

2.4 FERTILIZATION

Both fertilizers and manure promote the production of agricultural products substantially. The application results therefore often in the reduction of the radiocontamination of the crop if expressed per kg crop. The total uptake of radionuclides increases indeed, but the yield increases more than the uptake. Of course the method is most successful on soils which received no or limited fertilization before the contamination. On well fertilized soils it is difficult to increase the yield. Some effects may still be obtained by a dilution of the K and Ca content in the soil, resulting in a decrease of the Cs and Sr uptake, respectively. Some results are shown in Table 2.

TABLE 2. Reduction of root uptake of ^{137}Cs by fertilization [8].

Treatment	Relative improvement factor			
	Maize for silage	Beet root	Lupin	Peas
Control	1.0	1.0	1.0	1.0
Fertilizers, N 60, P 90, N 120 [a]	1.5	1.8	1.1	1.6
Lime, 1.5 times requirement	1.8	2.0	2.7	2.4
Manure, 50 ton/ha	1.8	3.7	1.3	1.9
Lime + fertilizers	1.8	2.5	2.7	2.5
Manure + fertilzers	2.6	1.8	2.7	2.4
Manure + lime + fertilizers	2.8	2.0	2.7	2.5

Treatment	Relative improvement factor					
	Maize for silage			Fodder beet		
	1988	1989	1990	1988	1989	1990
Fertilizers, N 60, P 90, N 120 [a]	1.5	1.3	1.2	1.8	2.1	2.7

[a] Fertilizer applications in kg/ha

In many countries, also in CIS countries, the best soils are used as arable land, leaving the other soils as pastures. As a consequence fertilization may be most successful on pastures. Very good effects are obtained by a combination of ploughing and fertilization. Most grass species are not rooting very deeply so that the contaminated layer does not have to be buried as deep as for arable crops. Prister [8] reports a sufficient depth of 18-20 cm for light soils, 30-35 cm for heavy and peat bog soils and 40-45 cm for drained swamp soils. Table 3 shows some amazing results. In general, -Prister reports this explicitly-, the reduction which can be reached is not more than a factor 2-4.

Reports on the successful application of K fertilizers are not limited to the CIS countries. Robison [21] reports a dramatic reduction of the ^{137}Cs content of coconuts on Bikini in the Marshall Islands. The soils were contaminated during the nuclear weapon tests in the early sixties. The mineral fraction of the soil consists for almost 100 percent of calcium carbonate; organic matter content is rather low and illite seems to be missing. The effect of the application lasts for about 5 years; thereafter the ^{137}Cs of the nuts starts to increase. A second application of K fertilizer results in a further decrease of Cs in the nuts. Some results are shown in figure 6. Coconut palms store large quantities of K in their trunk after an application of K. The explanation of the long term effect is therefore complicated and still under investigation.

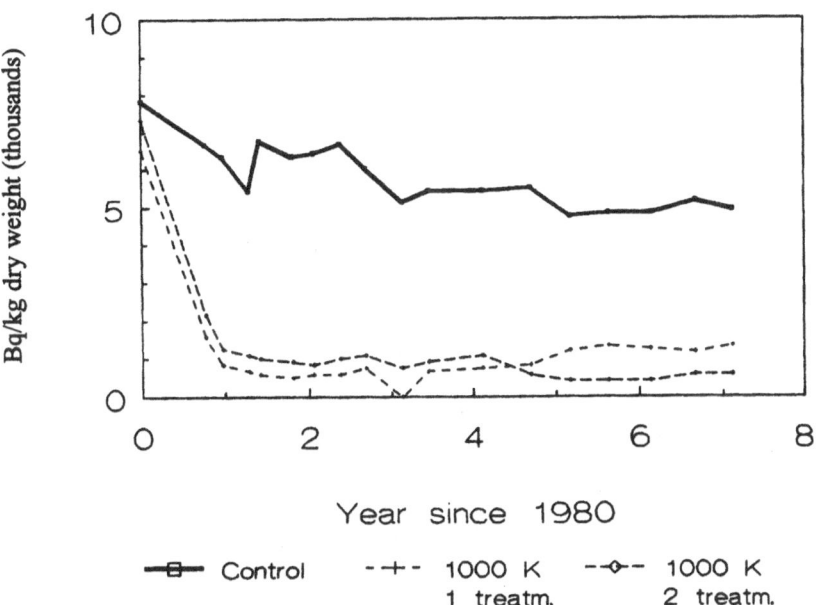

Figure 6. The contamination of coconuts on the island of Bikini. First treatments with K fertilizer in 1980, second treatment 1985 [21].

TABLE 3. Effectiveness of radical improvement of ^{137}Cs uptake in meadows on peat bog soils [8].

Treament	Transfer coefficient (Bq/kg dw/ kBq/m²)	Treatment	Transfer coefficient (Bq/kg dw/ kBq/m²)
Control	71	Ploughed + K120 + 1.5 dose lime	4.9
Ploughed	17	Ploughed + P90 + K120 + 1 dose lime	4.3
Ploughed +N60 + P69+K60	21	Ploughed + N90 + 1 dose lime	12
Ploughed +N60	39	Ploughed + N60 + P60 +K180 + 1 dose lime	12
Ploughed + N60 + P60	24	Ploughed + N120 +P60 + K120 + 1 dose lime	9.9
Ploughed + P90	30	Ploughed+N120 +P120 +K120 + 1.5 dose lime	11
Ploughed + K120	6.0	Ploughed + N60 +P150 + k60 + 1 dose lime	17
Ploughed + P90 + K120	5.3		

2.5 OTHER AMENDMENTS

Several non-fertilizer chemicals have been tried to reduce the uptake of radionuclides. To them belong various types of artificial ion exchangers. There is little doubt that ion exchangers limit the uptake of radionuclides; unfortunately they also limit the uptake of macro and/or micro nutrients. Specific ion exchangers may be developed, an example being Prussian blue. Ammonium-ferric hexacyano-ferrate (related to Prussian blue) applied in quantities of 1, 10 and 100 g/m^2 to a sandy soil reduced the uptake of Cs by ryegrass with a factor of 4, 25 and 225 respectively [22]. Many other compounds have been tried, among them alginates, alumino-silicates, phosphates, zeolites and humolite (mixture of zeolites, illites and excrements).

Very successful appeared to be sapropell. Sapropell is a sediment of natural lakes decomposed under anaerobic conditions. Addition to a soil stimulates microbial activity and adds mineral nutrients to the soil. Its effects are similar to fertilizers. There seem to be no problems with too high applications. Table 4 shows some results [8].

An overview of results which can be expected from all kinds of soil amendments was prepared at the REACT (Relative Effectiveness of Agricultural Countermeasure Techniques) meeting [23]. The Tables 5 and 6 are short versions of the tables in this overview.

TABLE 4. Reduction of the root uptake of ^{137}Cs by sapropell [8].

Treatment	Relative improvement factor			
	Maize for silage	Beet root	Lupin	Peas
Control	1.0	1.0	1.0	1.0
Sapropell, 100 ton/ha	1.8	4.5	3.0	1.8
Sapropell, 50 ton/ha + fertilizers	1.8	4.5	4.6	1.2
Sapropell,100 ton/ha + fertilizers	1.9	3.0	4.4	1.3
Sapropell, 150 ton/ha + fertilizers	2.1	3.0	2.2	1.2
Sapropell, 100 ton/ha + fertilizers + lime	2.1	2.5	1.5	1.5

	Maize for silage			Fodder beet		
	1988	1989	1990	1988	1989	1990
Sapropell, 100 ton/ha	1.8	1.9	2.2	4.5	2.3	2.9
Sapropell, 50 ton/ha + fertilizers	1.8	1.8	1.9	4.5	2.5	2.5
Sapropell,100 ton/ha + fertilizers	1.9	1.9	2.2	3.0	2.9	5.0
0Sapropell, 150 ton/ha + fertilizers	2.1	2.3	2.3	3.0	3.9	4.0
Sapropell, 100 ton/ha + fertilizers+lime	2.1	2.3	2.3	2.5	4.4	6.7

TABLE 5. Evaluation of soil-based countermeasures for reducing radiocaesium uptake by vegetation [23]

Counter measure	Reduction factor	Objective	Effectiveness	Comments
K fertilization.	up to 5	Increase of exchangeable K in soil. Decrease of Cs/K ratio in soil.	Effective only in soils with low pH.	Inexpensive. Standard agricultural technique. Max. 200 kg/ha /y.
Liming	up to 2	Increase of soil pH Increase of ion exchange capacity of soils.	Effective on soils with low pH.	Inexpensive. No effect on calcareous soils. Limit 3000 ton/ha.
Sapropell	up to 6	Decrease of Cs and K in soil solution.	Very effective.	Inexpensive. Easy to apply. Improves yield. Limited availability.
Alumino-silicates	up to 2	Increase of exchange capacity of soils.	Limited.	Rather expensive.

TABLE 6. Evaluation of soil-based countermeasures for reducing radiostrontium uptake by vegetation [23]

Countermeasure	Reduction factor	Objective	Effectiveness	Comments
Liming	up to 10, but 2-3 is more likely	Increase of soil pH. Increase of ion exchange capacity and exchangeable Ca in soils. Decrease of Sr/Ca ratio in soil	Effective on soils with low pH.	Inexpensive. No effect on calcareous soils. May cause desorption of adsorbed Sr because of preferential Ca adsorption on organic soils.
Organic matter, including sapropell	up to 5	Increase complexation of Sr. Decrease of Sr/Ca ratio in soil solution	Very effective.	Applicable in place of liming but not in conjunction with it. Org. matter may increase avail. Cs.
Phosphates sulphates and silicates	up to 10	Reduces availabilityof Sr by coprecipitation	Limited.	May reduce availability of other (essential) nutrients.

2.6 REMOVAL OF RADIOACTIVITY

It has often been proposed to remove radionuclides by developing a crop which takes up radionuclides very efficiently. It is probably just as unreal as the development of crops which do not take up radionuclides.

Leaching of radionuclides is another topic which appeared in many research programmes. In general leaching is much to slow to be effective. Perhaps Cs can be leached from organic soils if very high K and NH_4 applications are used. If successful, the question remains what to do with the drainage water which contains the Cs. The only effective way to remove radioactivity from a soil is to remove the soil. Politically it is an attractive method; it looks simple and settles the problem at once. In practice this appears not to be the case, the dominant problem is what to do with the removed soil.

Sophisticated methods have been described by Jouve *et al* [24]. A polymer foam is spread on the contaminated soil. The foam penetrates the contaminated layer, whereafter the gel polymerizes. The polymerized foam can easily be removed together with a minimum quantity of soil. Of course it works best on a bare soil which was flattened before. The same applies to a method which applies grass seed for the forming of a sod which can be removed easily by conventional machinery.

Vovk et al [20] give an overview of methods applied in the CIS after the Chernobyl accident. A number of studies concentrated on the stripping of the top layers of soils by conventional road equipment such as graders, scrapers and bull dosers. On experimental plots efficiencies of 80-100 % could be reached. In practice, results are less good. Scrapers belong to the best type of equipment; they can remove soil layers with a thickness varying from 10 to 35 cm and transport it over long distances without a need to load into other equipment. It avoids messing contaminated soil. In some cases it can be an advantage to prepare the ground for a scraper by smoothing it and to compact it by caterpillar tractors or rollers.

The removed soil has in some cases been brought to special waste depositories; in other cases large pits were used and the filled pit covered with clean soil. A problem which results from removing the top soil layer may be that also the fertile soil layer will be removed. In that case special techniques have to be applied to develop a vegetation cover. Despite the success of the method, removal can only be applied to limited areas where other methods can not be used, eg because of very high contamination levels or between buildings. An overview of decontamination factors which can be reached is given in Table 7 [20].

TABLE 7. Efficiency of various mechanical and technical methods of land decontamination [20].

Decontamination method	Equipment used	Decontamination factor
Removal of a soil layer of > 5 cm thickness	Auto scraper	26
	Auto grader	10
Removal of a soil layer of > 10 cm thickness	Bull dozer	6
Manual removal of soil	Spade	6
Cover with 15 cm clean soil	Auto scraper	20
Cover with 30 cm clean soil	Bulldozer + truck	20

2.7 CHANGE OF CROP USE

The application of other varieties, other crops or another agricultural system can significantly reduce the dose which is received from a contaminated area. Green vegetables can be replaced by cereals, pastures by arable land, food crops by industrial crops such as flax, while as a last possibility all agricultural systems can be replaced by forestry systems. From a soil science point of view the replacement may be not so easy as it looks. Many pastures are found on so called marginal soils. The soils are not suitable for other products. Reasons may vary; most common is that the soil moisture regime is not suitable. It may be too wet or too dry. Or the soil layer may be to shallow for arable land. Soils on which green vegetables are cultivated are excellent soils; nevertheless they may be too wet for many arable crops. Forests will never be a success on too shallow soils. The social and economical consequences of drastic changes of crop use are, however, much larger. Agriculture is a highly specialized enterprise with specialized machinery. A change of a production system has therefore to go together with large investments. Still more important is the employment situation. Change of land use will always go into the direction of a less intensive farming system, thus causing unemployment. Agrotechnical, economical and social constraints limit therefore the possibilities of the change of land use severely.

3. References

1. Mohrmann, J.C.J. and Kessler, J. *Water deficiencies in European agriculture* (1959) H.Veenman en Zonen, Wageningen, Netherlands.
2. Konoplev,A.V. and Bobovnikova, T.I. Comparative analysis of chemical forms of long-lived radionuclides and their migration and transformation in the environment following the Kyshtym and Chernobyl accidents (1990) in *Proceedings of Seminar on Comparative Assessment of the Environmental Impact of Radionuclides Released during Three Major Nuclear Accidents: Kyshtym, Windscale, Chernobyl*. CEC , DG XI, Luxembourg, 371-395.
3. Frissel; M.J., Jakubick, A.T., van der Klugt, N., Pennders, R.M.J., Poelstra, P. and Zwemmer, E.T.A. (1883) Modelling of the transport and accumulation of radionuclides of strontium, cesium and plutonium in soil. Technical Research Report no 101. Association Euratom ITAL, Wageningen, Netherlands
4 Strebl, F, Gerzabek, M.H., Karg, V. and Tataruch, F. [137]Cs behaviour in Austrian forest ecosystem (1994). Lecture presented at the VAMP meeting, IAEA, Vienna.
5. Tikhomirov, F.A., Shcheglov, A.I. and Sidorov, V.P. Forests and forestry: Radiation protection measures with special reference to the Chernobyl accident zone (1993). *The Sci of the Total Envir* 137, 289-305.
6. Van Veen, J.A. Het onzichtbare beheersen. (1994, in dutch). RU Leiden.
7. Frissel, M.J. and van Veen, J.A., eds, (1981) *Simulation of nitrogen behaviour of soil-plant systems*. PUDOC, Wageningen, Netherlands.
8. Prister, B.S., Perepelyatnikov, G.P., Perepelyatnikova, L.V. Countermeasures used in the Ukraine to produce forage and animal food products with radionuclide levels below intervention limits after the Chernobyl accident (1993). *The Sci of the Total Envir* 137, 183 -189.
9. Frissel, M.J. The adsorption of some organic compounds, especially herbicides, on clay minerals (1961) *Verslagen van Landbouwkundige onderzoekingen nr 67.3*, PUDOC, Wageningen, Netherlands.
10. Data reported by the IUR Working Group on Soil to Plant Transfer. Data are only partly published, all are present in the Workings Group databank. IUR Torenlaan 3, 6866-BS Heelsum, Netherlands.
11. Aleksakhin, R.M. Countermeasures in agricultural production as an effective means of mitigating the radioecological consequences of the Chernobyl accident (1993). *The Sci of the Total Envir.* 137, 9-20.

12. Shutov, V.N. Influence of soil properties on ^{137}Cs and ^{90}S intake to vegetation (1992). VIIIth report IUR Working Group Soil to Plant Transfer. 11-15.
13. Vetrov, V.A. Personal communication.
14. Salbu, B., Krekling, T., Oughton, D.H., Oestby, G., Kashparov, V.A. and Day, J.P. The Significance of Hot Particles in Accidental Releases from Nuclear Installations (1992) in *Proceedings of the International Symposium on Radioecology-Chemical Speciation-Hot Particles.* CEC, IUR, SCSR, Znoijmo, Czechoslovakia.
15. Petryaev, E.P., Sokolik, G.A. Ovsyannikova, S.V., Leynova, S.L. and Ivanova, T.G. Forms of occurrence and migration of radionuclides from the Chernobyl NPP accident in typical landscapes of Byelorussia (1990) in *Proceedings of Seminar on Comparative Assessment of the Environmental Impact of Radionuclides Released during Three Major Nuclear Accidents: Kyshtym, Windscale, Chernobyl.* CEC , DG XI, Luxembourg, 185-209.
16. Lembrechts, J. (1993) Personnel communication.
17 Handbook of Parameter Values for the Prediction of Radionuclide Transfer in Temperate Environments (1994) *IAEA Technical Reports Series no 364.* IAEA Vienna, Austria.
18. Prister, B.S. Agricultural aspects of the radiation situation in the areas contaminated by the southern Urals and Chernobyl accident (1990) in *Proceedings of Seminar on Comparative Assessment of the Environmental Impact of Radionuclides Released during Three Major Nuclear Accidents: Kyshtym, Windscale, Chernobyl.* CEC , DG XI, Luxembourg, 449-463.
19. Partly derived from a draft of an IAEA report on Agricultural Countermeasures
20. Vovk, I.F., Blagoyev, V.V., Lyashenko, A.N. and Kovalev, I.S. Technical approaches to decontamination of terrestrial environments in the CIS (former USSR) (1993). *The Sci of the Total Envir* 137, 49-63.
21. Robison, W. (1994) Lawrence Livermore Nat. Lab., Livermore, USA. Personnel communication.
22. Vandenhove, H. (1995) Personal communication.
23. Nisbet, A.F., Konoplev, A.V., Shaw, G., Lembrechts, J.F., Merckx, R., Smolders, E., Vandecasteele, C.M., Lönsjö, H. Carini, F. and Burton, O. Application of fertilisers and ameliorants to reduce soil to plant transfer of radiocaesium and radiostrontium in the medium to long term -a summary (1993). *The Sci of the Total Envir* 137, 173-182.
24. Jouve,A., Schulte, E., Bon, P.,and Cardot, A.L. Mechanical and physical removing of soil and plants as agricultural mitigation technique (1993). *The Sci of the Total Envir* 137, 65-79.

RADIOECOLOGY IN THE URALS; INVESTIGATIONS OF HYDROMORPHOUS SOILS

I.M. MOLCHANOVA AND E.N. KARAVAEVA
Biophysical station of the Institute of Plant and Animal Ecology
Urals Division, Russian Academy of Sciences
ulitsa 8 Marta, 202
CIS-620219 Yekatarinburg, Russia

1. Introduction

Fundamental science as a basis of radioecology and the requirements of life situations appeared to be beneficial for the development of radioecology. The latter comprises agricultural radioecology, radioecology of forest ecosystems, and, more generally, continental radioecology studying radioecological processes in biogeocenosis of land and inland water reservoirs [1]. Radioecological investigations of the soil-vegetative cover received an essential place in the body of studies because the soil-vegetative cover is the first shield against radioactive substances coming to the earth's surface from the atmosphere.

Finally, soil becomes a main depot of radionuclides in the terrestrial environment. Soil is a fine and subtle shell of the biosphere, mostly inhabited and full of life and, at the same time, very sensitive to injury caused by radiation from radioactive contamination. At present, man is studying many processes of the behaviour of radionuclides within the soil-vegetative cover, as well as mechanisms of their accumulation by living organisms and their transportation via food chains, depending on ecological and physicochemical factors. Peculiarities of the behaviour of radionuclides in various geochemical landscapes and individual biogeocenoses are also investigated. Fast development of nuclear industry and nuclear power engineering necessitates a multiaspect evaluation of the environmental behaviour of radionuclides.

2. The Institute of plant and animal ecology and its biophysical station

The theoretical bases for the advancement of radioecology were established by V.I.Vernadsky [2]. Other important scientists are B.B.Polynov [3] and A.I.Perelman [4] for biogeochemistry of landscapes and V.N.Sukachev [5] for the doctrine of biogeocenosis. The fundamentals of earlier genetic soil studies were laid down by V.V.Dokuchayev. The first radioecological investigations in our country were carried out in the fifties under the supervision of N.V.Timofeev-Resovsky [6] at the Institute of Plant and Animal Ecology of the Urals Branch of the Russian Academy of Sciences. Propagating the ideas

F. F. Luykx and M. J. Frissel (eds.), Radioecology and the Restoration of
Radioactive-Contaminated Sites, 103–114.
© 1996 *Kluwer Academic Publishers.*

of his great predecessors and considering the experience acquired during centuries of observing harmful environmental impacts of industrial waste, Timofeev-Resovsky asked for a complete study of all possible effects of the development of the nuclear industry on the biosphere.

3. Scientific approaches

The author's methodological approach to solving the main radioecological problems of the soil-vegetative cover consists of two complementary methods:
1) experimental techniques and
2) system analysis of natural objects.

The principle of the experimental technique is analytical reduction, separating a complicated natural system into a sequence of simple ones, in which causality relations are revealed and studied. The essentials of this approach were founded and successfully developed by A.A. Titlyanova, a collaborator of Timofeev-Resovsky in the fifties.

3.1 THE SYSTEM APPROACH

The system approach, the essentials of which were founded by V.V.Dokuchayev, is used for studying a system of several interactive objects. It proceeds from the point of view that the objects in a system are individual but interrelated at the same time and, due to their interaction, produce common results. For example, in a radioecological investigation of a landscape, a distinction is made between parts such as watersheds, slopes, river valleys and swampy low grounds, which are localized on relief elements contiguous to a sink.

Whatever the method of analysing radionuclide migration in the soil-vegetative cover, the specificity of the investigations proceeding from the achievements of modern soil science should be observed. The concept specifies that the migration of microquantities of radioactive elements in soil has apparently a non-steady state character and, because they are microquantities, depends not only on the radionuclides own properties, but also on the properties of non-isotopic carriers.

3.2 THE EXPERIMENTAL APPROACH

In the experimental approach, the factors and mechanisms governing the migration ability of radionuclides in each link of the environment are being studied. In agreement with the systems analysis approach, the soil-vegetative cover is separated into such simple links as soil-solution, soil-plant and soil-vegetative cover of natural biogeocenoses.

3.2.1 The soil-solution
The solubility of all compounds of a soil depends on the liquid-to-solid phase ratio. A substantial part of the tests on the soil-solution system was performed, using a standardized liquid-to-solid phase ratio. This simple system allowed to estimate the

effects of several ecological and physicochemical characteristics of a soil (alkali-acidic conditions, concentration of isotopic and non-isotopic carriers, artificial complexing agents, water-soluble organic substances, soil moistening regime, temperature conditions) on the mobility of radionuclides, defined by their degree of absorption and stability of retention by soil.

3.2.2 Sorption

It was found that the rate and the degree of sorption of microquantities of radionuclides depend primarily on the intrinsic chemical properties of a radionuclide but depend only poorly on the soil properties. Variation of alkali-acidic conditions of the medium in a wide pH-range does not affect the sorption of non-hydrolysable radionuclides. The sorption of readily hydrolysable elements decreases with the conversion of ionic forms into colloidal ones. As a rule, non-isotopic carriers of radionuclides, when present in the solution, reduce radionuclides sorption by the soil. The mechanism of this phenomenon is either associated with the processes of co-sedimentation and sorption of radionuclides with colloids of macro-carriers or determined by competitive relations with the macro-elements, which are chemical analogues of the radionuclides. A majority of radionuclides, in the form of intercomplex compounds with artificial complexing agents and in the presence of water-soluble substances, is poorly absorbed by soil. Their mobility in soil depends, to a large extent, on the level of the water content. Tests revealed a higher migration ability of ^{90}Sr compared to ^{137}Cs and ^{144}Ce for various levels of water content in soil. The role of different forms of soil moisture (gravitational, capillary-adsorbed and crystalline) on radionuclide migration in soil was evaluated in a series of special experiments. They showed that at a low solid-to-water ratio, both Sr and Cs are present in capillary adsorbed water; at a high ratio, Sr is present in the gravitational water (Table 1).

TABLE 1. Distribution of radionuclides over soil moisture forms (%) as a function of the solid-to-liquid phase ratio.

Soil	Solid-to-liquid phase ratio	Gravitational		Capillary-sorptive	
		^{90}Sr	^{137}Cs	^{90}Sr	^{137}Cs
Soddy-podzolic	0.3	9.2	1.0	90.0	99.0
	1.0	84.5	2.0	15.5	98.0
Soddy-meadow	0.3	5.0	1.0	95.0	99.0
	1.0	76.4	9.5	23.6	90.5

3.2.3 Soil-vegetation studies

It appeared that in soils where radionuclides are firmly retained, they are to a lesser degree taken up by plants. A relatively high mobility of ^{59}Fe, caused by the presence of poorly sorbable colloidal forms, and of ^{90}Sr, caused by its poor retention by the soil, results in a relatively high uptake of these radionuclides by plants. The radionuclides, present in a soil as complex compounds are accumulated by plants to a greater extent than those introduced in the form of ordinary salts.

A study of radionuclide migration in a more complex system, i.e., in a natural forest, showed that the above-surface mass of herb plants contains less than one percent of the total amount of radionuclides in the system and the roots retain approximately one to two orders of magnitude more.

Investigations carried out in contiguous plots of a mountain-forest landscape in the Southern Urals and in a tundra landscape in the far north revealed that ^{90}Sr has a greater ability for migration than ^{137}Cs, both in the soil itself and in the landscape as a whole.

4. Studies on fallout

In the middle of the sixties, the main source of radioactive contamination of the soil-vegetative cover was global fallout from the atmosphere [7]. Distribution processes in space and time are influenced by climatic conditions, soil properties, type of vegetation and local geomorphology [8,9].

The aforementioned investigations, which started in the middle of the sixties, showed that the main quantity of ^{90}Sr and ^{137}Cs in all landscape elements was retained by the vegetative cover and the upper soil layers. However, ^{90}Sr, as expected, shows a greater migration ability in all landscape elements. This is indicated by the change of radionuclide content in the soil-vegetative cover in different areas and/or different values of Sr/Cs ratios. The changes were less than one times the stored amount in eluvial areas of the tundra and Southern Urals, while the ^{90}Sr content was 2-3 times higher than that of ^{137}Cs in radionuclide accumulating areas.

It is interesting to look at the distribution of ^{90}Sr and ^{137}Cs in geochemically contiguous areas of a particular landscape during the period with a stable fallout rate, i.e., in the eighties and nineties. Non-violated areas of the landscape, located in the Central Yakutia, the Middle Urals and the Southern Urals, were investigated. Twenty years after the nuclear weapon moratorium, ^{90}Sr was more or less uniformly distributed in the vegetative cover, forest litter and top 5 cm of soil layer. The ^{137}Cs content of the vegetative cover and soil appeared to be, on average, 4-5 times higher than that of ^{90}Sr. It also appeared that, in all soils studied, ^{137}Cs is accumulating in the well-decomposed forest-litter layer, which is the most humus-rich part of the A horizon. Calculations show that 40-80% of the total content of ^{137}Cs in soil is retained by this layer.

During the same period, the values of the Cs/Sr ratio for soils of various landscapes in the Central Yakutia were close to their ratios in fission products (Table 2); the amounts of ^{90}Sr and ^{137}Cs in the soil profile were 1.8-3.0 and 2.7-3.7 kBq/m², respectively.

TABLE 2. Radionuclide content in the soil-vegetative cover in the period of stable
radioactive fallout from the atmosphere, kBq/m2

Region	Topographical element	^{90}Sr	^{137}Cs	Cs/Sr
Central	Watershed	1.8±0.6	2.7±0.6	1.5
Yakutia	Slope	1.7±0.2	2.3±0.6	1.4
	Slope foot	3.0±0.5	3.7±0.8	1.2
Middle Urals	Plakor zone	0.7-1.5	3.0-6.6	4.0-6.1
Southern	Watershed	1.5±0.3	6.8±0.9	4.5
Urals	Slope	1.4±0.3	4.6±0.6	3.3
	Slope foot	1.2±0.2	4.9±0.8	4.1
	High moor bog	2.0±0.4	2.9±0.4	1.4

The Cs/Sr ratio is high for the Middle and Southern Urals for all topographical elements,
except for high-moor bog.

5. Mining and tailings

At the initial stages of the nuclear fuel cycle, mining and enrichment of uranium ore are
the main sources for the increase of radionuclide concentrations in the environment.
This can be illustrated by the behaviour of U, Ra and Th during a geological survey in
Southern Yakutia.

Uranium ore reserves were prospected in that area. A great volume of rocks, rich in
uranium and its decay products, were taken to the surface from adits and stored. For a
long time, waste dumps underwent weathering of their rocks and both chemical elements and radionuclides scattered around. This resulted in the formation of zones adjacent to the storage dumps with an increased radionuclide content in organisms. Thus, U
and Ra concentrations in soil samples exceeded the reference level by 100 to 1,000
times. The concentration of these elements in representative wood and herb plants growing
in the polluted areas is from 3 to 70 times higher than in similar plants species in a
reference area. The highest uranium content -up to 18 g/kg of ash- was detected in moss
and lichen. The ability of this type of vegetation to accumulate radionuclides was reported in [10-12].

6. The Beloyarsky nuclear power plant (BNPP)

The use of nuclear power may cause of release the radionuclides into the environment
Even the release of strictly controlled quantities may, in the long run, create areas with
an increased radionuclide content. Since 1978, the authors are engaged in radioecological
investigations in the vicinity of the Beloyarsky Nuclear Power Plant (BNPP), named
after I.V.Kurchatov, and put into operation in 1964.

6.1 TERRESTRIAL CONTAMINATION

A 30-km zone surrounding the BNPP was examined to estimate the contribution of aerosol releases from the BNPP to the radioactive pollution of the soil-vegetative cover. Sample plots locations were chosen taking into account the wind rose and the main direction of the exhaust gases. The data obtained show that the aerosol releases from the BNPP do not contribute significantly to the contamination by ^{90}Sr and ^{137}Cs of the adjacent overground ecosystems. The concentration of ^{137}Cs in the soils of the 30-km zone is, as a rule, about 4 kBq/m^2 and is near to the mean background values typical for the Middle and Southern Urals. The accumulated amount of ^{90}Sr in the soil-vegetative cover of the 30-km zone, around 1-1.5 kBq/m^2, is practically independent of the distance from the BNPP.

A detailed investigation of the vegetative cover within a distance of 3 km of the BNPP showed a relatively low content of radionuclides in consumable mushrooms (55-58 Bq/kg of ^{137}Cs and 11-13 Bq/kg of ^{90}Sr) and higher contents in moss and lichen (180-900 Bq/kg of ^{137}Cs and 90-400 Bq/kg of ^{90}Sr). Although 10 times higher than in spores and flowering plants, their radionuclide concentrations do not exceed the concentration in moss and lichen from other regions of the country [13,14].

6.2 AQUEOUS CONTAMINATION

The BNPP discharges its low level liquid waste into the Olkhovskoye bog-river system. This system can be called " a trouble spot" and has been under investigation for many years. It is located in the 5-km sanitary zone and consists of the Olkhovskoye bog with adjacent swamp areas and the small Olkhovka river originating from the bog and discharging into the Pyshma river. There exists also an artificial channel (200 m long) between the BNPP and the bog for discharge water and city sewage outlets.

In the investigation, the content and distribution of ^{60}Co, ^{90}Sr, ^{137}Cs, ^{226}Ra, ^{232}Th and ^{238}U were determined in the principal components of the system and the contiguous soil-vegetative cover. The data available show that the main contaminants are ^{60}Co and ^{137}Cs and, to a lesser extent, ^{90}Sr. For many years, the mean concentrations of these nuclides in surface water of the Olkhovskoye bog itself and of the Olkhovka river appeared to be lower than the permitted levels established by the Radiation Safety Standards for fresh water. Yet the levels exceed from 5 to 75 times the radionuclide concentrations in the Pyshma river (Table 3). It is noticeable that the radionuclide concentrations at the inlet and the outlet of the bog are almost equal. Taking into account the fact that the quantity of water leaving the bog is 10 times higher than the quantity of entering water, it will be clear that additional quantities of radionuclides are somehow delivered to the open hydrographic network of the Olkhovskoye bog.

The high sorption capacity of the bog's bottom sediments causes them to be a main depot of radionuclides. The bog concentrates 90% of the nuclide amount entering the bog with waters discharged from the BNPP. They are primarily accumulated in the upper 50 cm layer of the peaty-slimy formations. The total amount of radionuclides in the Olkhovskoye bog grounds is approximately 7.4E12 Bq (200 Ci), which causes the bog to be a potential source of radionuclides.

An examination of areas close to the bog revealed a higher content of ^{60}Co and ^{137}Cs (the main contaminants of the bog-river system) in the near-to-bog soils and plants than in other bogs and plants. The ^{137}Cs contamination in those soils is one order of magnitude higher than in soils more than 0.5 km away. The vertical distribution of ^{90}Sr and ^{137}Cs in soil can be satisfactorily described by an exponential model composed of the sum of two or three exponential relationships. The number of exponents can be associated with the different amount of radionuclide retention by the soil-absorber complex. Our investigations show that, in the bog's soils and peaty sediments, ^{90}Sr is mainly present in an exchangeable form, while the vertical distribution of ^{137}Cs shows various forms according to depth, resulting in an enrichment of the lower layers by mobile Cs.

TABLE 3. Radionuclides in the water of the Olkhovskoye bog-river ecosystem, Bq/l.

Locations	^{60}Co *	^{90}Sr		^{137}Cs	
	1978-1986	1978-1986	1989-1991	1978-1986	1989-1991
Discharge channel	2.6(2.2-6.4)**	0.6(0.1-1.2)	0.7(0.3-1.2)	10.0(2.8-28.3)	9.0(2.4-15.6)
Olkhovskoye bog:					
beginning	3.3(2.6-7.6)	0.8(0.1-1.6)	1.3(1.0-1.6)	16.3(2.3-49.2)	8.4(1.8-10.5)
centre	2.8(1.5-10.1)	0.9(0.1-1.7)	1.2(0.7-1.2)	20.7(4.1-43.4)	17.0(9.9-24.1)
Olkhovka river:					
source	2.8(1.2-7.6)	0.8(0.2-2.5)	1.0(0.8-1.1)	16.0(9.7-35.0)	11.6(4.7-20.2)
mouth	1.3(1.1-2.6)	0.4(0.2-1.0)	0.5(0.2-1.0)	10.0(2.4-16.0)	7.9(4.1-17.9)
Pyshma river:					
upstream, 0,5 km	0.1(0.1-0.2)	0.1(0.1-0.2)	0.08(0.03-0.1)	0.2(0.1-0.4)	0.1(0.08-0.15)
downstream, 0,5 km	0.1(0.1-0.3)	0.2(0.1-0.3)	0.35(0.04-0.5)	1.3(0.1-4.4)	0.5(0.16-0.9)

* During 1989-1991 ^{60}Co levels were below the detection limit.
** Annual variation.

7. The Mayak Area

The authors of the present report were engaged in a radioecological examination of both the soil-vegetative cover in the 30-km zone of the Chernobylskaya nuclear power plant accident and the territory of the Eastern Urals Radioactive Trace (EURT) and the flood plain landscapes of the Techa river. A methodological approach, employing a combination of experimental and comparative geographical methods of investigation, allowed to collect materials, characterizing contamination levels and distribution of radionuclides of the soil-vegetative cover, and revealed mechanisms of radionuclide migration in soil profiles and mesorelief elements contiguous to a sink [12].

7.1 DESCRIPTION OF THE AREA

The territory of the regions studied comprises large water arteries and lakes with an adjacent floodplain and often swampy banks. These different hydromorphous conditions resulted in the development of different soils [15]. A gley horizon developed as a result of periodic moistening and flooding in combination with a high phreatic level. It is a beneficial medium for the decomposition of anorganic substances and the formation of readily movable organic compounds of a ligand type which contributes to a more intensive migration of radionuclides in the soil. An important role in the vertical migration of radionuclides in hydromorphous soils is played by the downward gravitational water flux and upwards capillary flux.

7.2 THE "EURT" AREA

The authors tried to identify peculiarities of the behaviour of ^{90}Sr and ^{137}Cs in this region. Investigations were focused on the determination of ^{90}Sr and ^{137}Cs concentrations and their migration in flood-plain soils.

Therefore, they examined shore areas of the lakes Bolshoi Sungul and Chervyanoye, both stretching along the central axis of the EURT, and areas of the flood plain of the Iset river. These areas were contaminated as a result of the accident in Kyshtym in 1957. In the shore zones, soil cuts were made close to the bank, at a distance of 2 to 5 m, and usually in poorly drained lowerings. Furthermore, soil cuts were made at a distance of 150-200 m from the shoreline in plots with a relatively high, well-drained relief. Those plots are representative of arable, formerly ploughed and pasture lands. In this area, soddy-meadow and dark-grey forest soils prevail.

A related study included the spatial distribution and migration of radionuclides contiguous to the sink of landscape plots in the flood plain of the Iset river. A suitable geochemical profile was found at 2.5 km west from the central axis of the EURT. This profile, with a length of 450 m, encompasses a terrace of the flood plain and stretches to a former river bed in the central flood plain. There are differences of 5 m in height. Varieties of the hydromorphous soils include layered alluvial soddy soils in the shore zone and soddy meadow soils in the terrace of the flood plain.

Results are shown in Figure 1. A distinction is made between eluvial and accumulative plots. An analysis of the data showed that, in the shore zone of the Bolshoi Sungul lake, the ^{90}Sr concentration was 1.7 times higher than in the watershed in general. Mathematical data processing by comparison of radionuclide values of the contiguous plots showed that the differences were highly significant (Student's criteria $t_{ex} = 3.9$ at $t_{0.05} = 2.6$). The increase of ^{90}Sr concentration in the excessively moistened soils of the shore zone can be explained by two processes: migration of ^{90}Sr which is stored in the solid and liquid surface sink and an additional seasonal flood supply from the water of the reservoir contaminated in 1957. The ^{90}Sr concentration in the shore zone soils of the Chervyanoye lake was also 2 times higher than that in the watershed plots, though the significance of this observation is not high (Student's criteria $t_{ex} = 2,8$ at $t_{0.05} = 2,6$). Furthermore, the ^{90}Sr

concentrations in various plots of a geochemical profile in the flood plain of the Iset river are quite equal, i.e., 6-7 kBq/m² .

For ^{137}Cs, differences are revealed only in the shore zone of the Iset river. The maximum concentration of ^{137}Cs was found in the alluvial layered soils of the flood plain in a former river bed. Earlier, it was shown that about 80% of the ^{137}Cs content in the soil cover of the EURT is supplied from some unknown sources [10].

Another conclusion from Figure 1 is that both ^{90}Sr and ^{137}Cs penetrate to a greater depth in flood plain soils with excessive moistening compared to the never flooded terrace plots of the Iset river. In the soddy meadow soil of the terrace, 90% of the radionuclides are accumulated in the upper 10 cm, while in the alluvial layered soil of the former river bed, the maximum concentration of ^{137}Cs is shifted to a depth of 10-20 cm and that of ^{90}Sr to 20-40 cm.

7.3 THE TECHA RIVER AREA

From 1949 to 1952 76 million m³ of effluent water with a total beta activity of 100 PBq (2.75 MCi) were discharged into the Techa river, due to technological imperfections in the production and storage of fission waste [20].

The pollution of water and sediments by liquid radioactive waste of the Techa river contaminates the soils in the flood plain. The flooding of those soils by river water results in the long term in stagnant water at low-lying sites. This causes the flood plain to be a severe source of secondary contamination.

To study the characteristics of the behaviour of ^{90}Sr and ^{137}Cs in the soil of this flood plain, soil cuts were made in the river bank and in that part of the central flood plain which is flooded seasonally. Distances from the river bed were 5-10 m and 40-50 m, respectively. The periodicity of the sedimentation during flooding is clearly shown in the alluvial soils. The maximum depth of sampling was 30 cm and, as a rule, reached the phreatic level.

Figure 2 shows the contamination levels of both soils. In soils from the river bank near the mouth of the river (sampling site B, 175-190 km from the discharge point), the concentration kBq/m² of ^{90}Sr was 3 times higher and that of [137]Cs 2 times lower than the concentrations in sampling site A (130-160 km from the discharge point). The fact that the ^{90}Sr concentration increases in the soils of the river bank near the mouth of the river is in agreement with the usually observed accumulation of salts in flood plains of rivers. The general trend is a gradual increase in the downstream direction. This fact reflects a greater migration ability of radiostrontium than of caesium in wet soils. A study of the ^{137}Cs distribution of bottom sedimentation in the downstream direction shows that it can be described by an exponentially decreasing function [18].

Some areas with a soil-vegetative cover in the central flood plain showed a higher accumulation of ^{90}Sr and [137]Cs (up to 600 kBq/m²) than in the river bank (Figure 2). This kind of accumulation occurs probably during the long period of water stagnation.

There is a steep decrease of total Sr and Cs concentrations in the soils of the central flood plain between inlet point and river mouth; a factor 25 for Sr, a factor 10 for Cs.

112

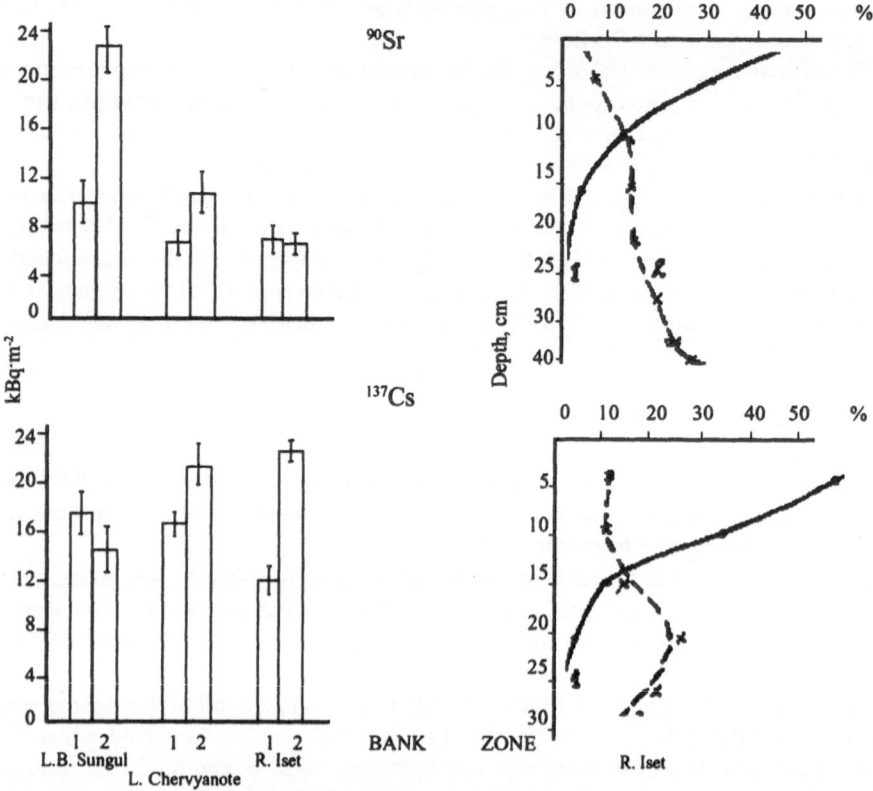

Figure 1. Radionuclides in interconnected landscape components soil
1 - eluvial, 2 - accumulative

A reduction of radionuclide concentration with distance is more vivid when data on upstream and downstream radioecological systems are compared. In this case, the difference in radionuclide concentrations may reach 3 orders of magnitude [19].

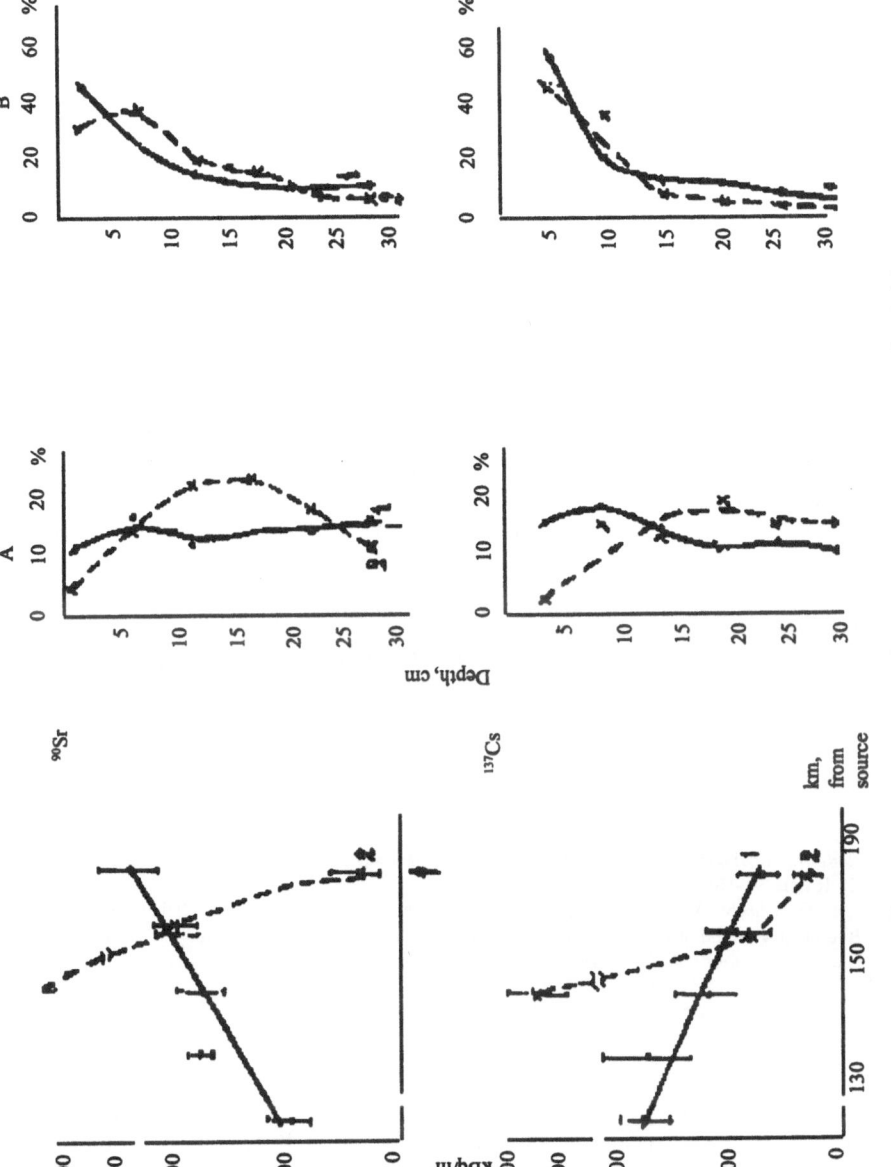

Figure 2. Radionuclides in the Techa flood plain soil (1 - bank flood plain, 2 - central flood plain) distance from source, km: A - 130-160; B - 175-190

114

8. References

1. Kulikov N. V., & Molchanova I. V.(1975) Continental Radioecology (Soil and Freshwater Systems) *M: Nauka*, 182 .
2. Vernadsky V.I. The Biosphere (1926)*L. Naych. Chim. izdat.*, 146 .
3. Polynov B.B.(1956) Selected Studies. Ed. Turin I.V., Saukov A.P. - *M: Ac. Sc. USSR - Izdat.*,751.
4. Perelman A.I.(1966) Geochemistry of Landscape. *M: Mysl'*,391 .
5. Sukachev V.N.(1966) The Basic Notions of the Biogeocenoses and General Tendency in their Study. The Program and Methods of Biogeocenological Investigations. - *M: Nauka* , 7-19.
6. Timofeev-Resovsky N.V.(1969) Some Problems of Radiation Biogeocenology, Summary of Published Works Presented for Defence of the Degree of D.Sc. (Biology). - Sverdlovsk.
7. Izrael Yu.A.(1979) Ecology and Environmental Monitoring. - *L: Gidrometeoizdat,*.
8. Pavlotskaya F.I.(1974) Migration of Radioactive Products of Global Fallouts in Soils. *M: Atomizdat* .
9. Tjurjukanova E.B.(1979) On Biological Cycles of Radionuclides. Modern Tasks and Problems of Biogeochemistry. - M: Nauka, 143- 150.
10. Ermolayeva-Makovskaya F.P., Litver B.J.(1978) Lead-210 and Polonium-210 in the biosphere, *M. Atomizdat.*
11. Kulikov N.V., Molchanova I.V., Karavaeva E.N.(1990) Natural Heavy Radionuclides in the Biosphere: Migration and Biological Effects on Habitat and Biogeocenoses. Edited by R.M. Aleksakhin. N. Nauka (1990).
12. Kulikov N.V., Molchanova I.V., Karavaeva E.N.(1990) Radioecology of Soil Vegetative Cover. - Sverdlovsk: Ur. D. Ac. Sc. USSR. .
13. Nifontova M.G., Kulikov N.V.(1977) On Accumulation of Sr-90 and Cs-137 by Lichens in Natural Conditions, *Ecologia* N 3, 93-95.
14. Nifontova M.G. , Kulikov N.V.(1981) On Accumulation of Sr-90 and Cs-137 by Some Species of Low Plants in the Surroundings of Beloyarskays NPP in the Urals, *Ecologia,* , N 6, 94-97.
15. Kovda V.A.(1973) Essentials of soil study, *M: Nauka,* **2.**
16. Aarkrog A., Dahlgaard H., Karavaeva E.N., Kulikov N.V., Myttenaere C., Molchanova I.V.,Nielsen S.P., Pozolotina V.N. Polikarpov G.G., Frissel M., Foulquier L. & Yushkov P.I.(1992) On the long-lived radionuclide content in soils and wood plants in the territory of the nuclear accident in the South Urals, *Ecologia*, N4, 50-55.
17. Resolution of the Committee on Assessment of Ecological Situation in the area of the "Mayak" enterprise production activity. Radiobiologia, (1991) **31**, N3.436-452.
18. Chukanov V.N., Trapeznikov A.V., Ekidin A.A., Yozhakov A.V., Lisovskikh V.G., Trapeznikova V.N. , Yarmoshenko I.V.(1994), A Complex ecological investigation of the rivers Techa, Iset and Miass, *Radiatsia, Ecologia i Zdorovie*, Srednii Ural. Sbornik nauchnykh trudov. Chast 1. Ekaterinburg. Ur.D.Ac/ Sc. Russia. 91-106.
19. Berzina I.G., Chechiotkin V.A., Khotuleva M.V,. Melnichenko N.A., Blohin V.I. , Scheinker I.G.(1993) Radioactive contamination of biological objects and natural media at the Musljumovo settlement (Cheliabinsk region). *Radiation biology. Radioecology,* **33**, N.2(5),747-759.

A MODELING APPROACH TO REMEDIATION OF FORESTS CONTAMINATED BY RADIONUCLIDES

W.R. SCHELL AND I. LINKOV
Department of Environmental and Occupational Health
Graduate School of Public Health
University of Pittsburgh, Pittsburgh, PA 15261, USA

1. Introduction

The extensive radionuclide contamination resulting from the Chernobyl Nuclear Power Plant accident triggered international political recognition of a shared global responsibility for the safety of all nuclear installations. The specific focus of this paper, contaminated forest ecosystems, has caused increasing concern in many European countries due to the enhanced radioactivity accumulation by forests. If forests are exposed to fallout from a nuclear-related event, forest products become contaminated resulting in possible exposure to man far from the contaminated site. This exposure can occur through the manufacture and use of forest products such as construction materials, paper, food harvested from forests, etc. Certain groups of the population, such as hunters, reindeer, berry and mushroom gatherers can have particularly large intakes of natural food products. Forestry workers have been found to receive a radiation dose up to 3 times higher than other groups of the population in the same area [1].

Contaminated forests are often given a low priority in the decontamination decision-making process, given the high complexity of forest ecosystems (i.e., the large mass of organic matter involved, layered structure of soil and plants, interaction with wild animals, etc.) and relatively low food production. On the other hand, the long-term impact of contaminated forests can be quite significant since forests have been found to be an efficient reservoir for radionuclides [2-6]. A wide variety of possible countermeasures for radiologically-contaminated forests has been proposed [1,7,8]. They range from non-destructive measures, such as the restriction of forest activity, to radical effects including tree defoliation and deforestation. Even though it is generally admitted that the decision made about the cost and applicability of a given countermeasure should be based on the net benefit from the specific action, the methodology to evaluate remedial action for a given situation has not yet been developed. At present, the experimental data used in developing a database is ambiguous and difficult to interpret. More importantly, such considerations do not address the optimal timing for remedial actions and the dynamics of the ecosystem.

Policy decision-makers need scientific guidelines for countermeasures, particularly since the significance of some forests as sources of radioactivity to man has increased. However, knowledge about radionuclide transport and fate in forest ecosystems is limited. Recent progress in this area has been summarized in publications by IAEA-VAMP and

115

F. F. Luykx and M. J. Frissel (eds.), Radioecology and the Restoration of
Radioactive-Contaminated Sites, 115–136.
© 1996 *Kluwer Academic Publishers.*

CEC-REACT Working Group [9-11]. The feasibility and effectiveness of different countermeasures for contaminated forests is qualitatively discussed in several papers. Only a few attempts have been made to approach this problem in a quantitative manner. For example, Lehto [12] illustrated the application of decision analysis techniques for mitigation purposes to a hypothetical nuclear accident in forested areas.

The present study illustrates the application of a generic dynamic model - FORESTPATH [3, 13] - to the quantitative evaluation of several possible remedial actions. The philosophy is that the model could be used as a predictive tool to guide remediation and regulation of the environmental contamination. The proposed methodology facilitates the dynamical evaluation of the system following the contamination and the applied interventions. The radioactivity accumulation in forests can be evaluated for different contamination scenarios at any time following the remedial action. The model can be developed further for evaluation of the human radiation dose and cost-benefit analysis of remedial policies.

2. Development of the generic model

2.1 RADIONUCLIDE TRANSFER IN FOREST ECOSYSTEMS

Forest ecosystems can be contaminated by atmospheric deposition of radionuclides as a result of nuclear accidents, weapons testing and by the disposal of radionuclides on the forest floor or subsurface. Following deposition onto the plants or disposal into the ground, a continuous process of radionuclide redistribution begins. Radionuclides are removed by several redistribution processes (such as washoff, throughfall, leaching and resuspension) which have characteristic rates depending on the type of forest, season, specific radionuclide, etc. and which define the radionuclide residence half-times in forest compartments.

The rate of removal of radionuclides from the forest litter, which is an important component of many forest floors, depends on several factors including temperature, moisture and type of foliage. Litter, along with lichen and moss, provide an important temporary storage reservoir for radionuclides which can accumulate for many years. Organic matter from leaf fall, lichen, moss, fungi, branches, etc. is the source of nutrients for many animals which are part of the food chain leading to man.

The hydrological cycle and plant uptake primarily determine the fate of a radionuclide in forest soil. Runoff leads to horizontal radionuclide resuspension and depends on site-specific topography. For a flat homogeneous generic forest, runoff and horizontal radionuclide resuspension are less significant. Leaching to deep soil horizons and then to the ground water are very slow processes in undisturbed forest soils. Half-times for trace element removal in the watershed have been found to be in the range of thousands of years. Thus, the deep soil compartment can be considered as a radionuclide sink for the time scale of several hundred years. The radionuclides in forest soils can be present in labile and fixed forms. The radioactivity deposited initially constitutes the labile pool, which is subjected to leaching, root uptake and absorption by the non-labile pool. Materials in the non-labile pool are not absorbed by roots nor leached into the deep soil.

To be incorporated into plants, the radionuclide must be transported through the organic layer into the root zone. It can be deduced from a literature review that the available experimental methodology and present state of knowledge on forest plant uptake constitutes a weak base for quantitative modeling. The usual approach for evaluating radionuclide uptake by plants is to use transfer factors (TF), which are defined as the ratio of radioisotope concentration in plants (Bq g^{-1} dry or wet weight) to that in soil at a defined depth (Bq g^{-1} dry or wet weight). Unfortunately, the application of TF is valid only in the case of equilibrium in the ecosystem [14]. It takes several decades for radionuclide fluxes to reach equilibrium in forest ecosystems. The uptake half-time, which depends on the chemical composition of the soil, the chemical form of the radionuclide, the biomass and biological species, can be readily incorporated in dynamic modeling.

The general pattern of radionuclide cycling in forest ecosystems can be perturbed due to seasonal changes as well as a result of animal and human activities. Although these perturbations can be significant in some forest ecosystems, for the purpose of the model developed in this paper they are considered as secondary processes for the generic forests and are not incorporated.

2.2 BUILDING A MODEL: STEPS REQUIRED

To build a radiological assessment model, information is required which allows prediction of a "health effect" resulting from some source of radionuclide input. The pathway to man is:

Concentration of Radionuclide $\rightarrow$ Exposure $\rightarrow$ Dose $\rightarrow$ Health Effects In Environmental Media

The assessing of possible health effects requires knowledge of the processes which influence the radionuclide concentration at any time, leading to exposure dose and ultimately affecting man.

The basic steps in developing of a model are shown in Figure 1.

118

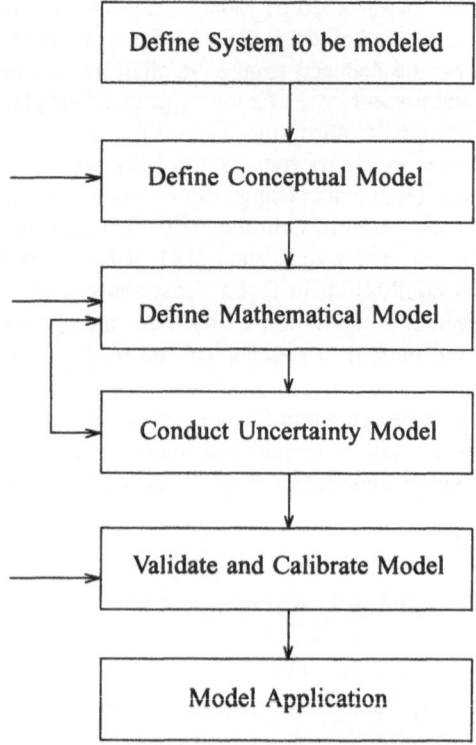

Figure 1 Steps required in model development

The Conceptual Model should be based on the available information for processes occurring in the ecosystem under study and can be pictured as a diagram. Figure 2 shows the comparison of water and radionuclide transport in a forest ecosystem.

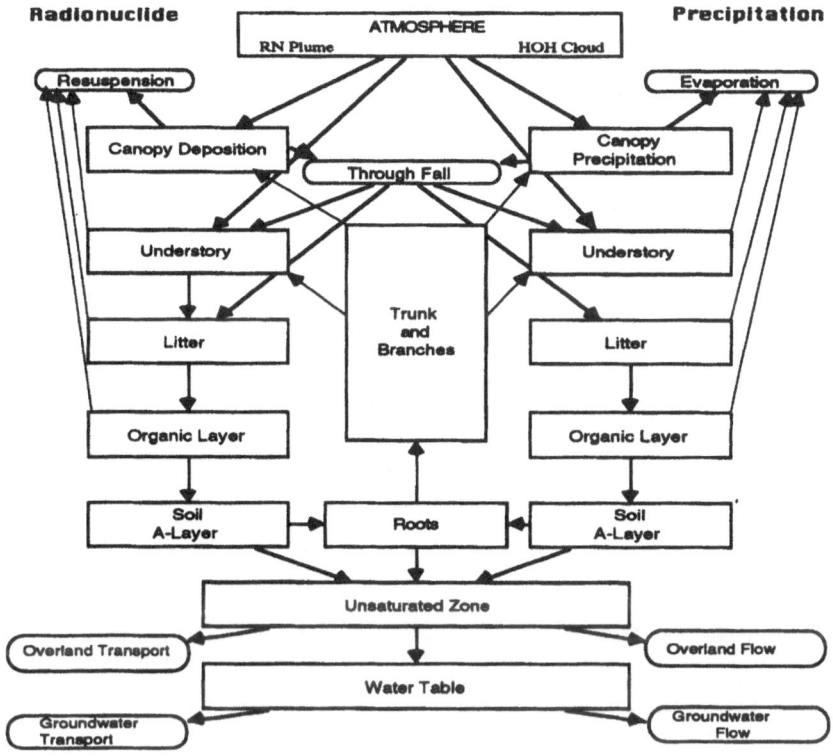

Figure 2. Radionuclide and Water Cycle in Forest Ecosystems after Nuclear Accidents.

This conceptual model for a radionuclide interacting with the forest must include:

a. Transport in the atmosphere
b. Interception by the forest canopy
c. Type of forest vegetation
d. Transport and relocation within the canopy
e. Interception by litter
f. Uptake and discharge by roots
g. Transport through the litter, humus and soil layers
h. Transport to the water table
i. Transport by overland flow to rivers and lakes

Developing a usable model for predicting the dose from radionuclide release and inter-action within the terrestrial environment requires an evaluation of the many independent and dependent parameters and variables. In addition, the dynamics of the process plays an important role. Two classes of model can be distinguished: dynamic (transient) and equilibrium (steady state). Dynamic models consider the time-dependent radionuclide quantities in various compartments, while equilibrium models consider data integrated over some time period. If equilibrium models are used, at least three time periods for the ecosystem must be considered, namely short term-weeks to months, intermediate term-months to a year, and long term-several months to many years . Table 1 shows the time when parameters related to the forest ecosystems become important.

TABLE 1. Relevant parameters for modeling the forest environment at the different times: Short term-weeks to months; Intermediate term-months to year; Long term-several months to many years.

Parameter	Short	Intermediate	Long	Comments
Deposition rates	X			
Resuspension rates	X	X	X	
Infiltration process rates			X	
Surface run-off rates	X	X	X	
Annual precipitation rates	X	X	X	Actual precipitation during deposition
Evapo-transpiration rates		X	X	
Mineral soil leaching rates			X	
Organic soil leaching rates			X	
Adsorption /desorption, kd		X	X	
Foliar deposition rates	X			
Litter understory and decomposition rates			X	
Mineral uptake, cycling and distribution rates			X	
Organic humus layer retardation rates		X	X	
Soil retardation rates			X	
Root transfer coefficient rates		X	X	
Plant-animal gut transfer rates		X	X	
Season-Input precipitation rates	X	X	X	
Foliar interception factor	X			
Leaf absorption rates	X			
Regional climatic conditions	X	X	X	
Microbiological activity rates		X	X	

Given the complex time related processes in contaminated forest ecosystems, dynamic models can provide more detailed information on the radionuclide transport, but they are usually more difficult to use. It is recommended that a comprehensive conceptual model be developed and then simplified depending on the objectives of a particular study. For example, Figure 3 presents a simplified generic conceptual model. The con-ceptual model then should be developed into a mathematical model by formulation of equations which describe the processes. Modern software can sometimes facilitate this stage, especially in the case of compartmental models. Figure 4 shows a conceptual representation of the FORESTPATH model using Stella-II software. The Stella-II framework has an underlying system of differential equations, which is represented by objects. For example, rectangles represent sink compartments for radionuclides and integrate all incoming and outgoing material flows whose directions are marked by unfilled arrows. At this stage, the boundary conditions and model coefficients should be determined.

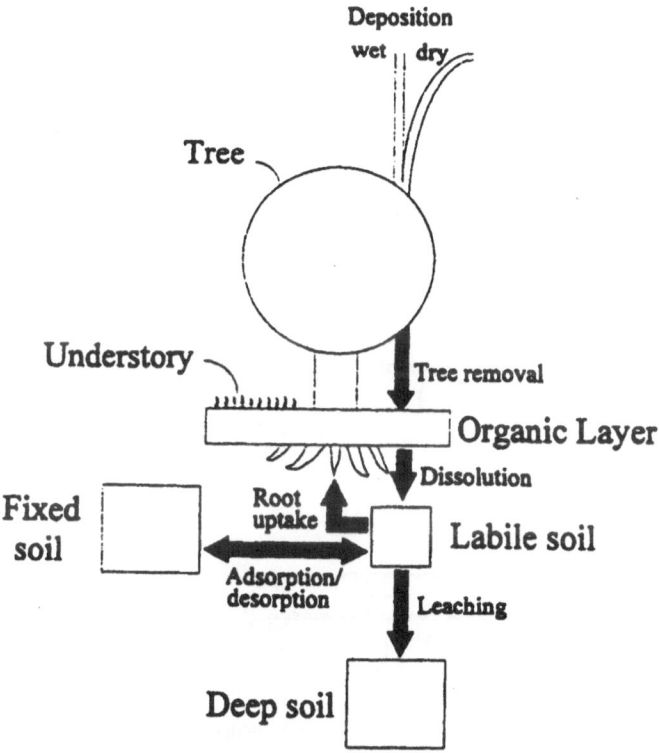

Figure 3. Generic Conceptual Model For Forest Ecosystem

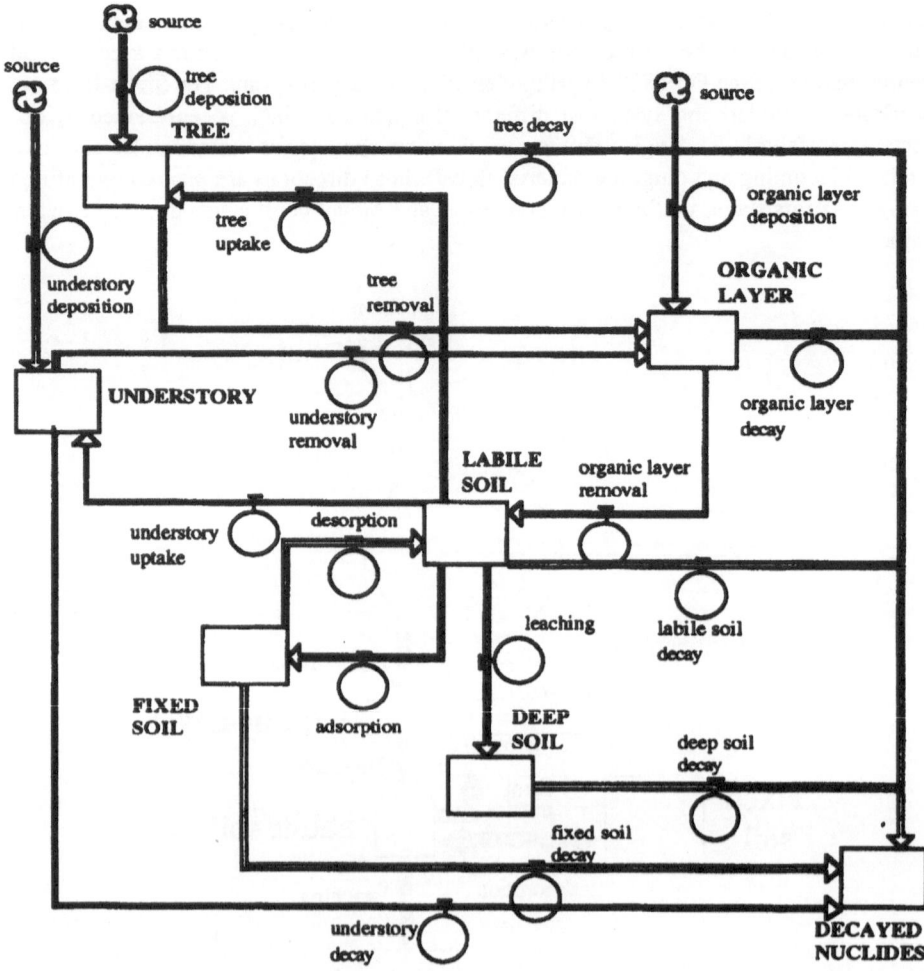

Figure 4. Conceptual Presentation of the FORESTPATH Model

The mathematical model can be analyzed deterministically or probabilistically, depending on the structure, reliability desired, and the data available. The deterministic approach requires single value data for input variables. These single values can be identified through their range and a mean and/or median value chosen. The final dose can be obtained as the solution of the model equations and is also a single value, i.e., without an indication of the uncertainty surrounding it. The probabilistic approach is used when it is not possible to assign precise values to the parameters characterizing the ecosystem but can be described by a probabilistic distribution of values. The possible model outputs would also be represented by a distribution of values. In this case, the Monte-Carlo method can be a very powerful tool to analyze uncertainties in the model predictions. Using this method, values for the input parameters are generated randomly. Each parameter appears with the frequency suggested by its probability distribution. For each combination of input parameters, the output of the model is recorded. By combining all possible outputs generated in this manner, the distribution of model outputs is synthesized. This procedure is often the basis of the so-called "Uncertainty Analysis." Thus, the task in using a deterministic or probabilistic approach then becomes a problem of identifying and collecting all relevant data and managing the information.

A distinction has to be made between the "Uncertainty Analysis," the goal of which is to analyze the consequences of the fact that the values of the input parameters within a specific ecosystem are never precisely defined and are best described by a probability distribution, and a "Variability Analysis," the goal of which is to analyze the consequences of the difference in values of input parameters across several ecosystems. For example, it would be appropriate to conduct "Uncertainty Analysis" for a specific site (such as the Chernobyl forests) using the Monte-Carlo technique. For the generic model, we are interested in the "Variability Analysis."

Different objectives are determined which can require generic or site-specific models. A generic model is designed to generalize the radionuclide cycling in a variety of similar ecosystem types. For example, in the following discussion we present the FORESTPATH model, which generalizes radionuclide cycling across different forest ecosystems and thus provides some information about various temperate forests. The input coefficients are based on an analysis of the existing data on radionuclide transport in forests and the values used correspond to averages over time, space and species. However, forest ecosystems vary significantly and depend on climate, soil types, plant diversity, etc. The available experimental data often are unreliable due to the limited number of experimental measurements. Thus, the use of the FORESTPATH input coefficients and model structure should be properly adjusted for application at a specific site.

2.3 FORESTPATH MODEL FORMULATION

FORESTPATH is a linear compartment model. Compartmental modeling implies the division of the system under study into N interconnected compartments. In a linear model, the rate of radioactivity change in each compartment is a linear function of the radioactivity in this and other compartments. The following set of first-order differential equations can, thus, be written:

$$\frac{dQ_i(t)}{dt} = \sum_{j:i \neq j} v_{ji} Q_j - \sum_{n:n \neq i} \mu_{in} Q_i + F_i(t) - \lambda Q_i$$

where
    $Q_i(t)$ is the nuclide radioactivity in the compartment i at time t (Bq x m^{-2}),
    V_{ji} is the rate of transfer from the j-th compartment to the i-th compartment (time^{-1}),
    μ_{in} is the rate of transfer from the i-th compartment to the n-th compartment (time^{-1}),
    $F_i(t)$ is the radionuclide inventory in the compartment i at time t (Bq x m^{-2} x time^{-1}),
and
    λ is the radionuclide decay rate (time^{-1}).

The major difficulty in compartmental modeling is in the determination of the rates of transfer, v and i. Prohorov and Ginzburg [15] approached the problem in the following way: if experimental data on Q_i and F_i are available for several times, t, then the system of equations (2.5.1) can be integrated for the determination of transfer rates. These values are then used for predicting the system evolution in later times. Aleksakhin and Mednik [16, 17] found a good agreement with experimental data using this approach. However, fitting the model with site-specific experimental data limits its application to different forest ecosystems. Another method for determining transfer rates for the forest ecosystems is to conduct a series of independent experiments aimed at the study of single processes (i.e., transfer between i and j compartments). However, this approach excludes other processes which run concurrently with the single process chosen. Nevertheless, we believe that recent experimental data permit an estimate of general transfer rates.

In the following discussion, the residence half-times for a particular compartment will be used to estimate the rates of the intercompartmental transfer. The residence half-times, τ_{ji}, are defined as the time needed for half of radionuclides injected to the i-th compartment to be transferred into the j-th compartment (we assume that other processes in the system do not affect the transfer between the i-th and the j-th compartments). Residence half-times are simply related to rate of transfer by the following expression:

$$\tau_{ji} = \frac{\ln 2}{v_{ji}}; \qquad\qquad \tau_{in} = \frac{\ln 2}{\mu_{in}}$$

The model FORESTPATH calculates a time series of inventories for a specific radionuclide distributed within the following six compartments: Understory, Tree, Organic Layer, Labile Soil, Fixed Soil and Deep Soil. A description of each compartment and its internal components is given in Table 2. The initial radionuclide contamination over an area is given from measurements. Radionuclide fluxes among the forest compartments (Figure 4) lead to its subsequent redistribution, which are calculated by FORESTPATH. The input coefficients used for generic forests and associated range of variation are given in Table 3. The model output values are converted to percentage of the total radionuclide inventory and can be further converted into concentration units for the purpose of assessing their impact on humans. The FORESTPATH formulation has the flexibility of including source terms from different processes (atmospheric deposition, disposal in soil, discharge to ground water, etc.) More details regarding the FORESTPATH model as well as its analytical solution can be found in Appendix 1.

TABLE 2. Components of the generic forest ecosystem.

Compartment	Components description
Tree	1. Leaves 2. Branches 3. Bole 4. Roots
Understory	1. Grass 2. Berries 3. Bushes 4. Mushrooms
Organic layer	!. Litter 2. Lichen 3. Moss
Labile soil	Reservoir of radionuclides which are subject to root uptake, leaching, and absorption to the non-labile reservoir (0.1-80 cm)
Fixed soil	Reservoir of radionuclides chemically bound in soil (0.1-80 cm)
Deep soil	Geophysical sink of radionuclides

TABLE 3. FORESTPATH model coefficients used as ^{137}Cs input in generic forests.

Parameter	Notation	Coniferous forest generic value	range	Deciduous forest generic value	range	Reliability
absorption half-time (years)	τ_{ah}	0.64	0.15-5	0.64	0.15-5	poor
tree biomass (%)	B_t	60	45-75	60	45-75	good
understory biomass (%)	B_u	20	5-35	20	5-35	good
desorption half-time (years)	τ_{da}	1.1	0.5-10	1.1	0.5-10	poor
interception fraction (%)	f	0.8	0.6-0.9	0.5	0.2-0.7	good
leaching half-time (years)	τ_{lc}	400	100-6000	400	100-6000	moderate
organic layer removal half-time (years)	τ_{or}	8	1-100	3	1-50	poor
radiation half-tine (years)	$\tau_{1/2}$	30.14	NA	30.14	NA	good
tree uptake half-time (years)	τ_{tu}	2	1-100	1	0.5-50	poor
tree removal half-time	τ_{tr}					
short (< 1 weeks)		3.6 d	1.4-5 d	3.6 d	1.4-5 d	poor
intermediate (< 1 yr)		80 d	21-175 d	25		moderate
long(>1 lyr)		3 yr	1-10 yr	0.72 yr	0.5-5 yr	poor
understory removal half-time	τ_{ur}					
short (< 2 weeks)		12 d	9-19 d	12 d	9-19 d	good
intermediate (< 1 yr)		32 d	8-64 d	32 d	8-64 d	moderate
long (>1 yr)		84 d	38-130 d	84 d	38-130 d	poor
understory uptake half-time (years)	τ_{uu}	8	1-1 00	1 0	1-1 00	poor

2.4 RADIONUCLIDE DISTRIBUTION AMONG FOREST COMPARTMENTS

An acute contamination case is presented by assuming that all deposition took place during the first week, as would occur in a major nuclear power plant accident. A 50-year FORESTPATH simulation for ^{137}Cs cycling in the deciduous and coniferous forests is presented in Figure 5 (generic parameters from Table 3 were used).

Several conclusions can be derived from the data. During the first fifty years after deposition, radioactive decay removes about 70% of radiocesium from active cycling in the forest ecosystem. The Deep Soil, which is the geophysical sink in our model, accumulates only a small fraction of the radionuclides deposited initially. The Organic Layer compartment holds the highest concentration of radiocesium (25%-85% of the initial radionuclide inventory over the 50-year period). Lesser amounts are associated with the Labile Soil and Fixed Soil compartments. Approximately 10% of the radionuclides are incorporated into both the Tree and Understory compartments, which are the main sources of human radiation dose. Thus, forests can be efficient reservoirs for radionuclides. It is clear from the data that the time at which the maximum ^{137}Cs accumulation occurs is different for deciduous and coniferous forests and even for different compartments within

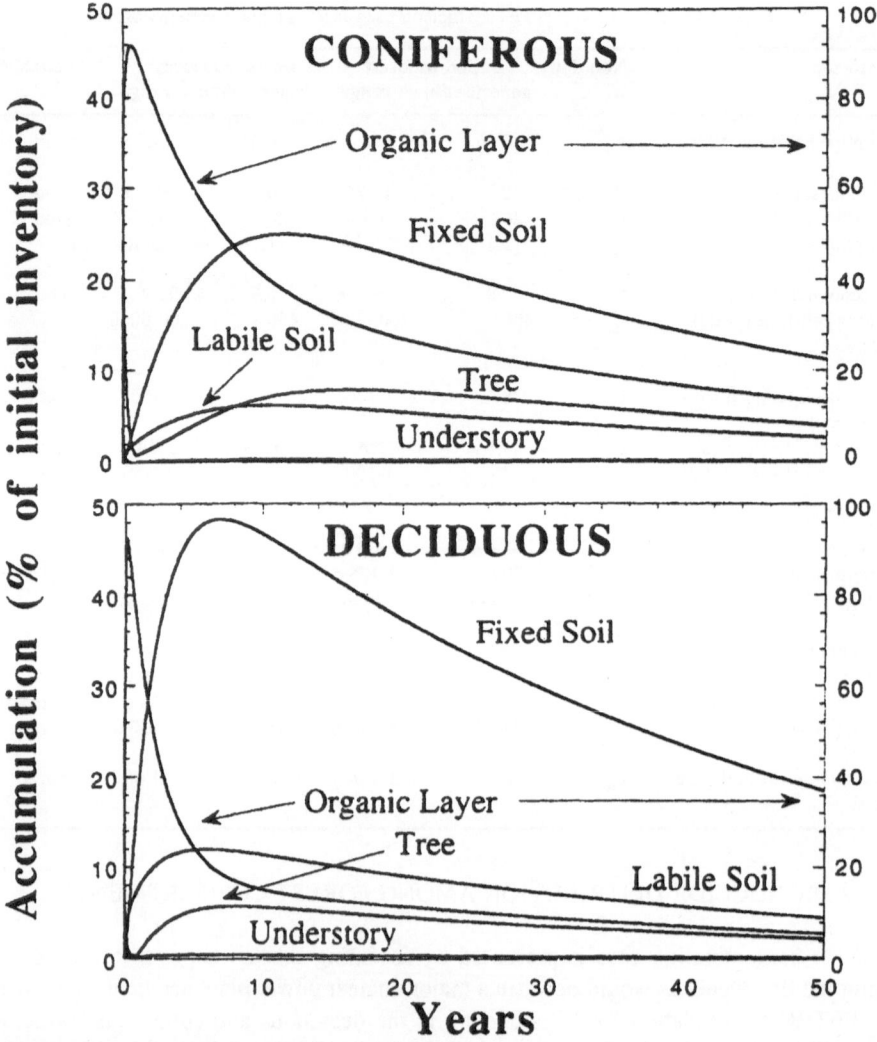

Figure 5. ^{137}Cs accumulated in generic forest model compartments over 50 years from an initial acute deposition (percent of initial inventory). Leaching to Deep Soil is negligible (not shown). The scale of the ordinate is 100% for the Organic Layer and 50% for all other compartments.

the same forest. In the case of deciduous forests, the Organic Layer reaches a maximum during the first few weeks after deposition. The amount of radionuclides in the Understory and Tree compartments increases immediately after deposition but decreases sharply during the first year, due to precipitation removal and leaffall. A second broad peak is observed in the Tree compartment about six years after deposition and is related to root uptake. The Fixed Soil and Labile Soil reach maximum contamination levels about four years after deposition, while the Deep Soil accumulates radionuclides continuously as a sink. In the case of coniferous forests, a similar behavior can be observed although the characteristic times are significantly longer. The quantity of radionuclides in the Understory and Tree compartments is larger for coniferous forests than for deciduous forests and reaches a maximum more than a decade after the initial deposition.

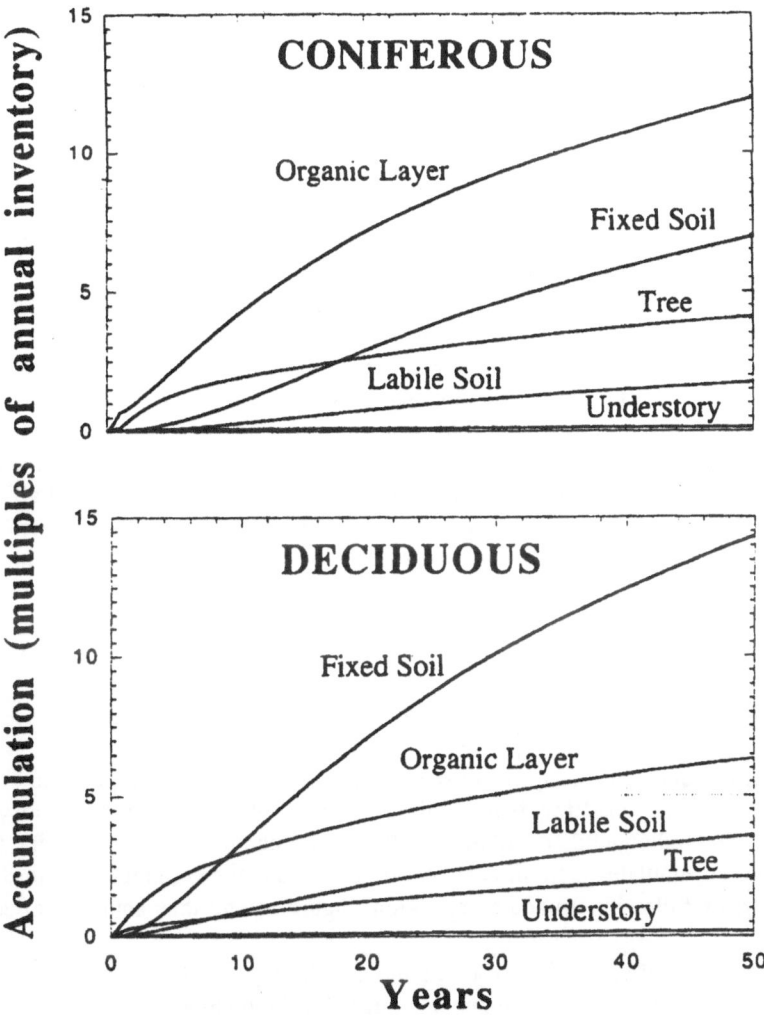

Figure 6. [137]Cs accumulated in generic forest model compartments over 50 years for a deposition with a constant rate in multiples of annual inventory (i.e., deposition, leaching from waste disposal sites, etc).

Figure 6 shows simulations with a constant annual rate of ^{137}Cs deposition on deciduous and coniferous forests (generic values from Table 3 were used). Such a situation would be typical of continuous low level emissions from nuclear power plants, nuclear waste disposal sites and redistribution by fires and dust. The radioactivity in all compartments increases with time and is much higher than the annually deposited inventory, i.e. tree contains about six times the annual radionuclide deposition, indicating that forests accumulate radionuclides. Moreover, more than half of ^{137}Cs accumulation occurs in the Understory, Tree and Organic Layer compartments, which contribute directly to the human radiation. Only a negligible part of ^{137}Cs reaches the Deep Soil.

The FORESTPATH model is designed to be used in radionuclide cycling in a variety of coniferous and deciduous forest ecosystems. Our literature review has shown that: (a) forest characteristics vary significantly from one ecosystem to another depending on the climate, soil types, plant diversity, etc., and (b) the available experimental data for environmental half-times have wide ranges due to the scarcity of measurements. Moreover, most of the measurements are not easily related to the FORESTPATH generic input parameters because, by nature, experimental measurements are taken at a site specific time and space; FORESTPATH requires parameter estimates which are averaged over time, space and species. To address this problem, variability analysis of FORESTPATH was conducted (Schell et al., 1995). We found FORESTPATH to be most sensitive to the variations in tree and understory uptake and organic layer removal half-times. FORESTPATH, when run with a representative set of values for these parameters, shows that the input parameters, on being varied from one ecosystem to another, can lead to different radionuclide cycling patterns. It is important, therefore, to narrow the ranges of variation of the input parameters for any site-specific model application.

3. Countermeasures for reducing radiation dose in forests

Organic layer removal is one of the most efficient countermeasures proposed to decrease the radiation dose resulting from radiologically-contaminated forest ecosystems [7, 8, 19]. The radionuclide distribution within forest compartments shows that the organic layer accumulates a significant portion of radionuclides (about 90% during the first two years following an acute deposition, Figure 5) and thus could be effectively treated.

Due to technical difficulties involved in removing the organic layer [8], we believe that no more than two thirds of the organic layer mass can be removed without encountering considerable problems. Moreover, the total removal of the organic layer is not thought to be safe for the health of the ecosystem as a result of the significant disturbance produced. Thus, we consider the following three scenarios for forest clean-up:
1) no action is taken,
2) about two thirds of the organic layer is removed within the first year following the release, and
3) about two thirds of the organic layer is removed 10 years after the release.

Figure 7 shows the resulting radioactivity accumulated in the Organic Layer and Tree compartments when each of these policies is applied to an acute deposition scenario. These two compartments are of particular interest since their contamination is proportional to the human radiation dose received. It is evident that the proposed intervention is beneficial especially if it is applied within the first several years following the accident. The efficiency of removing the organic layer at a later period (more than 10 years) is questionable because a significant portion of radionuclides would have migrated into the deeper soil compartments.

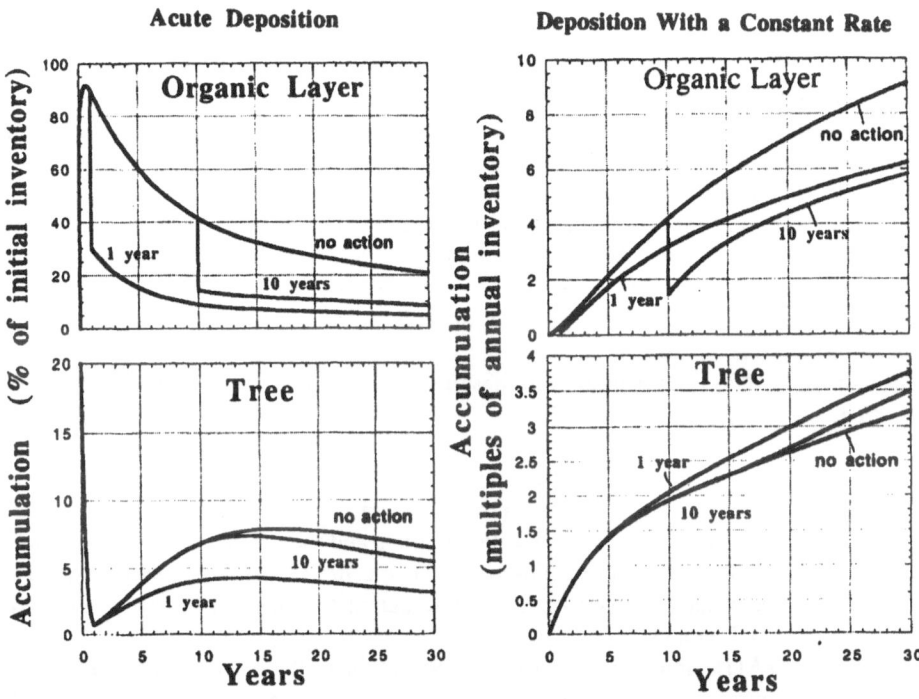

Figure 7. Effect of the Organic Layer Removal in Radionuclide Accumulation in Coniferous Forests.

Organic layer removal leads to a decrease in the residence half-time of radionuclides in this compartment. Radionuclides migrate faster through a depleted organic layer and thus can be more readily taken up by trees and the understory, resulting in an increase in their contamination levels. This factor can diminish the positive effects of this particular intervention. For example, in the situation where a constant rate of deposition occurs, organic layer removal would lead to an increase in the level of radioactivity in the Tree compartment as compared with a no-action policy (Figure 7). The application of this countermeasure in the later stages is questionable because the radioactivity decrease in the organic layer would have to be balanced against an increase in the tree contamination and the cost of the intervention.

4. Discussion

Model validation can be achieved by compiling the original literature values which describe a time-based series of radionuclide measurements in forest compartments after fallout deposition. Given the high uncertainty and variability associated with the residence times, only a qualitative comparison has been made at this stage of the FORESTPATH model development.

High uncertainty exists in the reported time series for radionuclide concentrations in different forest compartments. Many studies report that the Organic Layer is the major reservoir of radionuclides during the 15 years following the initial deposition [17]. This is consistent with the results of the FORESTPATH model (Figure 2). The available experimental data exhibits a gradual decrease in radionuclide concentrations in trees [20, 21]. Other data indicate a gradual increase Of ^{137}Cs in trees [18, 22, 23]. These variations may result from different nutrient concentrations in forest soils. In addition, climatic conditions and the variability of plant species found in the forest population affects the radionuclide cycling. This model allows one to simulate these types of outputs using different input environmental half-times. The FORESTPATH model shows that radionuclide cycling is faster in deciduous forests than in coniferous forests, which is a consequence of the biological processes operating in the two types of forest. For example, leaves in deciduous forests fall annually, while the lifetime of needles is between three and six years.

The generic FORESTPATH framework described in this paper can be easily adjusted for different site-specific applications [4]. For a given specific site, FORESTPATH can incorporate its particular radionuclide pathways by adding or changing compartments of the model. If available information on specific parameters or pathways is not sufficient, the generic FORESTPATH structure can be used. For these situations, the input parameters can be represented by probability distributions and their range of variation should be narrowed for better model predictions for the specific site. Analysis shows that it is important to determine the lower limit for input parameters in order to narrow the variation in the model outputs [3].

The FORESTPATH model may be used to evaluate the effectiveness of countermeasures. As an example, the effects of removing the organic layer from contaminated forests was considered and this intervention was found to be very efficient for the acute deposition scenario. Organic layer removal is often carried out by scraping and removing the upper layer of the forest floor. Technically, it may be difficult to remove only the organic layer without picking up soil from the upper mineral horizon, which can lead to a significant increase in the mass of generated waste (according to Vovk et al. [24], removal of 1 cm layer from mineral soil leads to generation of 100-150 tonne per ha of extra waste). To overcome this difficulty and to limit environmental damage and radionuclide resuspension, Arapis et al. [25] proposed the use of vacuum-type cleaners. Jouve et al., 1993 applied a polymer gel before commencing the operation and used turf-cutter machinery to remove the organic matter. Guillitte and Wildrodt [8] achieved an 80% reduction in radiocesium contamination by applying this technology in the Grand-Duchy of Luxembourg in 1988. For the best results, they recommended removing the organic layer immediately after the autumn leaffall of the deciduous forest, while Tikhomirov et al. [1] specified the optimal timing for this action to be during the first 2-3 years following the initial deposition. The FORESTPATH calculation suggest similar magnitude and optimal timing for the organic layer removal policy.

5. Conclusions and recommendations

A generic model FORESTPATH was developed to describe radionuclide pathways and dynamics in forests based on the current understanding of the physical, chemical and biological processes involved. Due to the different nature of biological transformations in deciduous and coniferous forests, two generic forest models must be considered. The model is designed to be flexible and simple in order to incorporate different radionuclides and contamination scenarios (atmospheric deposition~ waste disposal, etc.) as well as to evaluate human radiation dose.

The model uses only fourteen input parameters. It has been found that these parameters can vary from one ecosystem to another and have a high uncertainty due to the scarcity of experimental measurements. The results of simulations carried out using the FORESTPATH model show that forests are efficient sinks for radionuclides. The Understory, Tree and Organic Layer are the major sink compartments for radionuclides in the short and medium terms. Even several decades after a short-time radionuclide release, these compartments hold more than 50% of the total radionuclide inventory and can contribute significantly to the radiation dose to man. Constant annual deposition can lead to a significant accumulation of radioactivity in forest compartments. For example, 50 years after the initial deposition, the forest ecosystem exhibits the accumulated radioactivity more than 35 times higher than that released annually. The results of FORESTPATH modeling are in good qualitative agreement with existing experimental data.

Organic layer removal has been found to be an efficient countermeasure for the clean-up of radiologically contaminated forests. Simulations using FORESTPATH supported this empirical conclusion for the case of acute deposition and early intervention time. The deposition pattern and time of action are crucial factors for the clean-up efficiency. For deposition with a constant annual rate, the efficiency of organic layer removal is questionable since the clean-up increases the rate of radionuclide migration down to the root zone and thus results in a higher plant contamination.

The complex problem of a contaminated forest ecosystem requires use of a model to synthesize and analyze the properties of the entire ecosystem as well as to evaluate the ecosystem response to possible interventions. As we have shown, the same strategy can be beneficial or useless depending on the specific environmental conditions and time. It is evident that use of a dynamic model can facilitate the decision-making process in helping to design the efficient remedial policies. The future development of a model-based decision-support system on forest remediation requires: (a) human radiation dose evaluation, and (b) cost-benefit analysis of possible remedial policies. Social environment and consequences should be incorporated in the analysis. Advance decision science methodologies, such as multi-attribute utility theory and analytical hierarchy process could be used to facilitate this task.

6. Acknowledgments

The authors are very grateful to Professor B. Morel for the most fruitful discussions and ideas developed in the paper. We like to acknowledge Dr. Vives-Batlle for the critical review of the manuscript and Mrs. P.M. Schell for the editorial and technical assistance. This paper was prepared with the support of the U.S. Nuclear Regulatory Commission (NRC) under Grant No.NRC-04-91-093. The opinions, findings, conclusions and recommendations expressed herein are those of the authors and do not necessarily reflect the views of the NRC. Additional support was provided by NRC-NAS under CAST program.

7. References

1. Tikhomirov, F.A., Shcheglov, A.I., Sidorov V.P.(1993) Forests and forestry: radiation measures with special reference to the Chernobyl accident zone. *Science of the Total Environment* 137,289-305.
2. Schell, W.R; Berg, M.T.; Myttenaere, C.; Massey, C.D.(1994) A review of the deposition and uptake of stable and radioactive elements in forests and other natural ecosystems for use in predictive modeling, *Science of the Total Environment* 157153-161.
3. Schell, W.R.; Linkov, I.; Myttenaere, C.; Morel, B.(1995) A dynamic model for evaluating radionuclide distribution in forests from nuclear accidents. *Health Physics*,(in press).
4. Schell, W.R.; Linkov, I.; Rimkevich, V.; Chistic O.; Lutsko, A; Dvornik, A.M.; Zhuchenko, T.A.(1995) Model-directed sampling in Chernobyl forests: general methodology and 1994 sampling program. *Science of the Total Environment* (in press).
5. Myttenaere, C.; Schell, W. R.; Thiry, Y.; Sombre, L.; Ronneau, C.; Van der Stegen de Schrieck,(1993) J. Modeling of the Cs- 137 cycling in forest: recent developments and research needed. *Science of the Total Environment* 136,77-91.
6. Paajanen, A.; Lehto, J.(1992) Disposal of Radioactive Wastes from the Cleanup of Large Areas Contaminated in Nuclear Accidents. *Nordske Seminar- og Arbejdsrapporter 524.*
7. Guillitte, O.; Tikhomirov, F.A.; Shaw, G.; Johanson, K.; Dressler, A.J.; Melin, J. (1993) Decontamination methods for reducing radiation doses arising from radioactive contamination of forest ecosystems - a summary of available countermeasures. *Science of the Total Environment* 137, 307-314.
8. Guillitte, O.; Willdrodt, C.(1993) An assessment of experimental and potential countermeasures to reduce radionuclide transfer in forest ecosystems. *Science of the Total Environment* 137: 273-288.
9. Howard, B.J., Desmet, G.M., Eds.(1993) Relative Effectiveness of Agricultural Countermeasure Techniques-REACT. Proceedings of a workshop organised by the Commission of the European Communities DG XII-D-3 Radiation Protection Research DG XI-A- 1 Radiation Protection, Brussels, 1-4 October 1991. *Science of the Total Environment* 137 .
10. Desmet, G.M.; Janssens, A.; Melin, J., Eds.(1994) Forest and Radioactivity. A collection of papers presented at the seminar on the dynamic behavior of radionuclides in forests, Stockholm, Sweden, 18-22 May, 1992. *Science of the Total Environment* 157.
11. Desmet, G.M.; Nassimbeni, P.; Belli, M.(1990) Transfer of Radionuclides in Natural and Semi-Natural Environments, Proceedings of the Workshop organized by CEC. *Elsevier Applied Science*, London and New York.
12. Lehto, J., Ed.(1994) Cleanup of Large Radioactive-Contaminated Areas and Disposal of Generated Waste, *Final Report of the KAN2 Project. Tema Nord* .
13. Schell, W.R.; Linkov, I.(1995) Radiologically-contaminated forests: A modeling approach to safety evaluation and management, in: R. Lewis(ed) *Challenges and Innovation in the Management of Hazardous Waste.* Air and Waste Management Association (in press).

14. Ward, G.M., Johnson, J.E.(1985) Validity of the term Transfer Coefficient. Health Physics **50**,411-414.

15. Prohorov, V.M.; Ginzburg, L.R.(1972) Modeling the process of migration of radionuclides in forest ecosystem and description of the model. *Soviet Journal of Ecology* **2**,396-402.

16. Aleksakhin, R.M.; Ginzburg, L.R.; Mednik, I.G.; Prohorov, V.M.(1994) Sr-90 cycling in a forest biogeocenosis. *Science of the Total Environment* **157**,83-93.

17. Aleksakhin, R.M.; Ginzburg, L.R.; Mednik, I.G.; Prohorov, V.M.(1976) Model of Sr-90 cycling in a forest biogeocenose. *The Soviet Journal of Ecology* **7** ,195-202.

18. Mednik, I.G.; Tikhomirov, F.A.; Prohorov, V.M.; Karaban, R.T.(1981) Model of 90Sr migration in young birch and pine forests. *Soviet Journal of Ecology* 40-45.

19. Arkhipov, N.P.; Meshalkin, G.S.; Arkhipov, A.N.; et al.(1992) Measures (and their Effectiveness) to Improve the Radioecological Situation Given the Particular Features of the Contamination Caused by the Kyshtym and Chernobyl Accidents, in Proceedings of Seminar on Comparative Assessment of the Environmental Impact of Radionuclides Released during Three Major Nuclear Accidents: Kyshtym, Windscale, Chernobyl, EUR-13574. Luxembourg, 977-992.

20. Block, J.; Pimpl, M.(1990) Cycling of radiocesium in two forest ecosystems in the state of Rhineland-Palatinate. in Desmet, G. et al. *Transfer of Radionuclides in Natural and SemiNatural Environments,* Elsevier.

21. Maubert, H.; Duret, F.; Combes, C.; Roussel, S.(1990) Behaviour of the radionuclides deposited after the Chernobyl accident in a mountain ecosystem of the French southern Alps, in Desmet G. et al. *Transfer of Radionuclides in Natural and Semi-Natural Environments,* Elsevier.

22. Romanov, G.N.; Stukin, D.A.; Aleksakhin, R.M.(1991) "Peculiarities of Sr-90 Migration in the Environment," in Proceedings of Seminar on Comparative Assessment of the Environmental Impact of Radionuclide Released during Three Major Nuclear Accidents: Kyshtym, Windscale, Chernobyl, EUR-13574. Luxembourg, 421-436.

23. Yushkov, P.I., Chueva, T.A., Karavaeva, E.N.; Molchanova, I.V.; Kulikov, N.V.(1993) Contents Of 137+137Cs and 90Sr in the crown of birch trees in the first years after the accident at the Chernobyl nuclear power plant. *Soviet Ecology:* 70-74.

24. Vovk, I.F.; Blagoev, V.V.; Lyashenko, A.N.; Kovalev, I.S.(1993) Technical approaches to decontamination of terrestrial environments in the CIS (former USSR). *Science of the Total Environment* **137**,49-64.

25. Arapis, G.; Millan, R.; Iranzo, E.(1991) Comparison of the countermeasures taken for the recovery of rural areas contaminated by radioactivity, in Proceedings of Seminar on Comparative Assessment of the Environmental Impact of Radionuclide Released during Three Major Nuclear Accidents: Kyshtym, Windscale, Chernobyl, EUR-13574. Luxembourg 1057- 1069.

26. Jouve, A.; Schulte, E.; Bon, P.; Cardot, A.L.(1993) Mechanical and physical removing of soil and plants as agricultural mitigation techniques, *Science of the Total Environment* 137, 6579.

APPENDIX 1

Analytical Solution for FORESTPATH Model.

$$\frac{dQ_1}{dt} = -a_{11} Q_1 + a_{14}Q_4 + F_1 - \lambda Q_1 \tag{1}$$

$$\frac{dQ_2}{dt} = a_{21} Q_1 - a_{22}Q_2 + a_{23}Q_3 + F_2 - \lambda Q_2 \tag{2}$$

$$\frac{dQ_3}{dt} = -a_{33} Q_3 + a_{34}Q_4 + F_3 - \lambda Q_3 \tag{3}$$

$$\frac{dQ_4}{dt} = a_{42} Q_2 - a_{44}Q_4 + a_{45}Q_5 - \lambda Q_4 \tag{4}$$

$$\frac{dQ_1}{dt} = a_{54} Q_4 - a_{55}Q_5 - \lambda Q_5 \tag{5}$$

$$\frac{dQ_6}{dt} = a_{64} Q_4 - \lambda Q_6 \tag{6}$$

$$\frac{dQ_7}{dt} = \lambda(Q_1 + Q_2 + Q_3 + Q_4 + Q_5 + Q_6) \tag{7}$$

where

Q_1 is nuclide radioactivity in the Tree (Bq x m^{-2})
Q_2 is nuclide radioactivity in the Organic Layer (Bq x m^{-2})
Q_3 is nuclide radioactivity in the Understory (Bq x m^{-2})
Q_4 is nuclide radioactivity in the Labile Soil (Bq x m^{-2}) Qs is nuclide radioactivity in the Fixed Soil (Bq x m^{-2})
Q_6 is nuclide radioactivity in the Deep Soil (Bq x m^{-2})
Q_7 is radioactivity decrease due to the decay (Bq x m^{-2})
a_{ij} is the rate of intercompartment transfer (time^{-1}) and
F_i is the external source term for the compartment i (Bq*m^{-2} x time^{-1}).

For the case of radionuclide deposition, the external source term can be written in the following form:

$F_1 = I_o \times f \times B_t/(B_t+B_u)$
$F_3 = I_o \times f \times B_u/(B_t+B_u)$
$F_2 = I_o-F_1-F_3$

where I_o is the total quantity of radionuclides (Bq x m^{-2} x time^{-1})
       f is interception fraction (unitless)
       B_t is tree biomass (unitless)
       B_u is understory biomass (unitless)

Rates of intercompartment transfer can be defined as follows:

$$a_{11} = a_{21} = \ln2/\tau_{tr} \qquad\qquad a_{14} = \ln2/\tau_{tu}$$
$$a_{22} = a_{42} = \ln2/\tau_{or} \qquad\qquad a_{23} = a_{33} = \ln2/\tau_{ur}$$
$$a_{34} = 1\,n2/\tau_{uu} \qquad\qquad a_{45} = a_{55} = \ln2/\tau_{dS} \qquad (8)$$
$$a_{44} = \ln2/\tau_{uu} + \ln2/\tau_{tu} + \ln2/\tau_{ab} + \ln2/\tau_{lc} \qquad a_{54} = 1\,n2/\tau_{ab}$$
$$a_{64} = \ln2/\tau_{lc}$$

Let us consider the case of ^{137}Cs deposition. The Decayed Nuclide compartment, described by Equation 7, contains the radionuclides which have physically decayed in the system and do not influence radionuclide transport within the ecosystem. Using the environmental halftimes given in Table 4 and considering long term half-times for tree and understory residences, it is possible to rewrite the system (1-7) as follows:

$$
\begin{pmatrix} \dot{Q}_1 \\ \dot{Q}_2 \\ \dot{Q}_3 \\ \dot{Q}_4 \\ \dot{Q}_5 \\ \dot{Q}_6 \end{pmatrix} =
\begin{pmatrix}
-0.231 & 0 & 0 & 0.347 & 0 & 0 \\
0.231 & -0.087 & 1.386 & 0 & 0 & 0 \\
0 & 0 & -1.386 & 0.069 & 0 & 0 \\
0 & 0.087 & 0 & -5.972 & 1.351 & 0 \\
0 & 0 & 0 & 5.556 & -1.351 & 0 \\
0 & 0 & 0 & 0.002 & 0 & 0
\end{pmatrix}
\begin{pmatrix} Q_1 \\ Q_2 \\ Q_3 \\ Q_4 \\ Q_5 \\ Q_6 \end{pmatrix} -
\begin{pmatrix} 0.023 \\ 0.023 \\ 0.023 \\ 0.023 \\ 0.023 \\ 0.023 \end{pmatrix} +
\begin{pmatrix} F_1 \\ F_2 \\ F_3 \\ 0 \\ 0 \\ 0 \end{pmatrix} \qquad (M)
$$

It is possible to use the stationary equations ($dQ_i/dt=0$) to compute the asymptotic radionuclide distribution within the forest compartments. Naturally, eventually the forest compartments will be free of radioactivity due to physical decay. When the radionuclides decay slowly with respect to the characteristic radionuclide half-times within the system, it is reasonable to treat the radionuclides as if they were stable. That means that Equation 7 can be eliminated for the asymptotic calculations. Equation 7 can be excluded from the system because it is not coupled to Equations 1-6 and can be integrated directly once Q_1-Q_6 are known.

According to the FORESTPATH model, Deep Soil is the geophysical sink for radionuclides. All long-lived radionuclides finally accumulate in the Deep Soil compartment. From the previous discussion and Table 3, one can see that the rate of radionuclide transfer into Deep Soil is very slow, the corresponding half-time is about four hundred years and is much longer than other characteristic half-times. Hence, in studying the system dynamics over time scales of several decades, we can neglect the leaching into Deep Soil and consider Equations 1-7 without Equation 6, which describes the Deep Soil.

Given the above simplifications, one can solve the Equations 1-5 without the deposition pattern ($F_i=0$) to estimate the ratios Q_i/Q_k. The actual value of Q_i can be computed using the initial conditions. A stationary state solution for ^{137}Cs is presented in the Table 5 and discussed in Section 4. 1.

System $(M)_R$ can be solved to study transient radionuclide behavior using Mathematica 2.2 software. The eigenvalues for the system (M) for ^{137}Cs are

$$\lambda_1 = -0.023 \qquad \lambda_3 = -0.023 \qquad \lambda_5 = -0.25 - 0.062i$$
$$\lambda_2 = -2.043 \qquad \lambda_4 = -0.25 + 0.062i \qquad \lambda_6 = -3.028$$

The real parts of all the eigenvalues are less than zero. Two eigenvalues are complex conjugates and contain an imaginary part, i.e., the system tends to a stationary state with slow oscillations of ^{137}Cs concentration between the compartments. In fact, the above solutions do not characterize the system completely. Radionuclide concentrations cannot be negative. Thus, an additional set of inequality constraints must be added:

$$Q_i \geq 0$$

The fact that all the concentrations must be positive adds a constraint on the time evolution of the system. In this particular example, the non-negative concentration constraint is always fulfilled. The analytical solution obtained by integrating the system, M, corresponds to the time distribution of the radionuclide, for the condition that the parameters do not change with time. The general solution for the case of acute deposition is:

$$Q_i = \sum_{j=1}^{3} a_{ij} \exp(-\lambda_i t)$$

The coefficients aij can be found from the initial conditions using the radionuclide deposition patterns. A graphical representation of these analytical solutions is in a good agreement with the numerical calculation using Stella-II software (Figure 4). The small difference in the transient behavior observed comes from the fact that the analytical solutions represent the case in which environmental half-times are assumed to be time-independent while, for numerical simulation, tree and understory residence times are different for short, intermediate and long terms (Table 4). In this particular case, the concentrations remain positive over the entire time interval considered and, therefore, non-negativity constraints have no effect.

MODEL FOR THE EVALUATION OF LONG TERM COUNTERMEASURES IN FORESTS

M.J. FRISSEL¹, G. SHAW², C. ROBINSON³, E. HOLM⁴ & M. CRICK⁵
¹ *IUR, Torenlaan 3, NL-6866-BS Heelsum, Netherlands*
² *Centre for Analytical Research in the Environment. Imperial College at Silwood Park, Ascot Berkshire. SL5 7TE UK*
³ *NRPB Chilton, Didcot. Oxfordshire OX11 0RQ UK*
⁴ *Radiation Physics Department. Lund University Hospital. S-221 85 Lund, Sweden*
⁵ *IAEA P.O. Box 100. A-1400 Vienna, Austria*

1. Introduction

There are vast forests in Belorus, Russia and Ukraine which are severely contaminated because of nuclear accidents in Chernobyl or the Kyshtym region. The wood which can be harvested from these forests is an important source of income for these countries. Because high radioactivity levels in wood limit its applicability, and thus its price, it is worthwhile to evaluate strategies which can be applied to reduce radiocontamination. Moreover, forests form a source for numerous forest products such as mushrooms, game and berries. Finally a forest serves as an area for recreation.

2. Aim of model

This presentation describes a robust generic model which can be used for the development of a strategy to minimize the damage in a forest system which is caused by radioactivity. This model considers the various pathways which cause a radiation dose. The main pathways are:
- Ingestion dose from forest products.
- Inhalation dose from presence in a forest, both for public and workers.
- External dose from presence in a forest, both for public and workers.
- Doses which are received in industries which use contaminated wood as well as public which receives doses from wood, as e.g. timber.

The model considers all processes which might be important in a particular situation. The doses obtained are used to evaluate the effects of possible countermeasures on the various doses. Within this publication this is done by expert evaluation, in a similar model by the IAEA this evaluation is performed in a more objective way by attributing costs to radiation detriments, countermeasures and yields from a forest [1]. The roots of both models, the one used by IAEA and the one used in this publication, are the same.

F. F. Luykx and M. J. Frissel (eds.), Radioecology and the Restoration of Radioactive-Contaminated Sites, 137–154.
© *1996 Kluwer Academic Publishers.*

The main product of a forest is wood, therefore the concentration of radionuclides is a key variable. The production of wood is a long term process, in general the time between planting and felling covers 70 years. Although the model takes into account short lasting processes such as the direct contamination of bark and leaves it is not designed to model such short term processes, it concentrates on long term processes. For short term processes the reader is referred to Schell and Linkov [2].

3. Options for countermeasures in a forest system

There exist various ways to reduce the radiation dose which workers and public may receive in a forest and from forest products. They are listed in order of increasing environmental impact.

3.1 DO NOTHING, i.e. NORMAL OPERATION

The term 'do nothing' means not any application of countermeasures. It involves that maintenance and felling of trees follows the normal operation pattern. The wood which is harvested will be processed in a saw mill, it will be used as timber for the construction of houses, etc, it will be used in paper industry and as fuel in power plants. The population will use the forest for collection of berries, mushrooms, game and firewood. The most important use is probably recreation. Any countermeasure should be judged against this 'do nothing' or "normal operation" scenario.

3.2 REMOVAL OF FALLING LEAVES, NEEDLES AND LITTER

This method can successfully be applied during the first month after the contamination. Various reasons limit the applicability. During the first month the doses from short living radionuclides may be relatively high. Often there will be no equipment available because this is needed for urban areas. After a year 80 % of the radionuclides are already present in the litter layer and upper soil layer. Removal of this layers becomes very impractical, moreover it is destructive for nature. Experience has shown that removal of leaves, needles and top litter layer has never been applied on any more than an experimental scale.

3.3 LIMIT ACCESS TO THE AREA

Varies degrees are possible. In increasing order of impact they are: Picking of mushrooms not permitted; shifting of hunting season, banning hunting; picking of any edible material not permitted; collection of fire wood for domestic cooking not permitted; no access to forest for public; no access to forest for workers. The latter option has its limits. It is possible to stop thinning of forests, but the yield will get lost. Although the biological production will remain the same, a non-maintained forest produces no timber. Disease control, fire preventing and fire fighting has to be maintained always. A fire may cause serious secondary contamination.

More refined measures are possible, one might think on not permitting particular mushrooms species, while other ones are permitted. Differences between the uptake by different types of mushroom varieties are so large that this seems to be an attractive option.

3.4 EARLY FELLING OF TREES

The underlying hypothesis is that the concentration of radionuclides in wood in particular forests will increase with time. In such a situation it is recommendable to do felling earlier. This makes sense of course only if the forest is already harvestable.

3.5 DELAY FELLING OF TREES

All radionuclides decay with time, moreover they are leached downward which reduces external radiation. If it is possible to delay the felling without loss of wood products this seems an attractive option. The moment that income is generated is, however, also delayed.

3.6 REMOVAL OF TREES, SOIL IMPROVEMENT AND REPLANTING OF TREES

This option may be used for very severely contaminated areas planted with young trees. After removal of the young trees the soil may be deeply ploughed before replanting. The external dose will be reduced, also the radionuclide content in products such as mushrooms, berries etc will decrease. The yield of these products, in particular of mushrooms, will decrease too. The contamination of trees will not be influenced by deep ploughing. The method can be combined by -partly- removal of litter and understorey. Because the original trees are lost the method does not seem attractive.

3.7 CHEMICAL TREATMENT OF SOIL

The addition of fertilizers, lime, sapropel will reduce the concentration of radionuclides in forest products and wood. The effect is strongly nuclide dependent. The reduction of uptake is partly caused by a dilution effect in the soil, partly by a faster growth. In arable agriculture a decrease of the uptake with a factor 4-5 is often observed, in general the reduction of the uptake is not better than a factor two [3]. On well fertilized soils additional fertilization has hardly any effect. Also with trees a 50 percent reduction of the uptake of Cs has been observed after fertilization [4]. To maintain this effect re-fertilization from time to time may be necessary. Although limited dressings with lime and fertilizers will repay themselves because they increase the yield, repayment can not be expected for heavy dressings or dressings which have to be repeated to often. Knowledge on this topic for forest systems is limited.

3.8 APPLICATION OF SALT LICKS WHICH CONTAIN PRUSSIAN BLUE.

Prussian Blue limits the uptake of Cs by animals. It can be applied by adding it to Salt Licks which are placed in forests. The application can be very effective [5].

3.9 OPTIMAL CHOICE OF INDUSTRIAL APPLICATIONS:

Wood may be used as fuel in power plants, for paper production and as timber. Identified situations in which external radiation doses are received include [6, 7]:
- The ash of wood fuelled power plant contains an important part of the radioactivity which was originally present in the wood. The people who store the ash on a pile may receive a significant dose.
- For the manufacturing of wood it is treated which chemical solutions which dissolve not only the not wanted part of the wood, but also the majority of Cs. In particular because the leaching solution is re-used several times the solution accumulates Cs. The radiation dose which may be received near the storage tanks may be significant.
- Although paper, because of the leaching process described above, removes the larger part of the Cs from the pulp, paper still contains some Cs. People which are surrounded by paper made out of contaminated wood may therefore receive a dose from external radiation.

The use of adapted handling procedures, shielding, etc may be used to reduce doses received during this use.

3.10 SELECTION OF OTHER CROPS

This is a countermeasure which is successfully applied in normal agriculture. In silviculture there are hardly possibilities. One might consider the use of slowly growing species which produce wood of a higher quality.

3.11 REMOVE THE FOREST AND ABANDON THE AREA

In fact this option does not exist. Reforestation is a natural process which occurs rather fast. As said before, disease and fire control remain necessary.

4. Dose calculations for forests and wood based industry

4.1 PATHWAYS AND PROCESSES

For the dose calculations the following pathways or processes are considered:

4.1.1 Ingestion
Ingestion of:

Berries	Meat from roedeer
Mushrooms	Meat from other animals
Honey	Milk from marginal grazing and collecting hay in forest

4.1.2 Inhalation
Inhalation of:
Saw dust in a forest during thinning by workers.
Saw dust in a forest during thinning by public.
Saw dust in a forest during felling by workers.
(A separate calculation is required because felling occurs only in the year of harvest, while maintenance occurs continuously. The number of persons involved differs.)
Normal resuspended material in a forest by workers.
Normal resuspended material in a forest by public.
Resuspended material resulting from a forest fire by workers.
Resuspended material resulting from a forest fire by public.
Saw dust in a saw mill.
Ash dust resulting from the use of fire wood for domestic cooking.

4.1.3 External radiation
Presence in a forest of public.
Presence in a forest of workers because of maintenance, felling and fire fighting.
Presence on an ash pile of a wood fuelled power plant.
Presence in a pulp industry near tanks with wood/paper pulp.
Presence in a surrounding of paper which is produced from contaminated wood.
Presence in a saw mill.
Presence in a house made from contaminated timber.

4.2 UNITS AND PERIODS

The period over which an assessment has to be made should at least include one harvest of wood, otherwise it is not possible to consider the impact of contaminated wood products. The normal age at which a forest is harvested is 60-70 years. To be able to assess delayed felling the assessment period is set to 90 years. An assessment period includes thus always three periods: the period till harvest, the year of harvest and the remaining period up till 90 years.
The surface area is set to 1 km². This is not as easy as it looks. In every calculation the number of persons which receives their products from 1 km² forest has to be taken into account. For the consumption of mushrooms or berries this is straight forward. For the presence on an ash pile it is more difficult. How many persons are, in the year of harvest, present on the ash pile because of the production of wood on 1 km² forest ? In case the production of wood of 1 km² will be used for various purposes, as e.g. timber, paper and fuel, the number of persons for each application should be weighted according to the use.

4.3 THE ECOLOGICAL PART OF THE MODEL

Figure 1 shows the ecological part of the model. There are 9 dynamic compartments (the rectangles), usually the dimension is Bq/m². The compartments 'Bark' and 'Leaves' take into account the direct contamination from the atmosphere. They are only important during the first years after contamination. The contamination of newly formed leaves and bark is not separately accounted for. Of course the contamination of leaves and bark contribute to resuspension and external radiation, but both processes can be accounted for by including newly formed leaves and bark in the compartment "litter". Litter is therefore an important compartment. Of course the earlier mentioned compartments "Bark" and Leaves" contribute to "Litter". The flux R8 accounts for the production of new litter. The compartment 'Tot_soil' represents the combined contents of the compartments 'Organic matter' and both mineral soil compartments. There are two mineral soil compartments, 'Min_soil_surface' and 'Min_soil_deep', to have the possibility to include shielding of the gamma radiation by the upper soil layer. The depth of the 'Min_soil_deep' compartment starts at a depth of 5 cm. A very important compartment is 'wood'. Its dimension is Bq/kg. The variables 'berries', 'mushrooms'. 'roedeer', 'other game', 'milk' and 'honey' are directly related to the total concentration in soil.

4.3.1 The equations
The rate equations between most dynamic compartments, as indicated in Figure 1, are first order relations. Default values are supplied in Table 1. All nuclide dependent values refer to ^{137}Cs and are taken from [1]. R6 accounts for losses of radionuclides from the system. The 'loss accumulator' has no meaning, it is included for a material balance check only. The fluxes F1, F2 and F3 are not real fluxes, they only serve accounting purposes. The compartment 'Tot_soil' accounts for the total amount of radionuclides in the OM and both mineral soil compartments.

The litter production flux R8 and wood production flux R7 are controlled by a TF (transfer factor) relation. The dimension of TF is m²/kg (derived from Bq/kg product/Bq/m² soil). To maintain proper dimensions the litter transfer has to be multiplied by the litter production rate (kg/m²).

The radionuclide concentration in wood consists out off two parts: The nuclides in the newly formed wood and the ones in wood which is already present. It is assumed that only the concentration in "newly formed" wood is influenced by the concentration of a radionuclide in the soil and the TF. The concentration in "older" wood decreases only due to radioactive decay and biological decay [8]. For the time being the default value for biological decay is assumed to be zero.

The dimension of the wood compartment is in Bq/kg which makes it necessary to take into account the weight of the wood by a normalization relation. It is assumed that the production of wood (kg/m²) is constant with time [8]. This is a severe simplification but for the most relevant period -i.e. age of tree between 20 and 70 years- assumed to be sufficiently accurate. The normalization relation is therefore independent of the growth rate. The basic equation for the concentration in wood at time t, $C_{wood,t}$, is :

$$C_{wood,t} = (C_{soil,t} \times TF + C_{wood,t-1} \times A_{tree,t-1})/(A_{tree,t-1} + 1)$$

where $C_{soil,t}$ =   The concentration of radionuclide in the total soil at time t

$C_{wood,t-1}$ = The concentration in wood at time t-1

$A_{tree,t-1}$ = The age of the tree at time t-1

Note: The actual age of the tree has to be taken into account, not the time elapsed since contamination. The equation applies also to newly planted trees.

Implicitly the equation considers a time step of 1 year; in the model the integration occurs by Euler integration.

The first year after replanting the equation reduces to:

$$C_{wood,t} = C_{soil,t} \times TF$$

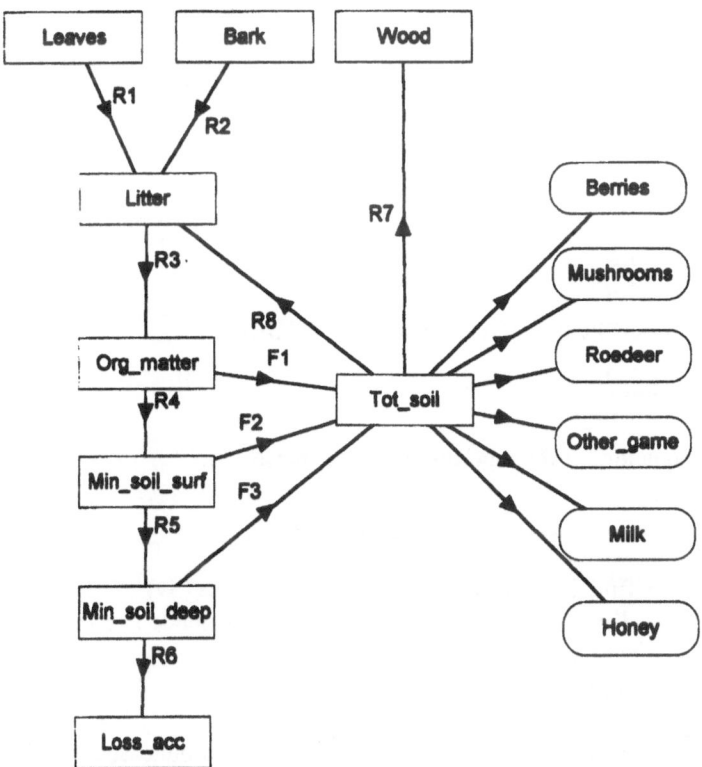

Figure 1. Radioecological part of assessment model. Rectangles are dynamic compartments. Ovals indicate compartments which are in immediate equilibrium with the indicated dynamic compartment.

TABLE 1. Transfer rate parameters of the dynamic part of the model [1]

Rate constant	a⁻¹	Rate constant	a⁻¹
R1 (leaves to litter)		R3 (litter to org. matter)	0.5
coniferous forest	0.345	R4 (OM to surface soil)	0.05
deciduous forest	0.95	R5 (Surface to deep soil)	0.07
R2 (bark to litter)	0.85	R6 (Deep soil to loss acc.)	0.01

Radioactive decay is accounted for in all dynamic compartments. These equations are not described.

The concentrations of radionuclides in the forest products are calculated with TF relations. All dimensions of the products are in Bq/kg and the TF's are in m^2/kg. Default values are provided in Table 2.

TABLE 2. Parameters for the ingestion dose pathway in a forest. (n.a.=not applicable) [1].

	Transfer Factor TF m^2/kg dw	Consumption kg/a fresh w	Persons involved number/km²	Fresh/dry ratio
Berries	0.035	2	50	0.2
Mushrooms	0.5	3	50	0.1
Roedeer	0.05	0.25	2	n.a.
Other game	0.02	0.25	2	n.a.
Milk	0.003	40	10	n.a.
Honey	0.05	0.1	2	n.a.
Wood	0.007	n.a.	n.a	n.a.
Litter	0.007	n.a.	n.a	n.a.

4.4 EQUATIONS FOR RADIOLOGICAL EXPOSURE

4.4.1 Ingestion dose

The general equation for the ingestion dose, per person, is:

$$E_{food} = \sum C_{food} \times H_c \times I_c$$

where
E_{food} = Time integrated committed effective dose (Sv)
$\sum C_{food}$ = Time integrated concentration of radionuclide in food (Bq.a/kg)
H_c = Dose coefficient (Sv/Bq)
I_c = Ingestion rate (kg/a)

Of course the calculation has to be performed for each food product and each radionuclide present. Default values are supplied in Table 2.

4.4.2 Inhalation dose

Two general equations for the inhalation dose are being used. The first one is based on the dust load directly, the second one on the concentration of radionuclides in the surface layer of soil and a resuspension factor. The first equation is:

$$E_i = H_i \times C_1 \times I \times D \times Res \times O \times A$$

The second one is :

$$E_i = H_i \times C_2 \times I \times R \times O \times P$$

where E_i = Committed effective dose or time integrated committed effective dose per person (Sv)

H$_i$ = Dose coefficient for inhalation (Sv/Bq)

C_1 = Concentration of radionuclides in inhaled material (Bq/kg or Bq.a/kg)

Depending on the pathway considered, the time integrated concentration or the concentration in the year of harvest has to be taken into account; the latter applications are labelled with a #. (From here on C might mean C or $\sum$ C).

C_2	=	Concentration in surface layer (Bq/m² or Bq.a/m²)
		For normal resuspension: Bark+leaves+litter
		For fire resuspension: Bark+litter+leaves+10% of organic matter
I	=	Inhalation rate (m³/h)
D	=	Dust load of resuspended material (kg/m³)
Res	=	Respirable fraction of dust (-)
R	=	Resuspension fraction (m⁻¹)
O	=	Occupancy (h/a)
A	=	Accumulation factor(-)
P	=	Probability (-)

To each of the processes listed in section 4.1.2 one of both equations applies, but not every symbol applies to each process. E.g. the accumulation factor A applies only to domestic cooking and takes into account the concentration ratio ash/wood. The probability P applies only to forest fires. For clarity, an overview of the parameters is given in Table 3.

TABLE 3. Overview of inhalation parameters [1].

First equation	H_i Sv/Bq	I m3/h	D kg/m3	Res -	O h/a	N persons/ha	A -
Thinning of forest, workers	4.6E-9	1.0	1.E-6	0.2	2,000	0.05	n.a.
Thinning of forest, public	"	"	"	"	100	100	n.a.
Felling of forest, workers #	"	"	"	"	2,000	20	n.a.
Activities in saw mill #	"	"	5.E-6	1	2,000	15	n.a.
Ash from domestic cooking #	"	"	2.E-6	1	4,500	15	50

Second equation	H_i Sv/Bq	I m3/h	R 1/m	Res -	O h/a	N persons/ha	P -
Normal resuspension, workers	4.6E-9	1.0	1.E-8	0.8	2,000	0.05	n.a.
Normal resuspension, public	"	"	"	"	100	100	n.a.
Resuspension forest fire, workers	"	"	1.E-3	0.05	100	20	1.5E-3
Resuspension forest fire, public	"	"	"	"	10	100	"

4.4.3 External doses

The external doses for all processes listed in section 4.1.3 are calculated with the same type of equation. In all calculations an 'aggregated dose conversion factor', A_g, is used which relates the total concentration in the forest system per unit of surface area to a dose. These A_g's are specially derived for forest systems [1].

The general equation for the external dose per person is:

$$E_e = C_{soil} \times A_g \times O \times F_{cor}$$

where E_e = Time integrated effective dose (Sv)
C_{soil} = Concentration in all forest compartments (Bq/m² or Bq.a/m²)

Depending on the pathway considered, the time integrated concentration or the concentration in the year of harvest has to be taken into account; the latter applications are labelled with a #. (From here on C might mean C or $\sum C$).

A_g = Aggregated transfer factor (Sv/Bq.m²)
O = Occupancy (h/a)
F_{cor} = Correction factor

A complication is that the A_g's for industrial applications are based on the situation observed in harvestable forests six years after contamination by Chernobyl fallout. In fact the dose due to industrial applications depends on the concentration of radionuclides in wood and not on the one in the soil. Consequently the A_g's have to be corrected by a factor, F_{cor} which corrects for this fact. The equation which is used is:

$$F_{cor} = C_{wood,t=t}/C_{wood,t=6}$$

where $C_{wood,t=t}$ = Concentration in wood at actual time t
$C_{wood,t=6}$ = Concentration in wood 6 year after contamination

To take into account the shielding caused by soil, the equation for the external dose is separately applied to the upper 5 cm zone and the zone below 5 cm. For this calculation the radioactivity which is present in the tree and humus is included in the upper 5 cm zone.

The values for the various parameters are provided in Table 4. The calculation for houses is based on the concentration of radionuclides in the year of harvest. A houses is, however, an enduring item. The dose is therefore integrated over the remaining assessment period. Radioactive decay is considered in this integration. Table 4 does not list occupancies or number of persons involved. They are already listed in Table 3.

TABLE 4. Parameter values for external doses [1, 9].

Pathway		A_e Sv/Bq.m²
Presence in forest, maintenance, worker		2.45E-12*
Presence in forest, public		"
Presence in forest, felling, worker	#	"
Presence on ash pile of wood fuelled power plant	#	7.E-11
Presence near pulp tanks in pulp factory	#	2.E-11
Presence in a saw mill	#	2.E-14
Presence in place surrounded by paper	#	2.E-14
Presence in houses constructed from contaminated timber	#	1.E-14

* For Cs deeper than 5 cm 0.9 E-12 Sv/Bq.m²

4.4.4 Collective doses

Collective doses are calculated by multiplying the number of persons involved by the personal doses.

4.5 RESULTS OF ASSESSMENT CALCULATION

The dose depends of course, besides on the processes described, also on the contamination density, the age of the forest at the time of contamination and the assessment period. The case study which is presented here applies to a situation which is at present relevant for forests in Belorus and Russia. They apply to a generic soil, a coniferous forest and an initial contamination of 1.10^6 Bq/m² (27 Ci/km²). The age of the forest at contamination is set to 35 years. The assessment period covers a period of 90 years and begins 10 years after the contamination. The normal felling of the forest occurs if the forest is 70 years old, i.e. during the 25th year of the calculation. In the 26th year a new forest is planted. The maintenance is constant during the 90 year assessment period.

The results of the assessment calculation are shown in the Figures 2-4. There is no difference between the Figures 3 and 4, only the scale of the Y-axis differs.

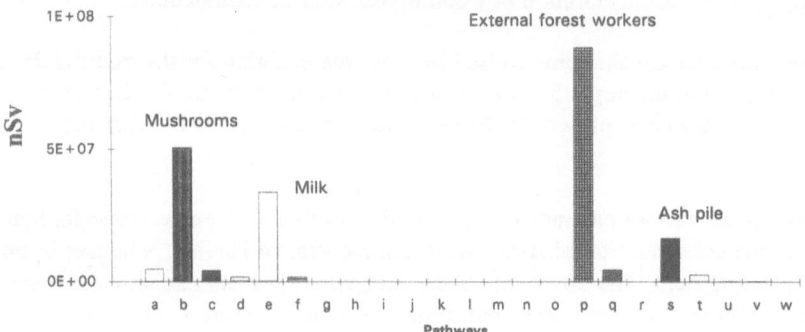

Figure 2. Effective dose per person per pathway (Table 5)
integrated over 90 years.

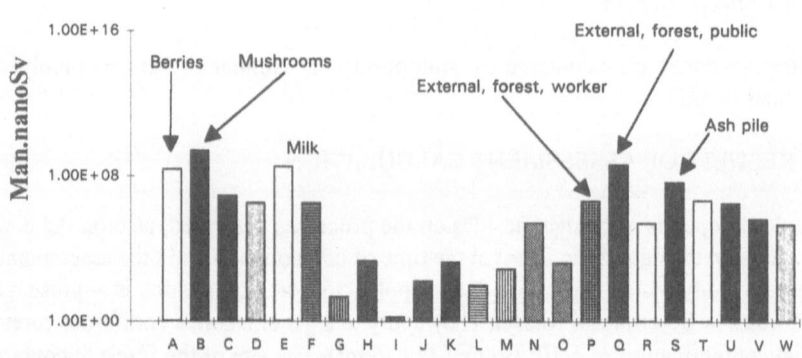

Figure 3. Collective effective dose per pathway (Table 5)
integrated over 90 years.

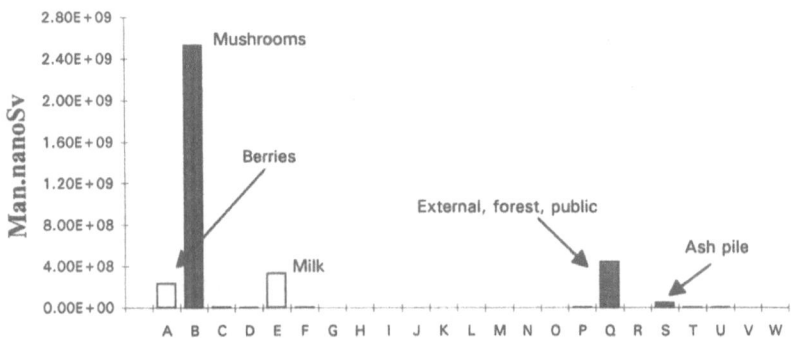

Figure 4. Collective effective dose per pathway (Table 5)
integrated over 90 years.

TABLE 5. Explanation of pathways indicated in Figures 2-4.

		Ingestion
a	A	Berries
b	B	Mushrooms
c	C	Roedeer
d	D	Other game
e	E	Milk marginal grazing
f	F	Honey
		Inhalation
g	G	Thinning of forest, dust, worker
h	H	Thinning of forest, dust, public
i	I	Normal resuspension, worker
j	J	Normal resuspension, public
k	K	Saw dust in sawmill, worker, year of harvest only
l	L	Cutting of forest, dust, worker, year of harvest only
m	M	Forest fire, resuspension, worker
n	N	Forest fire, resuspension, public
o	O	Cooking, ash dust, public, year of harvest only
		External radiation
p	P	Forest thinning, worker
q	Q	Forest, public
r	R	Forest felling, worker, year of harvest only
s	S	Ash pile power plant, worker, year of harvest only
t	T	Pulp industry, worker, year of harvest only
u	U	Timber, housing, public, year of harvest till end assessment period
v	V	Saw mill, worker, year of harvest only
w	W	Paper industry, worker, year of harvest only

5. Evaluation of results of assessment study

The first conclusion which can be drawn from the results shown in the Figures 2 to 4 is that a few pathways dominate. It is surprising that none of the inhalation pathways contributes significantly to the dose (pathways g-o, G-O). Of course, this applies to Cs, for Pu the result may be different.

The dominant ingestion pathways are the intake of mushrooms and milk. Of the collective doses, the intake of mushrooms dominates all other pathways. For the specified contamination of 1.10^6 Bq $^{137}Cs/m^2$ the individual time integrated dose due to the intake of mushrooms equals 47 mSv, for milk this value is 31 mSv. For the first year of the assessment period 1.3 and 0.9 mSv, respectively. The limitation of the collection of the most vulnerable mushroom varieties seems a useful countermeasure.

The dominant external process is the presence in a forest (p, q and Q). The dose of workers who are each year again in the forest (in the model those indicated by thinning) for the specified contamination is 120 mSv in 90 year and 3.9 mSv in the first year. Because the number of persons involved is low the contribution to the collective dose is comparatively low. It is worthwhile to note that no individual will work for 90 years in a forest. The external dose received on an ash pile is also significant, i.e. 25 mSv in the year of harvest. For all considered industrial processes the contributions by external radiation to the total dose are higher than the inhalation doses. It seems worthwhile to consider countermeasures which reduce the contamination of wood.

In a separate calculation the effects of three countermeasures are evaluated. The countermeasures are: fertilization, early cutting and delayed cutting. The fertilization is applied to the forest described before, and applied 10 years after the contamination. This forest can be harvested within the next 20 to 40 years. The fertilization is repeated from time to time so that the TF (transfer factor from soil to newly formed wood) is reduced by 20 %. The fertilization promotes the growth so that harvesting can be brought forward from 5 to 10 year. The result is shown in Figure 5.

The figure shows that the Cs content in the harvestable wood decreases indeed, because the harvest occurs some years earlier this positive effect is counterbalanced by less radioactive decay due to the earlier harvest.

Figure 6 shows the results of earlier or delayed harvesting. The curve indicated with 60 refers to a forest which was 60 years old at the time of contamination. The Cs content is still decreasing with time. A maximum will be reached 10 to 20 years from present. There is not any reason to postpone cutting .

The curve labelled with 40 refers to a forest which was 40 years old at the time of contamination. Its present age is 50. Harvesting may occur 20 to 30 years from present. Point B shows the concentration if the harvest is postponed to an age of 90 years. Sometimes this is possible, it reduces the contamination with a significant fraction.

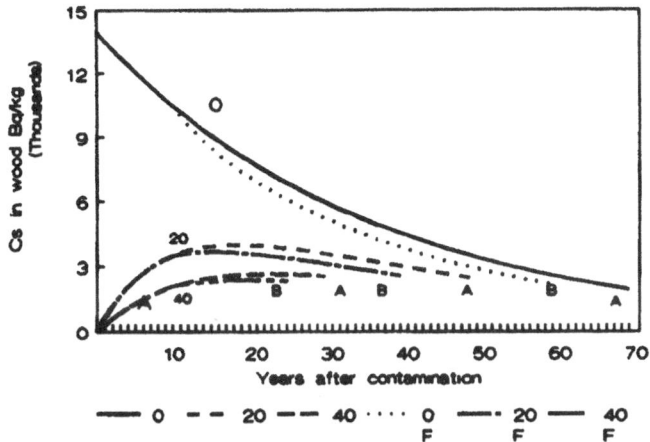

Figure 5. ¹³⁷Cs in trees as a function of the age of trees at the time of contamination and fertilizer applications (indicated with F below the age). The points A refer to the times of harvest of the control trees, the points B to the times of harvest of the fertilized trees. Initial contamination 1.10^6 Bq/m²

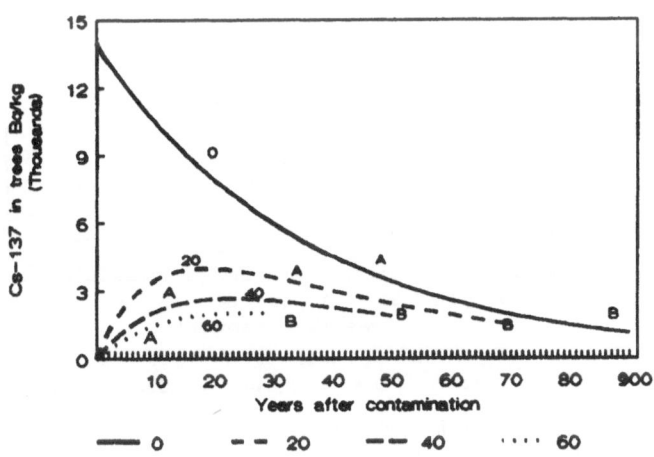

Figure 6. Effects of early and delayed harvesting as a function of the age of the tree at the time of contamination. The labels 0-60 refer to the age of the trees at contamination. The points A refer to normal harvest times. For the curve labelled with 60 that means in the very near future. Points B refer to a postponement of the felling to an age of 90. Initial contamination 1.10^6 Bq/m²

152

6. Sensitivity analysis

The number of parameters in the model is almost one hundred. But, because only a limited number of pathways/processes contribute significantly to the dose it is clear that the number of important parameters is much lower. Considering the dose equations it is clear that the number of persons involved, the TF of mushrooms and the amount of mushrooms ingested are dominant parameters. A sensitivity analysis with a special software programme [10] of the IAEA forest model [1] confirmed this expectation.

This analysis indicated a few other important parameters, namely the ones describing the leaching of Cs and the TF of newly formed wood.

In a separate calculation the effects of these parameters on the concentration of Cs in wood was studied. The effect of leaching is indicated by one single parameter: the percentage loss of the system per year. Figure 7 shows that the effect of the leaching rate is dramatic indeed. A leaching loss of 5 % per year is, however, unlikely. Also the TF of wood was varied. The default value is 0.014, calculations were made with values of 0.007 and 0.030. The results are dramatic again.

From experience with arable systems it is known that the leaching of radionuclides from a particular ecosystem is much better predictable than the TF. From studies on the TF factor of annual crops it is known that the TF depends very much on local conditions and type of crop. A conclusion might therefore be that the TF of wood is probably the most critical parameter of the model.

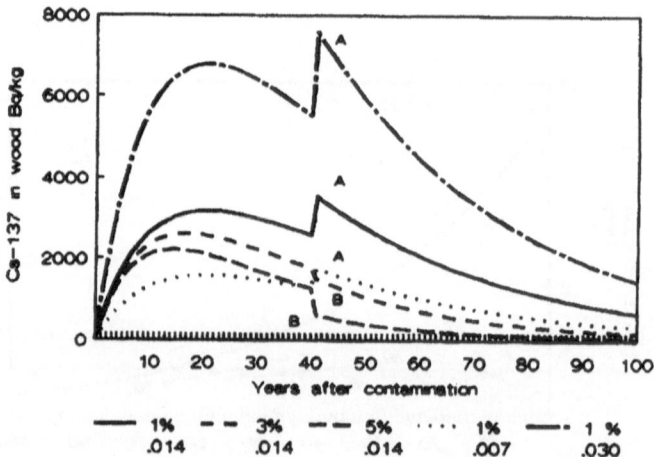

Figure 7. The concentration of Cs-137 in wood of trees as function of the soil-to-tree transfer factor TF (m²/kg) and leaching from the rooting zone (percentage per year). Age of tree at contamination 30 years. Age at harvest 70 year. The concentration in replanted trees is also shown.

Curve 1 TF = 0.014 (default) and leaching fraction = 1 % (default).
Curve 2 TF = 0.030 and leaching fraction = 1 %.
Curve 3 TF = 0.007 and leaching fraction = 1 %.
Curve 4 TF = 0,014 and leaching fraction = 5 %.
Curve 5 TF = 0.014 and leaching fraction = 3 %.

7. Conclusions

In this section the countermeasures described in section 3 will be evaluated on the basis of the results of the assessment study.

Ad 3.1 DO NOTHING i.e. NORMAL OPERATION

Does not apply.

Ad 3.2 REMOVAL OF FALLING LEAVES, NEEDLES AND LITTER

Removal of litter, etc. as soon as possible after contamination has an effect on all pathways. It is a very valuable method.

Ad 3.3 LIMIT ACCESS TO THE AREA

From the assessment calculations it appears that the dose due to the consumption of forest products is relatively high. The limitation of the consumption (of particular species) of mushrooms will decrease the dose considerably. The external dose is in some situations important, limiting access will reduce this dose.

AD 3.4 EARLY FELLING OF TREES

This option may be helpful to reduce the concentration of Cs in wood from forests which are at present 50 years of older. The concentration of Cs in these forests will increase the next 10 or 20 year.

Ad 3.5 DELAY FELLING OF TREES

This option may be applied to forests which are younger than 40 years. The concentration in wood will increase during the first 10 or even 20 years, but thereafter it will decrease below the present level.

Ad 3.6 REMOVAL OF TREES, SOIL IMPROVEMENT AND REPLANTING OF TREES

This option seems not attractive.

Ad 3.7 CHEMICAL TREATMENT OF SOIL

This option decreases the radionuclide concentration in wood indeed. The forest becomes, however, certainly harvestable earlier because of the fertilization. Decay of radionuclides will therefore be less. The overall effect will be low. From an economic point of view some fertilization seems attractive.

154

Ad 3.8 APPLICATION OF SALT LICKS WHICH CONTAIN PRUSSIAN BLUE.

In general the dose received from roedeer and other game will be low. There might be critical groups with a high consumption of game. Because of such groups the application of Salt Licks treated with Prussian Blue seems attractive.

Ad 3.9 OPTIMAL CHOICE OF INDUSTRIAL APPLICATIONS:

The doses are sufficiently high to take special precautions for particular processes.

Ad 3.10 SELECTION OF OTHER CROPS

Model does not apply.

Ad 3.11 REMOVE THE FOREST AND ABANDON THE AREA

Model does not apply.

8. Acknowledgements

The authors acknowledge with pleasure their colleagues Gerzabek (Austria), Dvornik (Belorus), Raito and Rantavaara (Finland) for the discussion which led to the derivation of the TF values for wood. They, as well as Pröhl, Wirth (Germany), Ianculescu (Rumania), Zabudko (Russian Federation), Luis Gutiérrez (Spain) and Schell (USA), contributed significantly to knowledge on radionuclide behaviour in forest systems and the derivation of parameter values and ranges.

9. References

1. IAEA. The IAEA model for aiding decisions on contaminated forests and forestry products. Vienna, Austria. (in preparation).
2. Schell, W.R. & Linkov, I. (1996) A modelling approach to remediation of forests contaminated by radionuclides. This issue.
3. Frissel, M.J. (1996) Soils role in the restoration of terrestrial sites contaminated with radioactivity. This issue.
4. Aro, L., Rantavaara, A., Raito, H. & Kaunisk, S. (1995) Personal communication.
5. Remez, V.P. The application of cesium selective sorbents for the remediation and restoration of radioactive contaminated sites. This issue.
6. Ravila, A. & Holm, E. (1994) Radioactive elements in the forest industry. The Sci of the Total Env. 157 339-356.
7. Ravila, A. & Holm, E. (1994) Assessment of the radiation field from radioactive elements in a wood ash treated coniferous forest in SW Sweden (In press).
8. Frissel, M.J. (1995) Personal communication.
9. Holm, E. Personal communication.
10. Crystal Ball: Forecasting and risk analysis for spreadsheet users. (1993) Decisioneering Inc., Denver, Colorado, USA.

MUSHROOMS, LICHENS AND MOSSES AS BIOLOGICAL INDICATORS OF RADIOACTIVE ENVIRONMENTAL CONTAMINATION

M.G. NIFONTOVA
Biophysical station of the Institute of Plant and Animal Ecology
Urals Division, Russian Academy of Sciences
ulitsa 8 Marta, 202
CIS-620219 Yekatarinburg, Russia

1. Introduction

At the time that the author started her investigations, it was already known that mushrooms, lichens and mosses are characterized by a number of properties which indicate them as potential biological indicators of radiocontaminants. These characteristics are wide distribution in nature, prolonged life time, eurytopness and a certain degree of tolerance. The specificity of the anatomical and morphological structure as well as the peculiarities of the physiology of these organisms favour their active accumulating function.

As a rule, radionuclide concentrations in lichens and mosses are higher than in herbaceous and woody-dwarf shrub vegetation. Mushrooms, lichens and mosses retain the radionuclides for a long time; the process of radionuclide accumulation is characterized by a cumulative nature.

2. Radionuclide concentrations in various ecosystems

2.1 RADIONUCLIDE CONCENTRATIONS CAUSED BY THE CHERNOBYL ACCIDENT

Over the period 1975-1985, samples of mushrooms, lichens and mosses have been collected in different regions which were subsequently affected by the accident in Chernobyl. The ^{90}Sr and ^{137}Cs concentrations in the collected samples did not exceed 2 kBq/kg (dry weight). As a result of the Chernobyl accident, the ^{90}Sr concentration increased by one to two orders of magnitude and that of Cs-134 and ^{137}Cs by two to three orders of magnitude. (Table 1). A mosaic-like pattern is observed for both radiostrontium and radiocaesium for different parts of the region studied (Figure 1).

F. F. Luykx and M. J. Frissel (eds.), Radioecology and the Restoration of
Radioactive-Contaminated Sites, 155–162.
© 1996 *Kluwer Academic Publishers.*

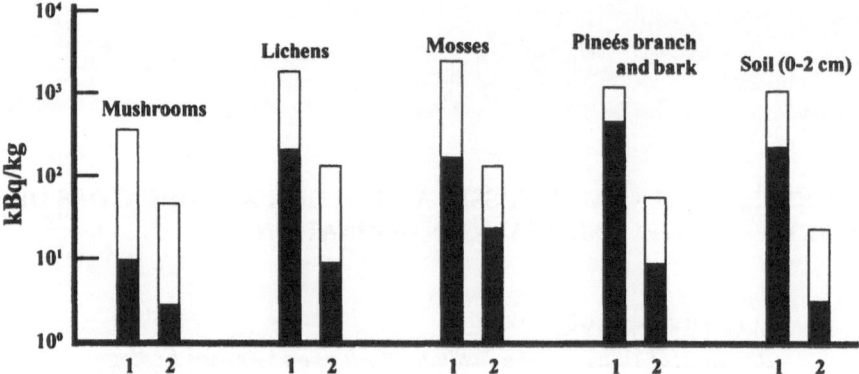

Figure 1. ⁹⁰Sr (dark) and ¹³⁴,¹³⁷Cs (white) concentration in vegetation and substrata of different areas near the Chernobyl Nuclear Power Plant.
1) 6 - 7 km south-south-west of the plant,
2) 18 km south   "   "   "

The accident in Chernobyl left practically unchanged the ⁹⁰Sr concentrations in lichens and mosses in the Middle Urals, but radiocaesium concentrations increased by a factor of 20-40. Over the subsequent years, a gradual decrease of the ¹³⁴Cs and ¹³⁷Cs content in lichens and mosses took place (Figure 2).

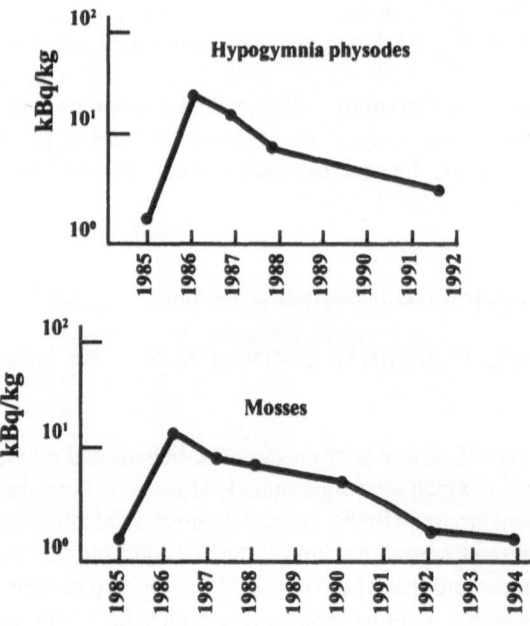

Figure 2. ¹³⁷Cs in lichens and mosses of the Middle Urals (mean weighted value).

TABLE 1. Radionuclide concentrations in mushrooms, lichens and mosses in the neighbourhood of Chernobyl Nuclear Power Plant, kBq/kg.

Species	^{90}Sr	^{134,137}Cs
Mushrooms		
Leccinum aurantianum	3.4±2.7	25.9±2.2
L. scabrum	12.0±3.5	889±165
Boletus edulis	19.7±7.0	306±88
Russula foetens	16.9±6.1	64.9±1.8
R. delica	56 9±5 2	193±12
Boletmus sp.	31 4±4 0	not determined
Lycoperdon sp.	1 8±0 2	27.6±2.2
Coltriciaperennis	93 6±33 4	229±22
Thelephoraterrestris	41.9±4.3	3
Cladina mitis		
C. arbuscula	152±75	1050±180
Hypogimnia physodes	8.8±0.8	100±7
Peltigera canina	10.2±1.1	155±24
Mosses	277±28	1460±570
Dicranum polysetum	120±35	2080.370
Polytrichum commune	55±11.5	860±270
Brachythecium sp.	230±70	2600±230
Pohlia nutans	290±60	1900±160
Plagiothecium sp.	27.1±2.7	118±5

2.2 RADIONUCLIDE CONTAMINATION CAUSED BY THE KYSHTYM ACCIDENT

As a result of the Kyshtym accident in 1957, the area of the Eastern Urals Radioactive Track became contaminated. At present, near the epicentre of the explosion the concentration of ^{90}Sr in mosses is 7.7 ± 0.8 kBq/kg (dry weight) and that of ^{137}Cs 4.9 ± 0.6 kBq/kg. The eastern part of the radioactive track crosses the boundary of the territory of the Kamensk region of the province of Sverdlovsk. The radionuclide concentrations in mosses of different taxonomic classes range from 0.5 to 1.4 kBq/kg for ^{90}Sr and from 0.5 to 1.1 kBq/kg for ^{137}Cs (Table 2).

TABLE 2.Radionuclide concentrations in mosses of different taxonometric classes in the area of the Eastern Urals Track, Bq/kg.

Species of mosses	^{90}Sr	^{137}Cs
Brachythecium collinum	1030±90	570±60
B. curtum	830±110	460±40
B. nelsonii	810±40	490±40
B. plumosum	800±160	400±90
B. populum	500±40	400±40
Cratoneurum commutatum	760±80	440±50
Climaclum dendroides	790±150	620±60
Leptodlctyum kochii	420±60	310±60
Mmum conferhdens	710±30	590±80
Myurella apiculata	630±70	590±80
Pleurozium schreberi	690±40	510±50
Plagiothecium laetum	810±79	690±50

158

An overview of the sampling sites in the Eastern Urals Radioactive Track is shown in Figure 3. The concentrations in moss cover per sampling site are shown in Figure 4. It appears that, more than 30 years after the Kyshtym accident, the area is still contaminated.

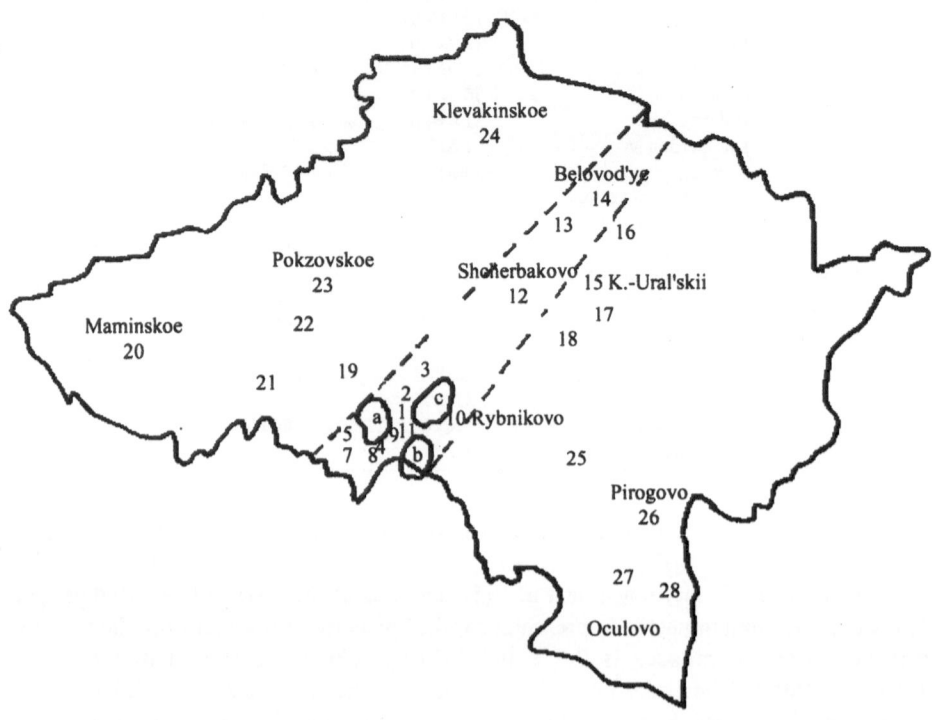

Figure 3.   Moss sampling sites within the territory of the Kamensk region.
- - - -   borderline of the zone of the Eastern Urals Radioactive Track (EURT).
a -   lake Tygish,
b -   lake Chervyanoe,
c -   lake Bol'shoi Sungul

2.3 RADIONUCLIDE CONTAMINATION FROM GLOBAL FALLOUT

The intensive exploitation of resources of the Yamal peninsula, its proximity to nuclear installations at Novaya Zemlya and the extremely scant information available on radiological conditions were the reasons for assessing the contamination of the soil and vegetation cover of this territory.

For the first time in this area, the radionuclide concentration in lichens and mosses of different taxonomic classes was determined. The ^{90}Sr concentration in vegetation ranges from 0.03 to 0.16 kBq/kg (dry weight) and that of ^{137}Cs7 from 0.05 to 0.41 kBq/kg. These levels are similar to those observed elsewhere from global radioactive fallout. Details are given in Table 3.

It appeared that the radionuclide concentrations in lichens and mosses in the far north decrease by a factor 3-7 compared to the maximum values registered during 1963-1965 (Table 4.). Over the same period, the ^{90}Sr concentration in the top 5 cm of soil decreased from 1260 Bq/m² (0.034 Ci/km²) to 630 Bq/m² (0.017 Ci/km²), and for ^{137}Cs from 1780 to 1150 Bq/m² (0.048 to 0.031 Ci/km²).

The main part of the radionuclides appeared to be concentrated in the soil and tundra litter (70-80%); only a small fraction of ^{90}Sr and ^{137}Cs is present in the herbaceous and shrubby vegetation (1.6-10.8%). The moss-lichen layer accumulates up to 20% of the total content of radionuclides in the soil and vegetation cover.

TABLE 3. Radionuclides concentration in lichens and mosses collected in the Yamal Peninsula, Bq/kg (dry weight).

Plant species	^{90}Sr	^{137}Cs
Lichens		
Cladina rangiferina	40±10	315±30
Cladina sylvatica	50±10	280±30
Cladina stellaris	40±10	150±30
Cladina arbuscula	70±10	220±30
Cladina mitis	50±5	320±30
Cladina uncialis	not determined	350-±50
Cladonia elongata	150±20	160±20
Cladonia amaurocraea	110±30	335±50
Cetraria chrysantha	30±5	180±30
Cetrarla hipatison	90±20	140±20
Cetraria islandica	40±5	280±90
Cetraria laevigata	30±10	50±10
Thamnolia vermicularis	120±20	140±10
Stereocaulon tomentosum	60±10	40±10
Mosses		
Dicranum spadiceum	160±40	170±10
Dicranum elongatum	110±15	270±20
Hylocomium splendens	130±10	120±10
Polytrichum commune	150±20	340±60
Polytrichum strictum	50±10	150±10
Polytrichum gracilis	80±20	90±20
Rhacomitrium lanuginosum	160±30	410±10
Sphagnum riparium	110±20	300±30
Sphagnum rubellum	120±20	220±30
Sphagnum squarrosum	30±10	120±20
Sphagnum wulfianum	60±10	80±20
Sphagnum majus	70±10	90±10
Sphagnum fuscum	130±10	330±50
Sphagnum balticum	150±10	360±90
Sphagnum lenense	120±30	260±50

TABLE 4. Radionuclide concentrations in the moss-
lichen cover of shrubby tundra of the Southern Yamal,
Bq/kg (dry weight).

Year	^{90}Sr	^{137}Cs
1963-1965*	110-370	1400-3200
1976	130-350	480-750
1988-1991	60-140	90-370

*) Data for lichen of the Yamal-Nenetzky Autonomous
District [1].

3. Conclusion

An extended discussion of the accumulation of radionuclides by mushrooms, lichens
and mosses is outside the scope of this publication. Certain features should, however, be
considered. The main fraction of radionuclides enters the fruit bodies of macromycetes
via the mycelium from the substrate soil, litter wood, etc. A positive correlation between
the accumulation in mushrooms and their concentrations in substrate has been observed
by various authors [2,3]. As a rule, an increase in radionuclide concentration in mushrooms is observed for some time after the fallout occurred [4-6]. The contamination of
lichens and mosses is of a global character [7,8]. The uptake of ^{90}Sr and ^{137}Cs by lichens
and mosses from atmospheric fallout occurs with efficiencies up to 100% [9] and the
distribution in lichens and mosses reflects the distribution in fallout [10-12]. The natural
elimination from mosses and lichens is a fairly inert process, the rate of which decreases
with time [1, 15].
A general conclusion of the investigations is that mushrooms, lichens and mosses are
suitable biological indicators for radioactive contaminants from accidental releases or
from global fallout. These organisms can be successfully used for long-term radioecological monitoring.

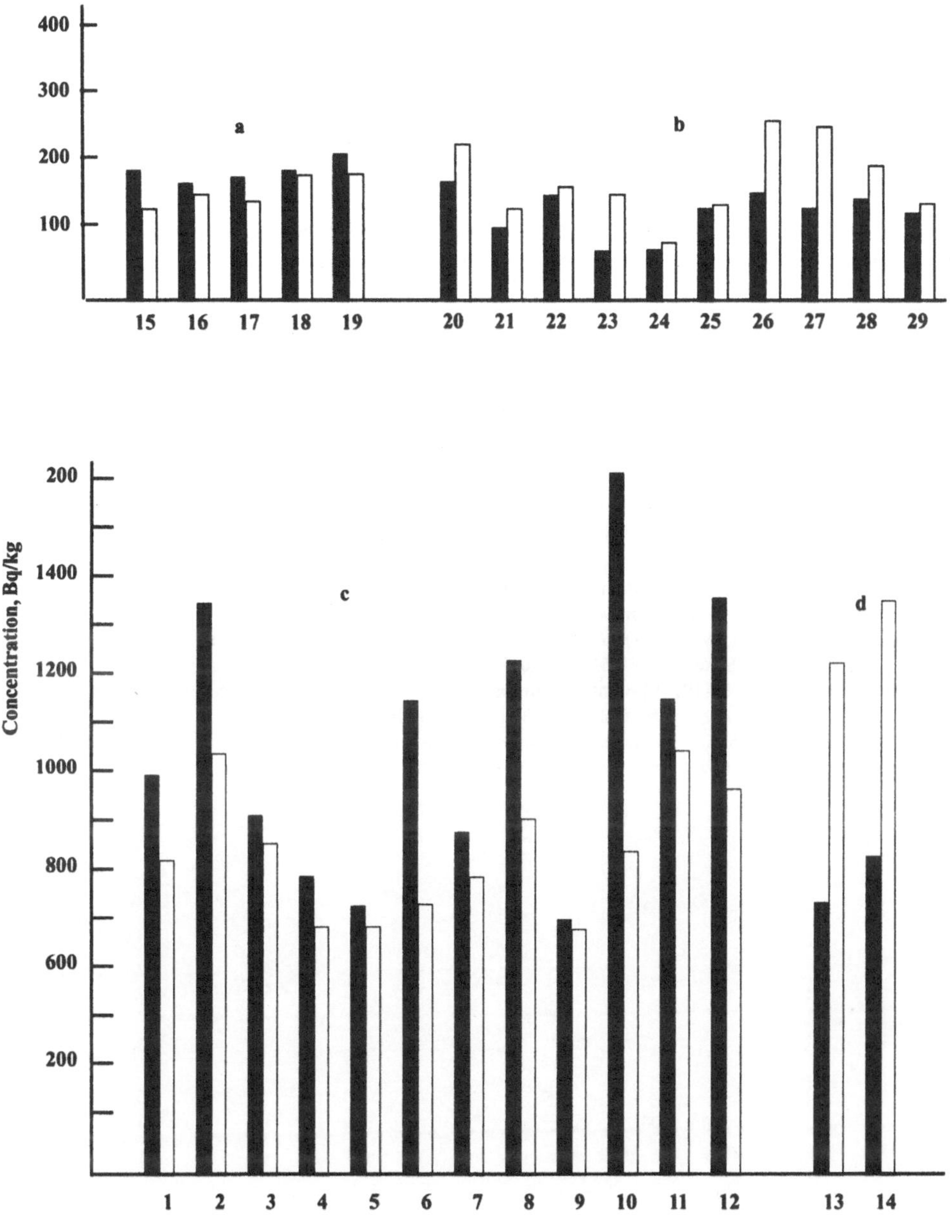

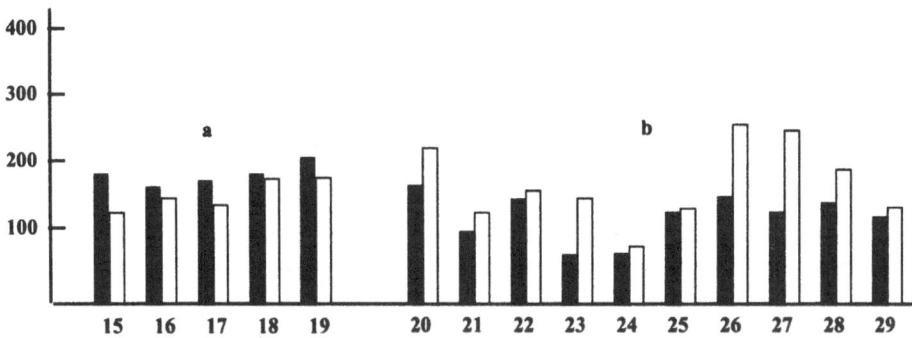

Figure 4. Concentration of ⁹⁰Sr (dark) and ¹³⁷Cs (white) in moss cover in the area of the eastern EURT zone

a,b   outside the zone,

c,d   inside the zone near the central axis of EURT in the neighbourhood of lake Bol'shoi Sungul (c) and at a distance of 18-25 km towards the northeast (d).

162

4. References

1. Troitzkaya M.N. & Ramazaev, P.V., (1971) Landscape radioecology of the far north. Contemporary problems of radiobiology. *Radioecology. Atomizdat.* M., p325 (in Russian).
2. Grueter, H., (1971)Radioactive fission product Cs-137 in mushrooms in W. Germany during 1963-1970, *Health Physics*, **20**, N6,655-658.
3. Bakken, L.R. & Olsen, R,A, (1990) Accumulation of radiocesium in fungi. *Can. J. Microbiol.*, **38**. N10,704-710.
4. Roemmelt, R., Hiersch, L. & Wirth, E. (1988) Uptake of Cs-134,137 by higher fungi in terrestrial ecosystems. Biological aspects of the cesium transfer. 4th Symposium Int. Radioecol. Cadarache " Impact accidents orig. nucl. envir." **1**, 151-161.
5. Oolbekkink,G. & Kuyper, T.W. (1989) Radioactive cesium from Chernobyl in fungi. *Mycologist*, **3** N1, 3-6.
6. Nifontova M.G., Aleksashenko.(1992) Sr-90 and Cs-134,137 content in mushrooms, lichens and mosses from the nearest zone of Chernobyl AEPS. *Russian J. of Ecology*, **23**, N 3.
7. Moseev, A.A. & Romzaev, P.V., (1975) Cesium-135 in the biosphere. *Atomizdat. M.*, 350 (in Russian).
8. Nifontova M.G., Kulikov N.V.(1977) On Sr-90 and Cs-137 accumulation by lichens under natural conditions. *Soviet Journal of Ecology*, **8**, N3.
9. Svenson, G.K. &Liden, K., (1965) The transport of Cs-137 from lichen to animal and man. *Health Physics*., **11**, 1393-1400.
10. Mattson, S., (1972) Radionuclides in lichen, reindeer and man. Lund. p230
11. Nifontova M.G., Kulikov N.V.(1990) The dynamics of distribution of strontium and cesium radionuclides within the components of terrestrial ecosystems in the zone of Beloyarsk NPS. *Ekologiya*, N 3.
12. Nifontova M.G.(1995) Radionuclides in the moss-lichen cover of tundra communities in the Yamal peninsula. *The Science of the Total Environment*, **160-161** .
13. Liden, K.., Gustafsson, M., (1967) Relationships and seasonal variation of Cs-137 in lichen, reindeer and man in Northern Sweden. 1961-65. Oxford, Pergamon Press. 108-115.
14. Martin, I.B., Koranda, I., (1971) Recent measurements of cesium-137 residence times in Alaskan vegetation, *Radionuclides in Ecosystem,*. Oak Ridge, Tennesseee. **1**, 108-115.
15. Nifontova, M.G., (1995) Radionuclides in the moss-lichen cover of tundra communities in the Yamal Peninsula. *The Science of the Total Environment*, **160-161**, 747-752.

ECOLOGICAL BASES AND PROBLEMS ASSOCIATED WITH THE IMPLEMENTATION OF AGRICULTURAL REMEDIAL ACTIONS

VANDECASTEELE C.M.[1], BURTON O.[2] & KIRCHMANN R.[3]

[1] Radioecology Laboratory, Radiation Protection Department,
SCK CEN, Boeretang 200, B-2400 Mol (Belgium).
[2] Radioecology Laboratory, Physics Department,
Faculty of Agronomic Sciences, B-5030 Gembloux (Belgium).
[3] Radioecology Laboratory, Botany Department,
University of Liège, Sart-Tilman, B-4000 Liège (Belgium).

1. Introduction

In the event of a major nuclear accident, involving the dispersion of radioactive material and a widespread contamination of the environment, the first concern of the responsible authorities will be to assess the radiological consequences for the population living in the affected areas, so that essential protective measures can be taken without delay. The main purpose of these measures is to limit the doses received from external and internal radiation to an acceptable level.

If the estimations carried out lead to the conclusion that acceptable levels could be exceeded, several actions can already be taken before (if sufficient time is available between notification of the release and arrival of the plume to allow for preventive actions) or during the passage of the radioactive cloud. These actions, reviewed and discussed by Willrodt [1] , aim to prevent or limit:

- the direct external exposure and contamination of populations to the radioactivity transported by the plume and the internal contamination by inhalation,
- the direct contamination of farm animals and
- the direct deposition of radionuclides on vulnerable plant products to be consumed in a near future by man or cattle.

They mainly consist in:

- evacuation or sheltering of inhabitants,
- housing of grazing cattle, sheep and goats, which will be fed with stored food and watered with well-water (no rain or surface water),
- closing of greenhouse windows and hotbeds,
- harvesting promptly all ripe crops, fruits and vegetables,
- covering with polyethylene foils uncovered feed and food store (hay, heaps of potatoes or beet roots on the field, silos, ...) as well as valuable vegetables and fruits to be harvested soon.

163

F. F. Luykx and M. J. Frissel (eds.), Radioecology and the Restoration of
Radioactive-Contaminated Sites, 163–178.
© 1996 Kluwer Academic Publishers. Printed in the Netherlands.

Other remedial actions can be applied after the passage of the radioactive plume, in order to limit the exposure of the population *via* external irradiation from radionuclides deposited on the ground, inhalation of resuspended contaminated particles and ingestion of contaminated food products. There are two time-scales in which such interventions can be envisaged [2] : the short-term when the radioactivity is still present on the surface of exposed material (plants, soil, roads, buildings, ...), and the long-term when radionuclides become distributed over the different compartments of the ecosystems.

- Short-term remedial actions consist essentially in removing the radionuclides from contaminated surfaces (washing of buildings, roofs, roads, ...) or eliminating contaminated material (vegetation, top soil and building materials). These techniques are generally very efficient and may reduce or prevent long-term problems, but they also generate huge volumes of radioactive waste.

- Long-term countermeasures are carried out once the radioactivity has become incorporated into soil, plant and animal components of the ecosystems. They aim to reduce the transferability of radionuclides between the various compartments of food chains, to increase the decorporation in contaminated living organisms or to process contaminated plant and animal products into cleaner by-products.

The decision whether or not countermeasures must be applied and the choice of particular remedial actions depend on radiological considerations, taking into account economic, ecological and social aspects [3]. The selection of potential remedial actions to be applied depends, however, greatly on the characteristics of the deposit (composition and speciation) and of the ecosystem considered. An adequate understanding of transfer paths and mechanisms is thus essential to intervene most efficiently on the most critical pathways.

The most important pathways of radionuclides in agricultural systems, reviewed by Vandecasteele *et al.* [4], are shown in figure 1.

The contamination of vegetation by most radioactive elements injected and dispersed into the atmosphere arises from two main processes: by direct deposition on aerial parts of plants or by indirect contamination when radionuclides, deposited onto the soil, are absorbed by the root systems together with water and nutrients. In a similar way, radionuclides present in irrigation water reach plants by direct deposition on aerial parts (sprinkling) or via soil and root absorption. Gaseous elements like ^{14}C and ^{3}H (as water vapour) penetrate the plants through the stomata and are incorporated into organic constituents by photosynthesis and other metabolic processes. Contamination of animals and animal products results from lung absorption of soluble inhaled radionuclides and from ingestion of contaminated soil, feed and water.

The following sections will consider the main steps of the radionuclide pathways and discuss for each of them possible remedial actions.

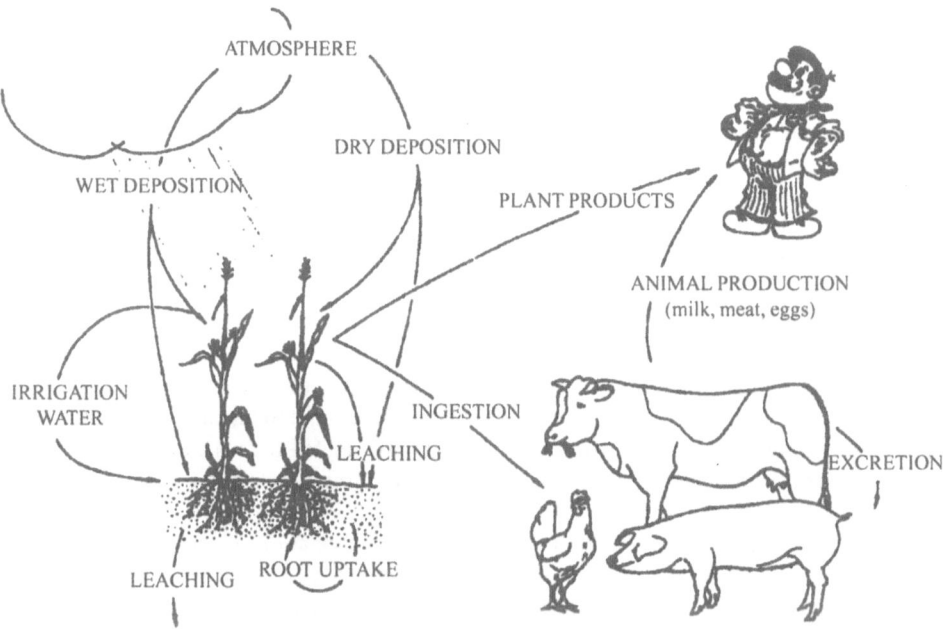

Figure 1: Main pathways for radionuclides to man in continental agricultural food chains.

2. Direct contamination of the vegetation

During the growing season, direct contamination of the plant aerial parts by dry and wet deposition is the first step in food chain contamination. It generally gives rise to higher contamination levels in plant products that indirect contamination (root uptake). In the case of short-lived nuclides like ^{131}I, a sufficient delay before consumption will provide clean products. Waiting is not a problem if the release happens in the early season; when radioactivity is deposited close to harvesting, storing as dry, frozen or sterilised by-products can be envisaged. For long-lived radionuclides, elimination of plant material by harvesting as soon as possible after the deposit (to avoid leaching by rain and other field loss mechanisms) is relatively straightforward and may be a very efficient way of decontaminating the environment, specially if the fraction of the deposit intercepted by the green mass is important. The harvested biomass must be considered as a radioactive waste and burning it in *ad hoc* designed installations can be recommended. Use of this material for production of bio-methane or ethanol can also be envisaged but will be restricted to a small fraction of the material to be treated due to the probably limited capacity of available fermentors. Due to the problem raised by the huge volume of waste produced by harvesting, ploughing in contaminated plant material appears as a very simple way to get rid of it, ensuring that radioactivity does not immediately enter the food chain and reducing external radiation from soil surface; but, this countermeasure may increase the long-term problems.

It is, however, not always recommended to eliminate the contaminated vegetation. If the accident happens in the spring, when stored feed for cattle is exhausted, the only forage available may be the contaminated standing grass. In these conditions, it should be recommended to house the animals and feed them with forage cut above the usual cutting level, since pasture grass exhibits a higher accumulation of deposited radioactivity in its lower parts than in the upper ones. Also some radionuclides are very poorly translocated from organs contaminated by a deposit early in the growth season to edible parts harvested later in the year; the edible parts (e.g. grains of crops) could be expected to exhibit contamination levels below the intervention levels. For instance, translocation of ^{90}Sr, ^{144}Ce and ^{106}Ru into the grains of cereals is minimal if deposition takes place in the early stage of development, while other nuclides such as ^{65}Zn, ^{55}Fe, ^{137}Cs, ^{60}Co and ^{54}Mn are easily translocated in the plant [54]. Middleton [5] reported that up to 50% of the caesium deposited on the leaves of potato plants may be transferred to the tubers but only 0.01 % of the strontium applied to aerial parts of the same plants migrates from the leaves into the tubers. Similarly, in wheat plants contaminated before ear emergence, 5 to 10 % of the caesium and only 0.1 % of the strontium initially retained by the plant was found in the grain at maturity.

Finally, if the culture has to be maintained, sprinkling irrigation, when available, can help to leach out a fraction of the radioactivity intercepted by the aerial organs. The leaching efficiency is highest if irrigation is started immediately after the deposit and decreases as the time before washing increases [6]. Monovalent radionuclides also appear to be more susceptible to leaching than di- or polyvalent elements which are more easily sorbed on leaf surfaces. Moreover, a light continuous drizzle could be more efficient in leaching than a large quantity of water spread over a shorter period [7].

3. Indirect contamination of the vegetation

Indirect contamination occurs after radionuclides have been deposited on the ground. These processes require the passage of the radionuclides from the soil into the plant by root absorption. They depend not only on the element characteristics, but also on the physiological properties of the plant roots and on soil processes.

One of the key properties of soils is their ability to adsorb ions and to immobilise them, to different extents, on the solid phase. The soil colloids (clay minerals and organic matter), that adsorb ions from the soil solution, contain a high specific density of predominantly negative charges. They attract mostly positive ions from the soil solution to their surfaces, where those ions exchange with others, already present at exchange sites. The ability of a soil to adsorb ions is proportional to the density of exchange sites and is expressed by its cation exchange capacity or CEC (in meq/100 g of soil or of a specific fraction of the soil solids). Values reported for the CEC vary depending on the clay type, ranging from 3 to 15 for kaolinite, 10 to 40 for illite, 80 to 150 for montmorillonite; the CEC of humic compounds is even higher, from 300 to 500 meq/100 g. Adsorption of anions, although limited, also occurs on the fewer positive sites present on the surface of clay minerals (especially illite and kaolinite clay types), on iron and aluminium hydroxide colloids bound to clay or on clay and organic matter by calcium bridges.

The affinity of ions for binding sites depends on their physico-chemical properties. It is higher for trivalent than for monovalent cations; divalent cations are intermediate. For cations of the same valence state, the binding affinity is inversely proportional to the hydrated radius of the ion. Adsorption of ions is a reversible process and an equilibrium tends to be achieved between the concentration in the soil solution and the sorption complex.

Binding of ions on the soil solid phase delays or prevents their leaching in percolation water to below the rooting zone. Thorium and some of the light rare-earth elements are so tightly bound to the solid matrix (clay and iron oxides) that they leach at a rate as low as 10^{-9} per year [8]. Caesium deposited by weapon test fall-out in the early sixties, disappears from the agricultural sandy soil in the vicinity of Mol (Belgium) at a rate of about 1 to 5 % per year, both by leaching and removal by harvesting [9].

The soil is also a dynamic system. Its properties are acquired and modified with time due to the joint actions of natural factors (variations of temperature and humidity, erosion, ...) and farming practices. Hence, through these mechanisms, the chemical form of radioelements, their sorption on exchange sites and their localisation in the soil profile may change with time.

When soluble radionuclides are deposited on the ground, they will be dissolved by rain, irrigation water or soil moisture and will migrate into the soil. They can adsorb on the sorption complex by exchange processes, precipitate as hydroxides, sulphides, carbonates or insoluble oxides, form complex with organic molecules or remain in the water phase in ionic form [10]. For example, in neutral and alkaline soils, rare-earth elements and others like yttrium, thorium, zirconium or niobium are precipitated as hydroxides and carbonates while strontium precipitates as carbonate. Elements like K and Cs may be trapped and immobilised between the lattices of illite type clay minerals. The reversibility of such a binding is very poor and the elements bound at these sites can only be removed by alternations of drying and re-wetting or of freezing and thawing.

How radionuclides behave and how they are partitioned between the different pools, largely determines their availability to plants and the time they will remain in the rooting zone.

Roots absorb their nutrients from the soil solution. The soil solution is thus continuously depleted from its solutes by root uptake, but it is also continually replenished from the soil solid phase. The solid phase constitutes a reservoir of nutrients (as well as of pollutants introduced in the system) which are made available through exchange reactions between the solid and liquid phases.

The main physical factors affecting the absorption of nutrients by roots are:
- ionic concentration in the water solution: at low concentrations (< 1 meq/l), as is the case for fission products released into the environment, the absorption rate is generally proportional to the concentration in the water phase;
- chemical properties of ions: ions with low valence are in general more easily absorbed than those with higher valence; the relative rate of absorption of several elements is the following:
  $^{89-90}Sr \gg {}^{131}I > {}^{140}Ba > {}^{134-137}Cs, {}^{106}Ru > {}^{144}Ce, {}^{91}Y, {}^{147}Pm, {}^{65}Zn, {}^{95}Nb > {}^{239}Pu$ [55];

- pH and Eh which effect the solubility of some elements (precipitation and dissolution reaction) and strongly influence the K_d values [11];
- ionic interactions, antagonisms and competition between elements, both for adsorption on soil sorption complexes and for root uptake.

A reduction of indirect contamination of the vegetation can be obtained by several means [12, 13]:
- removal of the contaminated soil surface layer
- (deep)-ploughing to reduce the contamination of the upper soil layers and subsequently reduce the uptake by shallow-rooted plants such as grass and vegetable;
- addition of fertilisers: phosphorus to insolubilise strontium as phosphates, calcium (lime) to compete with strontium for plant uptake, potassium (potash) to compete with caesium;
- addition of chelating agents to bind the radionuclides in a form unavailable for plant uptake or in soluble forms that will be leached below the rooting zone;
- cropping deep-rooted plants which exploit the soil below the contaminated layers (e.g. alfalfa, corn);
- cropping plants used for grain production (grains generally exhibit lower transfer factors than other plant organs), seed production, fibre or oil production, for cattle feeding;
- use of agricultural lands for alternative purposes: e.g. planting forest for timber production.

Among the various countermeasures aiming to reduce the indirect contamination of plants, removal of a more or less thin layer of top soil, as long as the deposited radioactivity is still concentrated at the surface, appears as a very effective method to prevent the entry of radioactivity into food chains. It has also the advantage to reduce the resuspension of radioactive particles and the long-term aftermath. Different methods have been developed and, for some of them, tested under real conditions: removal of top soil by scrappers, bulldozers, manual excavation or vacuum cleaners [14], trapping of surface soil particles in poly-urethane foams spread on the soil surface [15] or by the roots of turfing plants [16]. However one main drawback of these techniques, in particular the first ones, is that they generate impressive volumes of waste (e.g. removing only 5 cm of surface soil on 1 km² produces 50,000 m³ of contaminated waste) and therefore should be restricted to small areas. An alternative to removing the waste from the site would be to place it in self-shielding piles on the field (the top 5 cm accumulated in a 3 m high flat top pyramid mount would occupy only some 3 % of the cleaned area) [17]. Lixiviation and resuspension from these mounts could be prevented by the use of water impermeable barriers.

Ploughing, despite of the potential risk of long-term problems, generally appears as the most practical and cost-effective remedial action [17]. Normal ploughing (to a depth of 20–30 cm) would immediately suppress any tendency to resuspension and greatly reduce external irradiation and root uptake by many plant species. The use of modified ploughs, with a skimmer attachment, allowing the placement of a discrete surface layer at the bottom of the furrow would be even better. Deep ploughing (to a depth of 1 m) is a much more effective countermeasure than normal ploughing. This technique offers the advantages that the contamination is placed deep enough so as not to be disturbed by subsequent conventional ploughing and out of the reach of roots of many crops; it also greater reduces external radiation but has the drawback that it requires special ploughs and tractors.

Application of extra quantities of potassium and phosphate fertilisers, lime and/or organic matter (straw, dung, green manure) affects the absorption of radioelements by plants: liming acidic soils reduces the absorption of strontium due to a competition with calcium and application of potassium fertilisers may have a similar effect on the caesium uptake. The reduction observed is highest in poor soils but can be negligible in well fertilised farm lands. Also no effect of liming can be expected on highly calcareous soils. It should also be noticed that any measure taken in the Chernobyl contaminated zones to increase the crop productivity generally lowered the transfer of radionuclides into plant products [18].

Increasing the overall soil CEC by addition of zeolites, clay minerals, sapropel (lake sediments with a high organic matter content) or large quantities of organic manure results in a higher fraction of radionuclides associated with the solid phase. Conversely, the radionuclide concentration in the soil solution is lowered, reducing proportionally the uptake by plants. Moreover, the existence on certain clay types (illite, vermiculite, ...) of specific binding sites, on which caesium is *quasi* irreversibly fixed, contributes to the reduction of radiocaesium availability for plant uptake over the long-term. However, the experience in liquidation of Chernobyl accident consequences has shown that these amendments had in practice quite limited efficiency; appreciable effects can be expected only in very sandy soils, poor in clay minerals and organic matter.

Addition to the soils of binding chemicals which will immobilise the radionuclides appears as an attractive option. Modified alumino-silicate or Prussian blue compounds such as ammonium ferric hexacyanoferrate (AFCF) are envisaged. Application of 1 g AFCF/m^2 on surface contaminated soils, under laboratory conditions, reduces the transfer to rye grass by a factor around 5. Confirmation of the AFCF effect was also reported after fertilising soils with manure from cows treated with this compound [19].

Land drainage in water-logged soils has been reported (2) to reduce considerably Cs uptake by plants. The invoked mechanism, although not yet proved, could be the higher ammoniac concentrations found in these soils, due to the prevailing anaerobic conditions. Under other conditions, decreasing the soil moisture content may increase the caesium uptake, possibly due to concentration effects.

Selection of crops can offer another solution for contaminated land recovery. Differences up to a factor 4 to 5 were reported between plant varieties so that a sound choice of the cultivated variety can bring a very significant improvement without disturbance of the farmers' habits. Even higher differences, up to 10, were observed between plant species but replacement of one crop by another can be a more disruptive option, especially if the new proposed crop requests completely different cultural practices and material, and/or is intended toor different use (oil, fibres plants, production of bio-methanol). When agricultural perspectives must be abandoned due to excessive contamination the food products, afforestation can be proposed as an ultimate solution, allowing at least 30 years before exploitation during which the radioactivity will decay.

4. Transfer to animals

Animal products such as meat, milk, eggs and their derivates, are important elements in most continental population diets. Therefore, the pathways and mechanisms by which radioactivity is transfered to and accumulated in animals (more particularly to domestic animals) deserve a comprehensive knowledge. In particular, milk being an essential item in young children alimentation, the transfer to cow (but also to sheep and goat) milk must be given the necessary emphasis.

Besides direct absorption through the intact or wounded skin, which can generally be neglected, the two routes of pollutants penetration in animals are:
- inhalation of gaseous compounds, aerosols and particles;
- and principally, ingestion of drinking water, food and soil particles associated with the vegetation grazed by the animal.

Ingestion of contaminated soil is generally neglected as a contamination pathway; however, if we consider that grazing animals commonly ingest soil particles up to 20 % of their dry matter daily intake, this may represent the predominant contamination source for elements which exhibit high K_d values and low soil-to-plant transfer [20].

4.1 CONTAMINATION BY INHALATION

In the case of inhalation, pollutants are transferred from the lungs to organs through the blood. Noble gases, poorly soluble in aqueous media, may be neglected as a source of contamination of animal organs and products. Iodine, however, is well absorbed.

Aerosols and particles penetrate to different extent in the lungs, depending on their size. The largest particles (with a diameter from 5 to 30 μm) are deposited in the upper parts of the respiratory system; smaller particles (diameter < 1 μm) penetrate down to the alveoles. Some of these particles are re-excreted by clearance mechanisms up to the throat and may then pass into the digestive tract. Rapid clearance with retention half-times of a few days or less have been reported for certain chemical forms of strontium and iodine; plutonium-oxide can be retained for much longer periods [21]. The fate of elements in the lungs depends on their solubility and on their ability to cross the lung barrier. Pollutants like plutonium are more readily absorbed by this route than from the gastro-intestinal tract (G.I.T.). The contamination of animal products by inhalation is generally insignificant (several orders of magnitude lower) in comparison with ingestion [22]. Actinides are a possible exception [23].

4.2 CONTAMINATION BY INGESTION

The absorption by animals of ingested pollutants depends on their chemical properties and chemical form, on the animal species and on the particular physiological characteristics of the animals [24].

4.2.1. Influence of chemical properties of the radionuclide.

Caesium, like other alkali metals, is up to 100 % ABSORBED THROUGH the G.I. tract in monogastric mammals and to a slightly lower extent in ruminants (60-80 %). Gastrointestinal absorption after oral administration of alkaline earth's varies depending on the element: in general, absorption is highest for calcium absorbed by both active transport and simple diffusion mechanisms, less for strontium absorbed by simple diffusion only (about 20 %) and a few percent for radium. For elements such as rare earths the epithelial cells lining off the gut walls are practically impermeable; orally dosed Pu is absorbed to a very low extent (less than 1 %) [24, 25, 26].

4.2.2. Influence of speciation.

The chemical speciation of a given radionuclide can modify its gastrointestinal absorption. Absorption of technetium as pertechnetate is higher than that of technetium bio-accumulated in plant material [27, 28, 29]. In contrast, bio-incorporation of Pu in plants increases its availability for gastrointestinal uptake [30]. Differences in accumulation rates due to chemical speciation are also noticeable for ingested tritium depending on whether it is administered as tritiated water or incorporated in various organic molecules which increase the ^3H incorporation [31].

4.2.3. Influence of the diet.

The character of the diet (fibrous content of diet, presence of clay particles ingested together with the forage, ...) can modify the availability of radionuclides in the G.I. tract. The crude fibre content of the forage fed to cows were reported to influence the uptake of caesium: transfer coefficients vary from 0.0025 for alfalfa and corn silage to 0.01 for mixed grain [32, 33]. Iron-deficiency in food increases the absorption of U, Np, Am and Cm in rats [34]. The same effect has been reported by many investigators for other non-ferrous metals that may share intestinal absorption routes with iron [35].

4.2.4. Species.

Food processing in the G.I. tract also differs between animal species. Ruminants are characterised by having four chamber stomachs. The first chamber (rumen) acts as a fermentative vat that receives partially chewed vegetation. The aliments are digested by rumen bacteria, yeast and protozoa. Through fermentation, carbohydrates are broken down into various carboxylic acids. These fermentation products, along with some peptides, amino acids and short-chain fatty acids, are absorbed into the bloodstream from the rumen fluid. The fermented rumen fluid along with symbiotic micro-organisms are then passed, *via* the reticulum, into the omasum and the abomasum and are subject to enzymatic digestion, similar to that observed in monogastric animals. The rumen provides an anaerobic, reducing environment (Eh = -400 mV) that can modify the chemical form of the ingested radionuclides (e.g. Tc administered as TcO_4^- is reduced, resulting in a lower bio-availability [27, 35]).

Fermentation in the stomach is not limited to ruminating animals. It is also found in other animal species in which the passage of food through the stomach is delayed, allowing the growth of symbiotic micro-organisms such as in the crop of galliform (chickenlike) birds.

Fermentation processes are also present in monogastric mammals and take place in the caecum. This organ is more developed in herbivorous species.

The activity of the salivary glands and the amount of saliva produced depends on the alimentary regime of the animal species: a high fibre diet promotes a high production of saliva (up to 16 L/d in cow compared with 1 to 2 l/d in horse). Together with minerals excreted in saliva , large amounts of absorbed radionuclides can be recycled into the G.I. tract. This has been shown for technetium [27, 35, 36] but might also apply to caesium and iodine.

4.2.5 Physiological factors.

The accumulation of radionuclides in mammals depends to a large extent on the age of the animal: transfer coefficients (ratio of the radionuclide concentration in an organ or in milk at equilibrium to the daily amount of the radionuclide ingested) are higher for young individuals than for adults. This may be explained by a higher permeability of the gastrointestinal tract, especially in new-borns, and by a higher metabolic activity in growing animals than in adults. Lacourly et al. [37] have shown that calves aged six weeks have transfer rates for radiocaesium 15 to 25 times higher than cows. Considerably more niobium is absorbed by lambs and piglets receiving [95]Nb orally soon after birth than by those receiving it after weaning [38]. A similar observation was reported by Sullivan [39] for absorption of Pu in new-born swine. Field measurements performed after the Chernobyl accident on the transfer of radiocaesium to sheep showed higher transfer coefficient for lambs than for their dams [40, 41, 42]. However, Cs concentration in tissues of foetus and new-borns at birth are lower than the concentrations measured in the corresponding tissues of their dam [43, 44].

4.3. DISTRIBUTION IN THE ANIMAL

Radionuclides absorbed at the G.I. tract level are transported by blood and distributed into the various organs and animal products. The distribution varies according to the physiological status of the animal as well as to the nature and chemical form of the pollutant. Radioiodine is known to accumulate in considerable amounts in the thyroid gland, which therefore exhibits a much higher transfer coefficient than all other organs, but is also found in the stomach mammary and salivary glands. Technetium, administered as pertechnetate, is an analogue of iodine and concentrates in the same sites. Caesium, like potassium, is distributed in soft tissues. Strontium, radium, plutonium and rare earths elements are preferentially accumulated in bones. Liver and kidneys are preferential storage sites for many pollutants.

The selective capacity of the epithelial cells of the mammary gland for secretion into milk rules the relative secretion and accumulation of various radionuclides into the milk. The transfer of radionuclides ingested by animals to their tissues or products (meat, milk or egg) is generally quantified by a transfer coefficient (F_f for meat and F_m for milk). These transfer coefficients, originally defined by Ward and Johnson [45], are namely based on a prerequisite that steady-state equilibrium is achieved between intake and excretion and secretion into milk. If their use as a predictive tool has obvious advantages, caution is however necessary because of its shortcomings. A collation of transfer factor data was provided by Ng [46].

4.4 EXCRETION

Unabsorbed radionuclides are rapidly excreted via the faeces. The absorbed fraction is progressively eliminated at rates depending on the turn-over rates of analogous/homologous elements or of compounds in the body which the radionuclides are associated with. This excretion occurs via urine, faeces (endogenous secretion) and animal products (milk and eggs). Retention of radionuclides in the body is generally described as the sum of exponential functions. Each exponential function can be related to a "compartment" (a more or less physically defined reservoir), characterised by two parameters: the initial capacity and the half-time. After oral dosing, the first compartment can generally be related to the radioactivity present in the G.I. tract; other compartments of increasing half-times represent fractions of radionuclides retained in the body in increasingly more inaccessible forms.

Excretion rates vary with chemical properties of the radioelement as well as the physiological characteristics of the target organ. For instance, the half-time of radiocaesium is longer in muscles than in organs such as liver or kidneys and tritium present in the organism as tritiated water is more readily excreted than organically bound tritium. The long-term retention of ^{65}Zn is attributed to the fraction of the zinc atoms trapped in the matrix or in the crystal lattice of developing bone during the period of intake [24]. Excretion rates also depend on the animal species. Stara et al. [24] suggested a linear relationship between the logarithmic value of the long-term retention of radiocaesium and the logarithm of their body weight. Except for pig, the plot of long-term retention in monogastric animals, including man, vs their life weight falls on the same line, indicating a proportional increase for caesium retention with increasing body weight; a similar relation is obtained for ruminants for which values fall on a separate, lower regression line.

4.4.1. Limiting contamination of animal products.

Available countermeasures applicable at the animal level were reviewed and discussed recently [47, 48, 49, 50, 51].

The first possibility to reduce the contamination level in animal products is undoubtedly the reduction radiocontaminated feed ingestion. This can be achieved for housed animal by providing them, exclusively or partly, with clean forage, from the previous year, when still available, or from remote, unaffected regions. For grazing animals, supplementing with clean fodder (hay, concentrates, ...) will contribute to reduce their radionuclide intake. Finally, cattle can be moved to non- or less contaminated pastures. Pasture management practices (fertilisation, grazing pattern and pressure, ...) can also contribute to lower the contamination levels in grazing animals.

A reduction of the contamination of animal products can also be obtained by reducing the gastro-intestinal availability of the ingested radionuclides. This can be achieved by the incorporation of additives in the animal feed. The same additives can also contribute to enhance the metabolic excretion of absorbed radionuclides.

Depending on the radioelement considered, alumino-silicates like bentonite, zeolite and vermiculite, stable potassium, calcium or iodine, charcoal and Prussian Blue (the most effective for Cs), ... can be used to reduce the contamination of mammals [24].

- The effect of spreading bentonites (single or repeated applications of 80 g/m^2) on pastures grazed by sheep has been investigated by Beresford et al. [40]; only the repeated treatment (every 2 d) was found effective in the reduction of the caesium transfer coefficient but was counterbalanced by a loss in animal body weight (18 % after 34 d) associated with a decrease in grass intake (39 %). It is, however, doubtful whether such a technique could be adopted in commercial agricultural practice because of practical difficulties in its application [40].
- One compound of the "Prussian blue's" group, AFCF, has been tested under practical feeding conditions after Chernobyl and proved efficient [11]. This compound can also be used in free ranging and wild animal, in the form of boli or salt licks.
- 5 % sodium alginate added to contaminated milk fed to young swine reduced by a factor of 6 the strontium content in the body [52]. The same authors reported a slight reduction in radiocaesium retention in swine under the same conditions. The transfer of strontium to milk can also be reduced by giving cows a feed containing 5-7 % Na-alginate [53]. This compound is however much less efficient in ruminant because of its polysaccharide nature; it is degraded to a large extent by the rumen microflora. Moreover, the proportion of alginates in feed can not be increased much above 5 %, since it then reduces the appetency of the animal for its feed.
- chelating agents of the group of amino-acetic acids are administered to enhance excretion of Pb, Cd, Mn and Hg in humans and have been experimentally used to enhance the excretion of ^{65}Zn and Pu. In the case of Pu, the best results were obtained with DTPA, followed in efficiency by DDETA. Zirconium citrate and phosphate compounds,such as hexametaphosphate, are also efficient when administered immediately [24].
- Increasing the fibre content of the ruminants diet may also help to limit caesium absorption [32, 33].

When the animals are used for meat production, a transition period could be forseen before they are slaughtered. Feeding of contaminated or less contaminated fodder during this period will allow them to become decontaminated by biological processes until an acceptable contamination level is reached. This approach has been applied in the UK for upland sheep: a considerable decrease in the radiocaesium concentration in muscle was obtained when lambs were brought from highly contaminated upland pastures to less (100 to 200 times) contaminated lowland pastures for fattening before being slaughtered [41].

5. Food processing

When confronted with unacceptably heavily contaminated plant or animal products, industrial or domestic food processing may provide an alternative for reducing the contamination levels in products consumed by humans.

The problem of ^{131}I contamination in food products may be solved very easily if it is possible to delay consumption to allow physical decay. This can be achieved by production of long shelf life products such as milk powder, cheese, chocolate, tinned foods, deep frozen meat and deep frozen soup concentrates.

Other techniques must be considered for long-lived radionuclides for which physical decay during storage cannot be envisaged:
- washing of vegetables can remove a fraction of the external contamination;
- grain, after removal of the external envelopes, can be used for preparation of white floor;
- milk can be treated by passage on exchange resins to remove strontium and possibly caesium;
- milk can also be used for the preparation of storable by-products in which the contamination is lower (butter, cheese, ...); it should be noted that acid precipitation of casein is more efficient than rennet in the case of Sr contaminated milk;
- some culinary methods of food preparation (such as washing and/or peeling vegetables, boiling meat instead of roasting, pickling, ...) can also be used to limit contamination ingested by man.

6. Conclusions

Many kinds of possible remedial actions, some scientifically sound, others of more empirical nature, have been developed and used to limit the contamination level in products consumed by man or animals. Although developed on a scientific basis, many measures may appear impracticable or too expensive to be used in real situations. As a general rule, remedial actions that can be performed with existing, common material and machines should be preferred to those requiring new technology to be especially developed for such applications.

Often a combination of countermeasures can be applied, simultaneously or successively, to ensure the best result.

The efficiency of countermeasures may vary depending on the specific conditions in which they are used, and can lead sometimes to undesirable side effects. For instance, excessive liming can lead to the precipitation of micro-nutrients and induce deficiencies in plants and animals, high fertilisation in semi-natural systems can induce severe ecological modifications of the ecosystems, ...

It must also be kept in mind that some countermeasures are not or hardly reversible (ploughing, deep-ploughing, ...) and, when they appear to be ineffective, it is very difficult to remedy the situation.

7. References

[1] Willrodt C. (1993), Agrotechnical countermeasures to be applied before and during deposition of radioactive fallout, Sci. Tot. Environ., 137: 21-29.

[2] Segal M.G. (1993), Agricultural countermeasures following deposition of radioactivity after a nuclear accident, Sci. Tot. Environ, 137: 31-48.

[3] International Commission on Radiological Protection (1991), Recommendations of the International Commission on Radiological Protection, Annals of the ICRP 21: 1-3, Pergamon Press, Oxford.

[4] Vandecasteele C.M., Zeevaert Th. & Kirchmann R. (1991), Factors influencing the transfer of radionuclides in agricultural food chains, in "Anticarcinogenesis and radiation protection" 2nd Ed., O.F. Nygaard & A.C. Upton Eds., Plenum Press, New-York, 181-199.

[5] Middleton L.J. (1959), Radioactive strontium and caesium in edible parts of crop plants after foliar contamination, Int. J. Rad. Biol., 1: 387-402.

[6] Kirchmann R., Fagniart E. and Van Puymbroeck S. (1966), Studies on foliar contamination by radiocaesium and radiostrontium, in Radiological concentration processes, Pergamon Press, Oxford - New York, 475-483.

[7] Tukey H.B. (1970), The leaching of substances from plants, Ann. Rev. Plant Physiol., 21: 305-324.

[8] Eisenbud M., Krauskopf K., Penna Franca E., Lei W., Ballad R., Linsalata P. and Fujimori K. (1984), Natural analogues for the transuranic actinide elements: an investigation in Minas Gerais, Brazil, Environ. Geol. Water Sci., 6: 1-9.

[9] Vandecasteele C.M., Fagniart E., Colard J., Culot J.P. and Kirchmann R. (1988), Transfer of radiocaesium deposited after the Chernobyl accident to agricultural plants, in Impact des accidents d'origine nucléaire sur l'environnement, CEN-CEA Cadarache, Vol. 1: D179-D187).

[10] Schulz R.K. (1965), Soil chemistry of radionuclides, Health Phys., 11: 1317-1324.

[11] Bair W.J. (1960), Radioisotope toxicity from pulmonary absorption, in Radioisotopes in the biosphere, Caldecott R.S. & Snyder L.A. Eds, University of Minneapolis, Minnesota, Comstock Publishing Associates, Ithaca and London.

[12] Baes C.F. III, Garten C.T. Jr, Taylor F.G. and Witherspoon J.P. (1986), The long-term problems of contaminated land: sources, impacts and countermeasures, Oak Ridge National Laboratory, Environmental Science Division, Publication 2593, ORNL-6146.

[13] Nisbet A.F., Konoplev A.V., Shaw G., Lembrechts J.F., Merckx R., Smolders E., Vandecasteele C.M., Lönsjö H., Carini F. and Burton O. (1993), Application of fertilisers and ameliorants to reduce soil to plant transfer of radiocaesium and radiostrontium in the medium to long term - a summary, Sci. Tot. Env., 137: 173-182.

[14] Marti J.M., Arapis G. & Iranzo E. (1990), Evaluacion de contremedidas para la recuperacion de suelo agricola, in "Environmental contamination following a major nuclear accident" Vol. II, IAEA Vienna, IAEA-SM-306/103: 111-127.

[15] Legrand B., Fache P., Hamoniaux M., Camus H. & Gauthier D. (1990), Premiers résultats expérimentaux du programme RESSAC sur les essais in situ de décontamination/fixation et études de migration des radionucléides dans les sols, IAEA Vienna, IAEA-SM-316/33: 507.

[16] Jouve A., Schulte E., Bon P. & Cardot A.L. (1993), Mechanical and physical removing of soil and plants as agricultural mitigation techniques, Sci. Tot. Environ, 137: 65-79.

[17] Sandalls F.J. (1990), Review of countermeasures used in agriculture following a major nuclear accident, in "Environmental contamination following a major nuclear accident" Vol. II, IAEA Vienna, IAEA-SM-306/44: 129-140.

[18] Kirchmann R. (1990), Agricultural countermeasures taken in the Chernobyl region and evaluation of the results, International Union of Radioecology, Report for Contract 88-ET-006 with CE/DGXI/A1.

[19] Hove K., Strand P., Salbu B., Oughton D., Astasheva N., Vasiliev A., Ratnikov A., Jigareva T., Averin V., Firsakova S., Crick M.J. & Richards J.I. (1995), Use of caesium binders to reduce radiocaesium contamination of milk and meat in Ukraine, Belarus and the Russian Federation, Int. Symp. on "Environmental impact of radioactive releases", IAEA Vienna 8-12 May 1995, Extended Synopses, IAEA-SM-339/153: 75-76.

[20] Zach R. and Mayoh K.R. (1984), Soil ingestion by cattle: a neglected pathway, Health Phys., 46: 426-431.

[21] Baes C.F III. and Sharp R.D. (1983), A proposal for estimation of soil leaching and leaching constants for use in assessment models, J. Environ. Qual., 12: 17-28.

[22] Hvinden T., Lillegraven A. & Lillesaeter O. (1964), Passage of a radioactive cloud over Norway November 1962, Nature, 202: 950-952.

[23] Zach R. (1985), Contribution of inhalation by food animals to man's ingestion dose, Health Phys., 49: 737-745.

[24] Stara J.F., Nelson N.S., Della Rosa R.J. and Bustad L.K. (1971), Comparative metabolism of radionuclides in mammals: a review, Health Phys., 20: 113-137.

[25] Coughtrey P.J., Jackson D. and Thorne M.C. (1985), Radionuclide distribution and transport in terrestrial and aquatic ecosystems: a compendium of data, A.A. Balkema, Rotterdam - Boston.

[26] Kirchmann R., Bell J.N.B., Coughtrey P.J., Frissel M., Hakonson T.E., Hanson W.C., Horrill D., Howard B.J., Lane L.J., Myttenaere C., Robison W.L., Ronneau C., Shaw G., Schell W.R., Van den Hoek J., Konoplyov A & Zezina N. (1993), Terrestrial pathways, in Radioecology after Chernobyl, SCOPE 50 Warner F. & Harrison R.M. Eds., John Wiley & Sons, Chichester, New-York, Brisbane, Toronto, Singapore: 101-177.

[27] Gerber G.B., Van Hees M., Garten C.T. Jr, Vandecasteele C.M., Vankerkom J., Van Bruwaene R., Kirchmann R., Colard J. and Cogneau M. (1989), Technetium absorption and turnover in monogastric and polygastric animals, Health Phys., 58: 337-343.

[28] Sullivan M.F., Garland T.R., Cataldo D.A. and Schreckhise R.G. (1979), Absorption of plant-incorporated nuclear fuel cycle elements from the gastro-intestinal tract, in Biological implications of radionuclides released from nuclear industries, IAEA Vienna, IAEA-SM-237/58, Vol.I: 447-457.

[29] Vandecasteele C.M., Garten C.T. Jr, Van Bruwaene R., Janssens J., Kirchmann R. and Myttenaere C. (1986), Chemical speciation of technetium in soil and plants: impact on soil-plant-animal transfer, in Speciation of fission and activation products in the environment, R.A. Bulman and J.R. Cooper Eds., Elsevier Applied Science Publishers, London - New York, 368-381.

[30] Sullivan M.F., Garland T.R., Cataldo D.A., Wildung R.E. and Drucker H. (1980), Absorption of plutonium from the gastrointestinal tract of rats and guinea pigs after ingestion of alfalfa containing Pu-238, Health Phys., 38: 215-221.

[31] Kirchmann R., Charles P., Van Bruwaene R. and Remy J. (1975), Distribution of tritium in the different organs of calves and pigs after ingestion of various tritiated feeds, Current Topics in Radiation Research Quaterly, 12: 291-312.

[32] Wilson D.W., Ward G.M. and Johnson J.E. (1969), A quantitative model of the transport of Cs-137 from fallout to milk, Environ. Contam. Radioact. Mater., Proc. Semin., Vienna.

[33] Eisenbud M. (1987), Environmental radioactivity from natural, industrial and military sources 3rd edition, Academic press Inc., Harcourt Brace Jovanovich Publ., New-york, 475pp.

[34] Sullivan M.F. and Ruemmler P.S. (1988), Absorption of U-233, Np-237, Pu-238, Am-241 and Cm-244 from the gastrointestinal tracts of rats fed an iron-deficient diet, Health Phys., 54: 311-316.

[35] Jones B.-E.V. (1983), Metabolism of technetium in goats, Int. J. Appl. Radiat. Isot., 34: 837-839.

[36] Helman J., Turner R.J., Fox F.C. and Baum B.J. (1987), ^{99m}Tc-pertechnetate uptake in parotid acinar cells by the $Na^+/K^+/Cl^-$ co-transport system, J. Clin. Invest., 79: 1310-1313.

[37] Lacourly G., Savy C., Lehr J. and Kirchmann R. (1971), Relations entre la contamination de la viande de bovin et celle du lait par le radiocesium, Health Phys., 21: 793-802.

[38] Mraz F.R. and Eisele G.R. (1977), Gastrointestinal absorption, tissue distribution and excretion of Nb-95 in newborn and weanling swine and sheep, Radiat. Res., 72: 533-536.

[39] Sullivan M.F. (1980), Absorption of actinide elements from the gastrointestinal tract of neonatal animals, Health Phys., 38: 173-185.

[40] Beresford N.A., Lamb C.S., Mayes R.W., Howard B.J. and Colgrove P.M. (1989), The effect of treating pastures with bentonite on the transfer of Cs-137 from grazed herbage to sheep, J. Environ. Radioactivity, 9: 251-264.

[41] Howard B.J., Beresford N.A., Burrow L., Shaw P.V. and Curtis E.J.C. (1987), A comparison of caesium-137 and 134 activity in sheep remaining on upland areas contaminated by Chernobyl fallout with those removed to less active lowland pasture, J. Soc. Radiol. Prot., 7: 71-73.

[42] Vankerkom J., Van Hees M., Vandecasteele C.M., Colard J., Culot J.P. and Kirchmann R. (1988), Transfer to farm animals (ruminants) and their products of Cs-134, Cs-137 and I-131 after the Chernobyl accident, in Impact des accidents d'origine nucléaire sur l'environnement, CEN-CEA Cadarache, Vol. 2: E111-E119.

[43] Howard B.J. and Beresford N.A. (1989), Chernobyl radiocaesium in upland sheep farm ecosystems, Br. Vet. J., 145: 212-219.

[44] Vandecasteele C.M., Van Hees M., Culot J.P. and Vankerkom J. (1989), Radiocaesium metabolism in pregnant ewes and their progeny, Sci. Tot. Environ., 85: 213-223.

[45] Ward G.M. & Johnson J.E. (1986), Validity of the term transfer coefficient, Health Phys., 50: 411-414.

[46] Ng Y.C. (1982), A review of transfer factors for assessing the dose from radionuclides in agricultural products, Nucl. Safety, 23: 57-71.

[47] Hove K. (1993), Chemical methods for reduction of the transfer of radionuclides to farm animals in semi-natural environments, Sci. Tot. Environ., 137: 235-248.

[48] Hove K., Strand P., Voigt G., Jones B.E.V., Howard B.J., Segal M.G., Pollaris K. & Pearce J. (1993), Countermeasures for reducing radioactive contamination of farm animals and farm animal products, Sci. Tot. Environ., 137: 261-271.

[49] Howard B.J. (1993), Management methods of reducing radionuclide contamination of animal food products in semi-natural ecosystems, Sci. Tot. Environ., 137: 249-260.

[50] Jones B.-E.V. (1993), Management methods of reducing radionuclide contamination of animal food products, Sci. Tot. Environ., 137: 227-233.

[51] Voigt G. (1993), Chemical methods to reduce the radioactive contamination of animals and their products in agricultural ecosystems, Sci. Tot. Environ., 137: 205-225.

[52] Van der Borght O., Colard J., Van Puymbroeck S. and Kirchmann R. (1966), Radiocontamination from milk in piglets (swine): influence of sodium alginate on the Sr-85/Cs-134 ratio of the body burden and on the comparative Sr-85/Ca-47 absorption, in Radioecological concentration processes, Pergamon Press, Oxford - New York, 589-593.

[53] Thompson J.C. Jr, Wentworth R.A. and Comar C.L. (1971), Control of fallout contamination in the postattack diet: in survival of food crops and livestock in the event of a nuclear war, Proc. Symp. Brookhaven National Laboratory, Upton, New York, Sept. 1970, B.W. Benson and A.H. Sparrow Eds., 566-595.

[54] Aarkrog A. (1975), Radionuclide levels in mature grain related to radiostrontium content and time of direct contamination, Health Phys., 28: 557-562.

[55] Nishita H., Romney E.M. and Larson K.H. (1961), Uptake of radioactive fission products by crop plants, J. Agric. Food Chem., 9: 101.

METHOD FOR EVALUATING RADIONUCLIDE INVENTORIES IN WATER RESERVOIRS APPLIED TO THE BELOYARSKOE ARTIFICIAL LAKE

A. TRAPEZNIKOV
Biophysical station of the Institute of Plant and Animal Ecology
Urals Division, Russian Academy of Sciences
ulitsa 8 Marta, 202
CIS-620219 Ekatarinburg, Russia

1. Introduction

At present there is on the territory of Russia and other countries of the globe a great number of surface waters suffering from radioactive contamination. They include cooling ponds of nuclear power plants (NPPs), water reservoirs, lakes and rivers receiving discharges from nuclear fuel cycle plants, during both normal operation and accidental conditions.

The assessment of the risk from the presence of radionuclides in these aquatic ecosystems has a high priority, since understanding the real hazard of radioactive environmental pollution for the environment is necessary.

2. Radioactivity inventory

2.1 CHARACTERISTICS OF LAKE BELOYARSKOE

As an example, the radionuclide inventory in the Beloyarskoe artificial lake, which serves as cooling pond for the Beloyarsk nuclear power plant (BNPP), has been evaluated. The Beloyarskoe artificial lake is situated in the Middle Urals, 60 km from the city of Ekatarinburg. It was created from 1959 to 1963 by the construction of a dam in the Pyshma river at 75 km from the river source. The length of the lake is 20 km, its width opposite the BNPP is 3 km and its surface is 47 km². The fairway depth of the Pyshma river is 15-20 m.

The Beloyarskoe lake is a hydrocarbonatecalcium reservoir characterized by a normal mineral composition and a low alkali content. The mineral composition of the lake water is practically the same in all locations [1]. The main bays and channels of the lake are shown on the map (Figure 1). The BNPP is located on the left bank of the lake. Liquid industrial effluents reach the lake via an industrial runoff channel. 700 m downstream from this outlet is the intake channel, which draws water from the lake to the BNPP's cooling system. Further down, at a distance of 2 km, is the outlet of the dis-

F. F. Luykx and M. J. Frissel (eds.), Radioecology and the Restoration of
Radioactive-Contaminated Sites, 179–186.
© 1996 *Kluwer Academic Publishers.*

charge channel for waste water from the BNPP's cooling system, which goes directly to Warm Bay.

As an example the ^{60}Co inventory was calculated. ^{60}Co is a typical radionuclide found in NPP cooling ponds. It is deposited in the main components of the Beloyarskoe lake: bottom sediments, water and aquatic plants.

2.2 WATER VOLUME

To calculate the ^{60}Co inventory in the water, one needs data regarding the total volume of water in the lake and the concentration of this radionuclide in various areas of the lake. Therefore, the lake was divided into three arbitrary zones: the lower, central and upper parts. The ratio between water volume and water surface area was estimated from depth measurements. The volumes of the lower, central and upper part of the lake ar 144.10^6 m^3, 120.10^6 m^3 and 1.10^6 m^3, respectively. The total volume of the lake is thus 265 million m^3. Hence, the radionuclide inventory in the Beloyarskoe lake was evaluated at appoximately 6.6 GBq. Table 1 shows that the radioactivity in the lower, central and upper part of the lake is respectively 56, 43 and 0.3 % of the total.

TABLE 1. ^{60}Co inventories in water of various areas of the Beloyarskoe Lake

Area of the lake	water volume, million m^3	^{60}Co concentration, Bq/m^3	^{60}Co inventory, MBq
Low	144	26	3,700
Central	120	28	2,880
Upper	1	21	21
The whole lake	265		6,645

2.3 SEDIMENTS

To calculate the ^{60}Co inventory in the sediments and soils on the bottom of the lake these soils and sediments were sampled in an offshore area at 100 m from the lake perimeter. Although this is not entirely correct, soils and sediments will hereafter simply be called sediments. Four main types of sediments were sampled up to a depth of 10 cm: sand, sandy-silt, overflooded soil and silt sapropel. Again the lake was divided into three main areas. In particular the area around the warm-water outlet in Warm Bay was carefully sampled. The contribution of each sediment was determined for every area (Table 2).

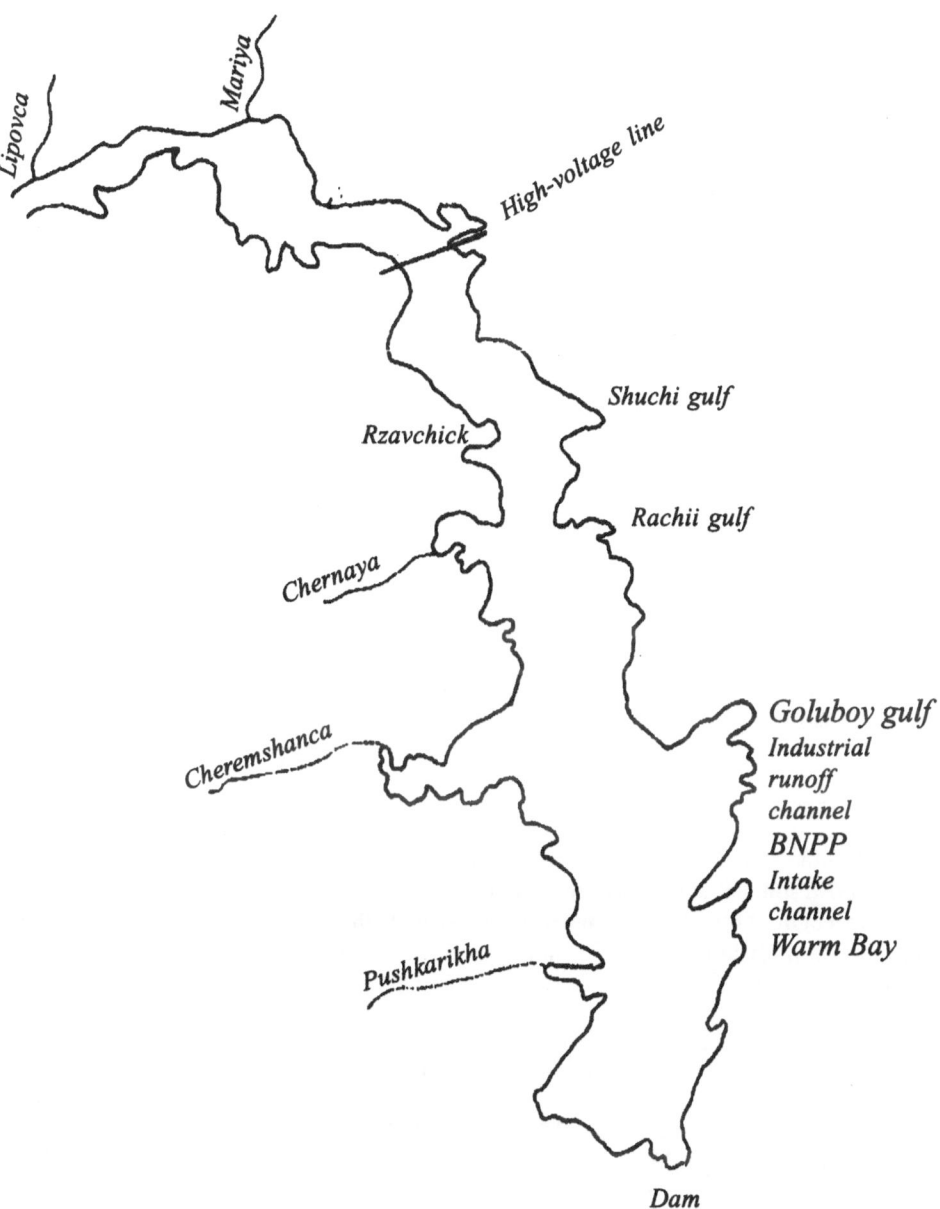

Figure 1. Beloyarskoe Lake

TABLE 2. Main types of sediments (%) in various areas of the Beloyarskoe lake*

Type of sediment	% of sediment types			
	Lower part	Warm Bay	Central part	Upper part
Sand	50	2	50	50
Sandy silt	10	none	10	none
Overflooded soil	35	48	35	35
Silt	5	50	5	15

* The data refer to the 100 m offshore area, excluding Warm Bay for which the data refer to the total area of Warm Bay.

This work was continued for several years. Radionuclide concentrations were averaged for each of the three parts of the lake, per type of radionuclide and per type of bottom sediment (Table 3).

TABLE 3. Average ^{60}Co concentration in main types of sediments in various areas of the Beloyarskoe lake

Type of sediment	^{60}Co content, Bq/kg fresh mass			
	Lower part	Warm Bay	Central part	Upper part
Sand	25.2	46.2	30.0	14.1
Sandy silt	38.1	none	25.2	none
Overflooded soil	93.6	2410.1	45.1	20.7
Silt	56.2	803.3	62.5	10.4

The importance of the sediments in the offshore area is decreasing in the order: sand, overflooded soil, silt and sandy-silt sediments. From the depth of the sediments sampled and the length of the offshore line one can calculate the total mass of bottom sediments for the 100 m offshore area in the various parts of the lake as well as their total amount in Warm Bay (Table 4).

TABLE 4. Total amount of sediments taken into account for the ^{60}Co inventory in the offshore area and the Warm Bay of the Beloyarskoe lake

Type of sediment	Sediment mass, 10^3 ton			
	Lower part	Warm Bay	Central part	Upper part
Sand	410	2	148	49
Sandy silt	32	none	23	none
Overflooded soil	87	24	63	21
Silt	12	25	9	9

Those values and the average ^{60}Co concentrations allow to calculate the radionuclide inventory (Table 5) which is approximately 90 GBq. The recent sediments in the deepwater zone of the lake consist mainly of silt.

TABLE 5. ^{60}Co inventory in the sediments of the offshore area and in Warm Bay of the Baloyarskoe lake

Type of sediment	Lower part	Warm Bay	Central part	Upper part	Total offshore area + Warm Bay
Sand	10.3	0.1	4.4	0.7	15.5
Sandy silt	1.2	none	0.6	none	1.8
Overflooded	8.1	57.8	2.8	0.4	69.1
Silt soil	0.7	2.0	0.6	0.1	3.4

The ^{60}Co concentration in the 10 cm layer of sediments in various selected parts of the deep-water zone was also determined. Results are shown in Table 6. The total amount of sediments in the deep water zone of the Beloyarskoe lake is shown in Table 7.

TABLE 6. Average ^{60}Co concentration in the sediments of various parts of the deepwater zone of the Beloyarskoe lake

Type of sediment	Lower part	Central part	Upper part
Silt	20.9	24.5	5.5

^{60}Co content, Bq/kg fresh mass

TABLE 7. Total amount of sediments taken into account in the calculation of ^{60}Co inventory in the deepwater area of the Beloyarskoe lake

Type of sediment	Lower part	Central part	Upper part	Total deep water zone
Silt ground	3,500	3,000	300	6,800

Sediment mass, 10^3 ton

From Tables 6 and 7, the total inventory in the bottom sediments of the deep water zone can be calculated; results are shown in Table 8.

The total quantity of ^{60}Co in all sediments of the Beloyarskoe lake amounts to 238 GBq (Table 9). The distribution across the lower part, central part, Warm Bay and upper part of the lake is 40, 34, 25 and 1 %, respectively.

TABLE 8. ^{60}Co inventory in the sediments of the deepwater area of the Beloyarskoe lake

| Type of ground | ^{60}Co, GBq | | | |
	Lower part	Central part	Upper part	Total deep water zone
Silt	74	73	2	149

TABLE 9. Total amount of ^{60}Co in sediments of the Beloyarskoe Lake (offshore and deepwater parts)

| Components of the Lake | Amount ^{60}Co, GBq | | | | |
	Warm Bay	Lower part	Central part	Upper part	Total Lake
Sediment	60	94	81	3	238

2.4 VEGETATION OF THE LAKE

To calculate the ^{60}Co inventories in the higher aquatic plants of the Beloyarskoe lake, several characteristics were determined: the biomass of plants per m^2 of surface area, the area occupied by the macro plant mass, their total biomass in the lake as well as the fresh-to-dry ratio fof the biomass.

Investigations concentrated on the dominant plants in the lake: Elodea canadensis, Ceratophillum demersum, Lemna minor, Potamogeton pectinatus.

Some important characteristics of these plants are shown in Table 10.

TABLE 10. Characteristics of aquatic plants of the Beloyarskoe lake

| Plant | Ratio fresh/dry | fresh mass of plants, kg/m^2 | Warm Bay Zone | | Beloyarskoe lake (excl. Warm Bay) | |
			Area occopied by plants, m^2	Fress mass of plants, total area, kg	Area occopied by plants, m^2	Fress mass of plants, total area, kg
Elodea canadensis	9.1	3.5	750	2,625	950	3,325
Ceratophillum demersum	12.6	2.7	800	2,160	250	675
Potamogeton pectinatus	8.1	1.2	20	24	160	192
Lemna minor	16.7	2.5	5,000	12,500	442,700	1,106,750
Potamogeton perfoliatus	9.6	2.1	300	630	1,200	2,520
Cladophora fracta	7.1	3.8	200	760	640	2,430

TABLE 11. Co-60 inventory in the main aquatic plants of the Beloyarskoe lake

Plant	Warm Bay			Beleyarskoe lake (excl W.B.)			Total inventory kBq
	Fresh mass of plants, kg	Av. Conc. ^{60}Co Bq/kg fresh mass	^{60}Co inventory, kBq	Fresh mass of plants, kg	Av. Conc. ^{60}Co, Bq/kg fresh mass	^{60}Co inventory, kBq	
Elodea canadensis	2,625	384.8	1,010	3,325	17.5	58.2	1,000
Ceratophillum demersum	2,160	403.3	871	675	18.5	12.5	900
Potamogeton pectinatus	24	32.3	0.8	192	11.6	2.2	3
Lemna minor	12,500	125.8	1,572	1,106,750	7.4	8,190	9,800
Potamogeton perfoliatus	630	47.8	30.1	2,520	8.8	22.2	50
Cladophora fracta	760	184.1	139.9	2,430	21.7	52.7	200

TABLE 12. ^{60}Co in the lake components

Component	Amount ^{60}Co, MBq	Distribution of ^{60}Co in the lake, %
Sediments	238,000	97.3
Water	6,645	2.7
Aquatic plants	12	0.005

Lemna minor dominates with an occupancy area of approximately 450,000 m^2 and a biomass of approximately 1,120 tons. Average ^{60}Co concentrations per species were determined, separately for both the total lake and Warm Bay. The total inventory in the aquatic plants of the lake is shown in Table 11.

The total amount is approximately 12 MBq.

2.5 TOTAL ^{60}Co INVENTORY

The larger part of ^{60}Co is present in the sediments of the lake, i.e., 238 GBq; the water contains 6.6 GBq and the aquatic plants only 12 MBq. The total amount of ^{60}Co in the lake is 245 GBq. The distribution over sediments, water and vegetation is 97.3, 2.7 and 0.005% respectively (Table 12).

186

3. References

1. Lubimova S.A. (1981) Hydrochemical conditions of the Beloyarskoe Lake // Radionuclide in the grounds and in the freshwater systems, Scientific Works of the Ecological Institute of the plants and animals., USC AS USSR (The Urals Scientific Centre of Academy of Science of the USSR).- Sverdlovsk, 43-46.

UP-TO-DATE STUDIES OF RADIONUCLIDES TRANSFER IN THE RIVER SYSTEM TECHA-ISET-TOBOL-IRTYSH-OB TO KARA SEA BY A JOINT INTERNATIONAL GROUP. *

A. AARKROG
Risø National Laboratory
P.O. Box 49, DK-4000 Roskilde, Denmark

1. Introduction

MAYAK is a nuclear facility established by the Former Soviet Union for the production of plutonium for military purposes. It is located 55°44'N 60°54'E, 70 km north of Cheliabinsk and 15 km east of the city of Kyshtym in the southern Urals. MAYAK covers an area of 90 km² and was operated by the former Ministry of Nuclear Power of the USSR. (At present Ministry of Atomic Energy of Russia).

The implementation of the nuclear programme in the Cheliabinsk region in the Ural involved contamination of the environment with long-lived fission products resulting from both accident conditions and the routine operation of the utility, particularly, in the initial period i.e. the late forties and early fifties.

During 1949-1952 liquid medium and low-level radioactive waste was thus disposed directly into the Techa river system.

The Techa river belongs to the Iset-Tobol-Irtysh-Ob river system. (Figure 1 and Table 1). In its upper region it passes the MAYAK nuclear installation. The liquid radioactive waste was released 6 km from the Techa source. During that period, 76×10^6 m³ of liquid radioactive waste was discharged, with a total activity of 100 PBq [1]. Most of the activity (95%) was discharged in 1950 and 1951. ^{90}Sr and ^{137}Cs contributed 11.6% and 12.2%, respectively. Now 45 y later, the environmental contamination from this source has decayed to 3.9 PBq ^{90}Sr and 4.3 PBq ^{137}Cs.

In September 1951, discharge of wastes was eliminated by introducing a cascade of reservoirs and by-pass canals. This permitted the removal of significant radioactivity from the open drainage network [2]; however, before this was accomplished, a large part of the flood plain and the bottom of the river, especially upstream, were contaminated. According to Nikipelov et al. [2], the major portion (about 99%) of radionuclides were deposited upstream Muslumovo (Figure 2) in the river floodland and bed. The Techa was withdrawn from economic use and the population of several villages was evacuated. Presently the Asanov swamps in the upper reaches of the river are the area of permanent

---

*   The Joint International Group consists presently (1995) of the following laboratories: Institute of Plant and Animal Ecology (IPAE, Russia), Institute of Biology of the Southern Seas (IBSS, Ukraine), National Radiological Protection Board (NRPB, UK) and Risø National Laboratory (RISØ, Denmark). The project is economically supported by INTAS (contract No. 94-1221).

F. F. Luykx and M. J. Frissel (eds.), Radioecology and the Restoration of
Radioactive-Contaminated Sites, 187–201.
© 1996 *Kluwer Academic Publishers.*

188

contamination of the Techa. Migration of radionuclides from the cascade of reservoirs and canals has been observed. This system contains 7.1 PBq ^{90}Sr and ^{137}Cs [1].
Most of the activity discharged to the Ob river system may thus still be found in the Techa river, in particular in the reservoirs. However, some of the activity must be found elsewhere, perhaps in the ocean via the Arctic Ocean.

Figure 1. Drainage area of the Ob river (2.9 × 10⁶ km²).

TABLE 1. Main hydrological characteristics of the Ob River System [3].

LOCATION (river)	Distance from discharge (km)	Mean width (m)	Mean depth (m)	Water flow (km³ year⁻¹)	Sediment transport (kg year⁻¹)
Nadirov Bridge (Techa)	49	37	2.1	~0.02	~1.66×10^6
Muslumovo (Techa)	78	20.5	1.1	0.06	~5×10^6
Verknaja Techa (Techa)	177	25.5	0.5	~0.14	~12×10^6
Pershinskoe (Techa)	214	78.5	0.7	0.18	~15×10^6
Zatechenskoje (Techa)	237	25.5	1.0	0.35	~30×10^6
Mekhonskoje (Iset)	356	82.9	1.2	2.57	-
Isetskoje (Iset)	480	131	2.6	2.48	0.05×10^9
Jalutorovsk (Tobol)	1,023	164	3.0	3.11	0.11×10^9
Tobolsk (Irtysh)	1,670	681	6.8	77.3	7.9×10^9
Salekhard (Ob)	2,542	2,894	13.3	421	15×10^9

Ongoing studies [4, 5] with participation of laboratories from the former Soviet Union and western countries aim to quantify the contamination of the Ob river system and of the Arctic Ocean [6] from the MAYAK activities. The present paper summarises the results of some of these studies.

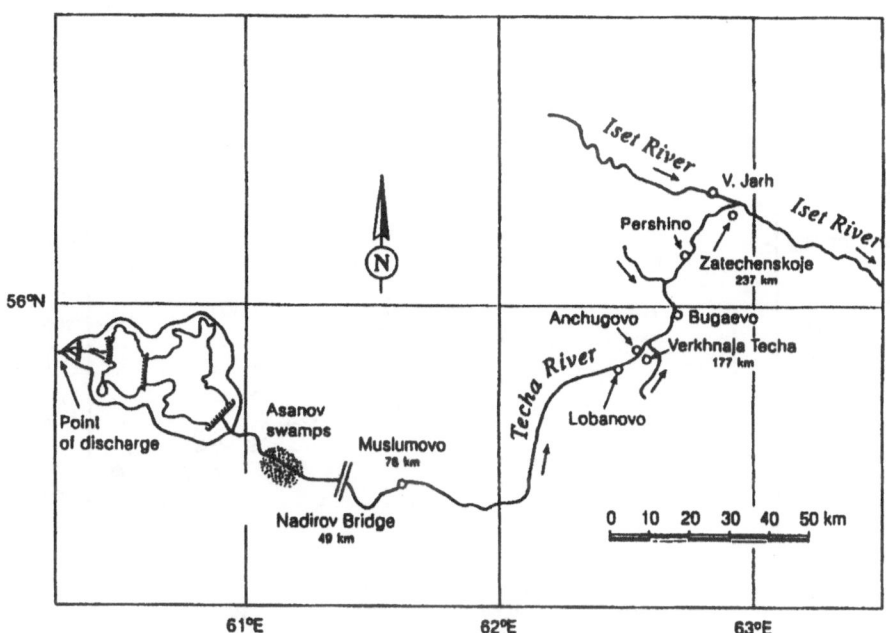

Figure 2. Techa River. (Distances from point of discharge)

2. Techa river studies

The Techa river source is in Lake Irtysh and joins the Iset river at the right bank (243 km long) (Figure 2). The drainage area is 7,600 km². Main hydrological characteristics are summarized in Table 1 [3]. Seasonal flooding occurs in April; maximal spring high flood is 2.2 m near the village Muslumovo and 4 m near the village Zatechenskoye. In winter, the upper stream freezes 4-6 d earlier than the lower stream and usually freezes through the bottom. Techa river runs a twisting course through a valley with low slopes and a width up to 2 km. The floodplain is open meadow, locally bushy. The landscape consists of hilly plains with groups of mixed forest. The river bottom near the banks is silt-sandy, in the middle of the river it is pebbly or sandy. Upstream of the village Muslumovo the river passes bogged areas. During 1949-1951 these areas accumulated significant quantities of radionuclides.

2.1 EARLIER STUDIES

Figure 3 shows how the annual ^{90}Sr mean concentrations in Techa river waters have varied at two locations since 1949, and in Table 2 the ^{90}Sr and ^{137}Cs concentrations at Muslomovo are compared. While the ^{137}Cs concentrations over about 40 years decreased by a factor of 40,000, the ^{90}Sr levels decreased by a factor of only 400. This reflects the higher and more rapid retention of ^{137}Cs in the river sediments and in the soil of the flooded areas.

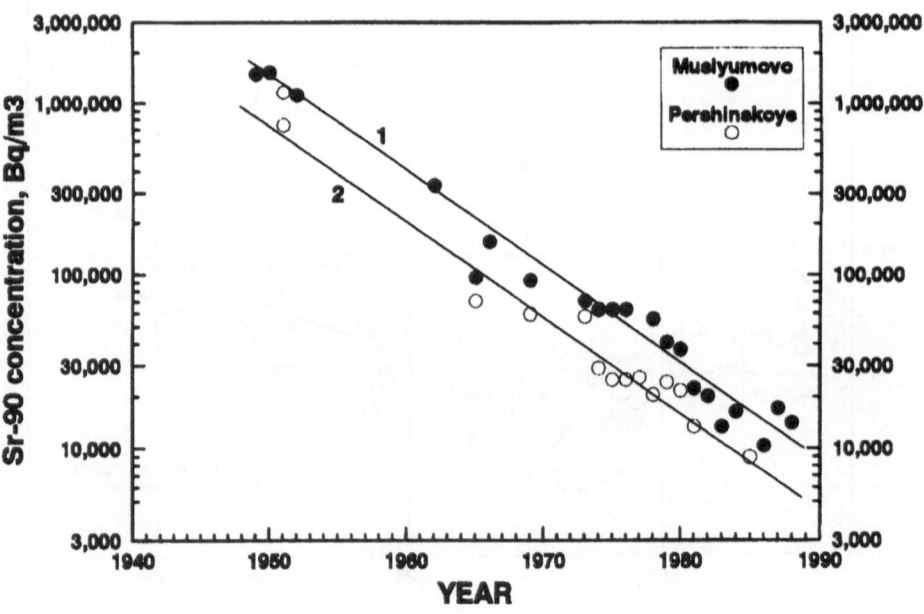

Figure 3. Mean annual ^{90}Sr concentrations in Techa river at Muslumovo and Perskinskoe since 1949[7].

TABLE 2. Annual mean water concentrations of ^{90}Sr and ^{137}Cs in Techa river at Muslomovo[8].

Year	^{90}Sr kBq m^{-3}	^{137}Cs kBq m^{-3}
1951	1,500	20,000
1962	350	150
1964	100	9
1973	75	1.5
1978	60	1.3
1983	13	0.9
1988	16	1.5
1990	4	0.5

*) Measurements by The Institute of Animal and Plant Ecology, Ekaterinburg and Risø National Laboratory [4].

2.2 RECENT STUDIES

The concentrations of ^{90}Sr, ^{137}Cs and ^{239,240}Pu (Table 3) in the Techa river water decrease with distance (x in km) from the point of discharge according to the exponential model: $y = e^{(a+bx)}$, where y is the radionuclide concentration in Bq.m^{-3}:

$$^{90}\text{Sr} \quad a = 8.62\pm0.05; \qquad b = -(0.0029\pm0.0003)$$
$$^{137}\text{Cs} \quad a = 7.03\pm0.13; \qquad b = -(0.0120\pm0.0008)$$
$$^{239,240}\text{Pu} \quad a = -(0.38\pm0.11); \qquad b = -(0.0074\pm0.0007)$$

The error term is 1 SD.

The equations show that at the discharge point, the ratio of ^{90}Sr to ^{137}Cs = 4.9. At Zatechenskoje, where the Techa river enters the Iset river, the ratio increased nearly an order of magnitude to 43. The annual water flow at the discharge point is estimated at $(2-3) \times 10^7$ m^3 y^{-1} and at Zatechenskoje it increased to $(3-4) \times 10^8$ m^3 y^{-1} [7].

When we pass from the point of discharge to the end of the river at Zatechenskoje, the amount of ^{90}Sr transported by Techa river increases from about 0.15 T Bq y^{-1} to about 1 T Bq y^{-1}, while the ^{137}Cs transport seems nearly unchanged (~0.02±0.01 T Bq y^{-1}). Thus, there seems to be a net transfer of ^{90}Sr to the river water during its passage through Techa river. The concentration of ^{60}Co was nearly constant in the Techa river water: 3.2±1.1 (±1SD) Bq ^{60}Co m^{-3}, which also suggests a supply of ^{60}Co.

From Tables 1 and 3, the following expressions for water inventories of radionuclides in the Techa river as a function of distance (x m) were derived:

$$\text{Bq }^{90}\text{Sr} \quad = \quad \int_{49,000}^{237,000} 5\cdot10^{10}\, x^{-1.1} dx \sim 2\cdot10^{10}\,\text{Bq}$$

$$\text{Bq }^{137}\text{Cs} \quad = \quad \int_{49,000}^{237,000} 4\cdot10^{14}\, x^{-2.1} dx \sim 10^{9}\,\text{Bq}$$

$$\text{Bq }^{239,240}\text{Pu} \quad = \quad \int_{49,000}^{237,000} 2\cdot10^{9}\, x^{-1.7}\, dx \sim 10^{6}\,\text{Bq}$$

TABLE 3. Radionuclides in Techa River water in July 1990 (relative counting error in %) [4].

LOCATION	^{90}Sr	Radionuclides (Bq m^{-3}) ^{137}Cs IPAE	^{137}Cs RISØ	^{239,240}Pu
Nadirov Bridge	5,000 (1)	620 (5)	500 (1)	0.52 (6)
Muslumovo	4,200 (8)	550 (1)	440 (1)	0.35 (10)
Verknaja Techa	3,200 (15)	155 (10)	156 (1)	0.174 (7)
Zatechenskoje	2,800 (32)	68 (2)	52 (2)	0.123 (6)

2.3 SEDIMENTS

The sediments in the Techa river contain relatively high ^{137}Cs levels upstream (Table 4), but the levels decrease with increasing distance from the discharge point. (Figure 4). This is best described by the following power function: $y = e^a x^b$ where x is the distance (km) from the point of discharge and y is the radionuclide concentration Bq kg^{-1}. The exponents a and b are given below with ± 1 SD:

^{90}Sr sand: $a = 23\pm2$; $b = -(3.7\pm0.5)$
^{90}Sr silt: $a = 12\pm3$; $b = -(1.2\pm0.6)$
^{137}Cs sand: $a = 28\pm2$; $b = -(4.5\pm0.4)$
^{137}Cs silt: $a = 33\pm1$; $b = -(5.1\pm0.2)$
^{239,240}Pu sand: $a = 14.4\pm0.3$ $b = -(3.0\pm0.1)$
^{239,240}Pu silt: $a = 22\pm2$ $b = -(4.2\pm0.3)$.

It should be emphasized that these equations are valid only between the points of observation, i.e. within 49-237 km from the point of discharge. The equations predict unrealistically high concentrations close to point of discharge.
From Tables 1 and 4, the following expressions for sediment inventories of radionuclides

as a function of distance (x m) were derived assuming a mean sediment layer of 150 kg m^{-2} of river-bed, corresponding to a mean sediment layer thickness of 10 cm:

$$\text{Bq } ^{90}\text{Sr} \quad = \quad \int_{49,000}^{237,000} 8 \cdot 10^{17} \, x^{-2.3} dx \sim 3 \cdot 10^{11}$$

$$\text{Bq } ^{137}\text{Cs} \quad = \quad \int_{49,000}^{237,000} 13 \cdot 10^{31} \, x^{-4.8} dx \sim 6 \cdot 10^{12}$$

$$\text{Bq } ^{239,240}\text{Pu} \quad = \quad \int_{49,000}^{237,000} 10^{24} \cdot x^{-3.9} \, dx \sim 8 \cdot 10^{9}$$

TABLE 4. Radionuclides in Techa River sediments in July 1990 (relative counting error in %)[4].

LOCATION	Type of sediment	Radionuclide (Bq kg^{-1} dry weight) ^{90}Sr	^{137}Cs IPAE	^{137}Cs RISØ	^{239,240}Pu	
Nadirov Bridge[a]	Peat	1,080 (21)	280,000 (51)	-	-	
	Sand and silt	2,050 (48)	141,00 (1)	25,000 (1)	42	±1
Muslumovo	Sand	1,230 (7)	4,000 (16)	3,900 (1)	47	(8)
	Silt	670 (28)	47,000 (12)	51,000 (1)	43	(6)
Verknaja Techa	Sand	39 (26)	150 (1)	159 (1)	0.40	(16)
	Silt	150 (40)	550 (11)	630 (1)	1.04	(16)
Zatechenskoje	Sand	22 (9)	34 (6)	13.1 (4)	0.18	(23)
	Silt	200 (36)	200 (2)	190 (1)	0.43	(20)

[a] Double determinations were carried out for transuranics at Nadirov Bridge; the error term is ± 1 SE

The calculation of radionuclide inventories in the Techa river does not include the amounts deposited between the discharge point and Nadirov Bridge, i.e. on the first 50 km of the rivers. Nor does it comprise what is retained in the floodlands along the river. Furthermore, it should be recalled that the sediment samples represent only the upper 10 cm of the sediments. Higher concentrations may be found below 10 cm; on the other hand, the mean thickness of sediments may be <10 cm, so we may have compensated for this in our sediment inventory calculation.

With regard to the activity deposited in the flood-lands, these amounts are estimated at 115-235 TBq ^{137}Cs and 79-189 TBq ^{90}Sr [7]. Of this, 5 TBq ^{137}Cs and 9 TBq ^{90}Sr are deposited downstream Muslumovo. If we assume proportionality between the inventories in floodlands and in river sediments, we could get a rough estimate of the deposition in the river sediments from the discharge point to Nadirov Bridge. We get 0.1 PBq ^{137}Cs and 2 TBq ^{90}Sr, i.e. an order at magnitude more than found downstream Nadirov Bridge in Techa River sediments.

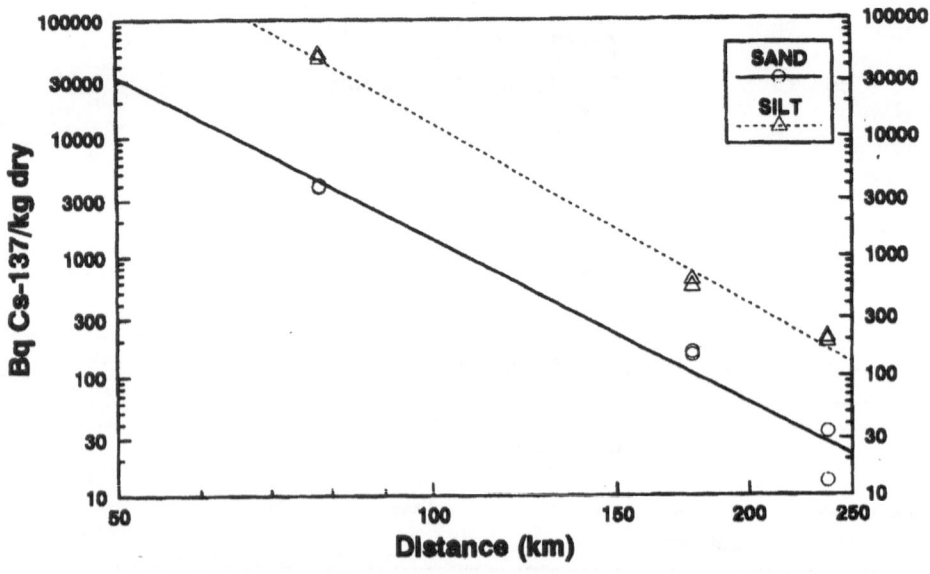

Figure 4. ^{137}Cs in sediment samples collected from the Techa River in July 1990. The concentrations are related to the distance from the outlet from MAYAK to the river.

If we include the floodlands, the inventories of Techa River become 0.3 PBq ^{137}Cs and 0.2 PBq ^{90}Sr. To these figures should be added a part of the 7.1 PBq retained in the cascade of reservoirs and canals previously mentioned, since this system also received radioactive waste also after the direct discharges to the Techa river stopped. We must therefore conclude that at least (4.3+4.6-0.3-0.2-7.1)= approx. 1 PBq of ^{90}Sr + ^{137}Cs has to be found elsewhere, *viz.* in the Ob river system and in the Arctic Ocean.

A considerable part of the radioactive disposal is contained within the flood plain of the Techa river. The density of contamination with ^{90}Sr and ^{137}Cs, in some of the flood plain areas (3-4 m from the river bank), exceeds the global fallout level by one or two orders of magnitude (Table 5). We used the flood plain of the Iset river near the V. Jarth village as a control location as it was considered to be uncontaminated from sources other than global fallout.

Whit regard to the radioecological situation within the central flood plain of the Techa (within 30-40 m from the canal) as seen at Zatechenskoje or at the right bank of the river near Pershino, the contamination levels decrease gradually with distance from the river, though the level at 30-40 m from the canal is still 10 times higher than the control. If the landscape is hilly (as at Bugaevo or on the left bank near Pershino) there is a geochemical barrier to the migration of the radionuclides. The total deposition, down to 30 cm and at 30-40 m from the canal, at these hilly locations, is higher than the deposition near the Techa river (Table 5).

The distribution of radionuclides in the soil profile varies as shown in Figure 5. At some locations strontium and caesium are distributed evenly. In other places, a maximum may be displaced towards the deeper layers, and at others, most of the radionuclides may be absorbed in the upper layers of the soil. The situation in the Techa flood plain is very dynamic. The following events occur periodically: (i) transference of small particles with spring high-waters and rain water; (ii) drift of soil in the river; (iii) covering of soil layers with river alluvium; (iv) vertical migration of radionuclides with subsoil waters. The majority of these processes promote the additional contribution of radionuclides to the river water.

TABLE 5. Radionuclides in the Soil of the Techa Flood Plain in July 1992 [5].

LOCATION Distance from Techa outlet (km)	River bank	Distance from the river (m)	^{90}Sr kBq m^{-2}	^{137}Cs kBq m^{-2}
v. Lobanovo	right	3-4	83	169
77	left	3-4	88	200
v. Anchugovo	right	3-4	230	235
65	left	3-4	96	145
v. Bugaevo	right	3-4	87	88
48		30-40	610	470
	left	3-4	230	167
v. Pershino	right	3-4	122	94
27		30-40	34	55
	left	3-4	290	155
		30-40	380	123
v. Zatechenskoje	right	3-4	230	83
3		30-40	25	58
v. V. Jarh	left	3-4	3.4	5.9
9				

196

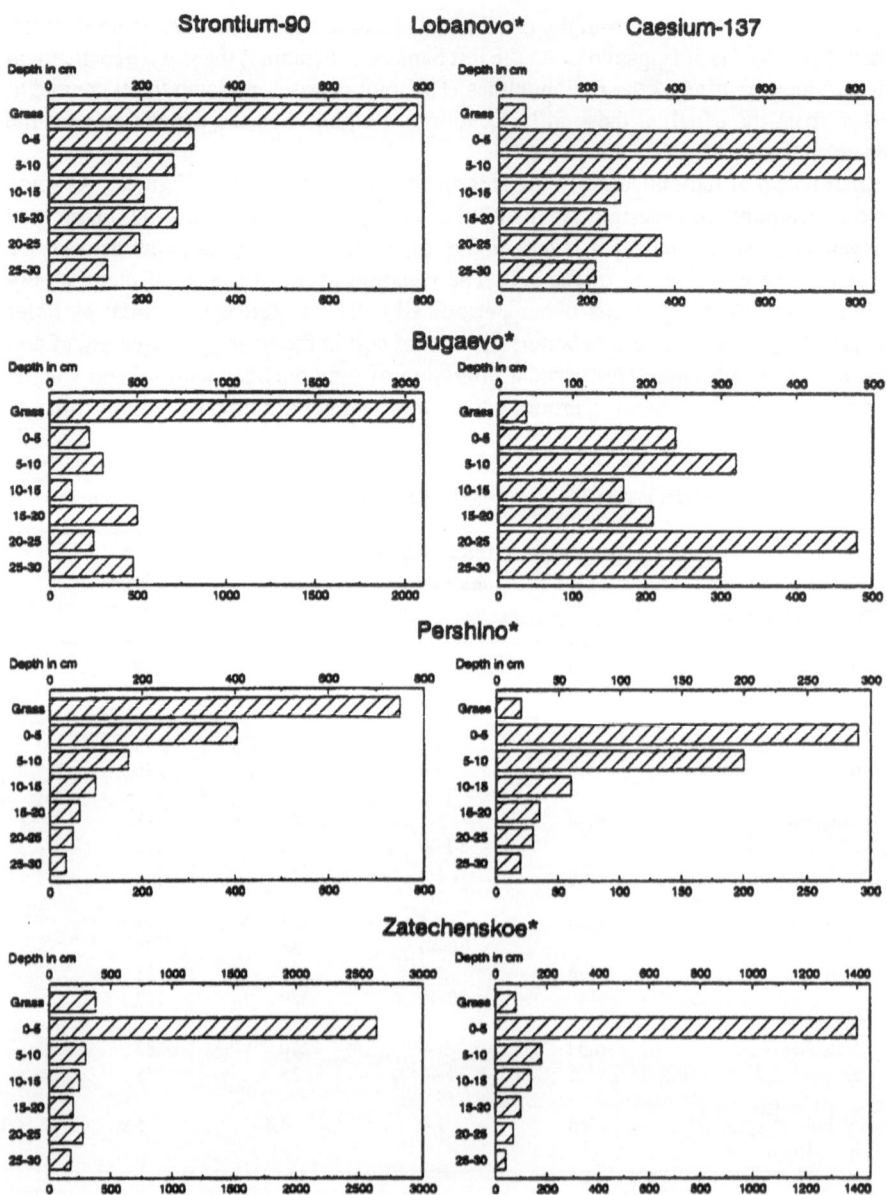

Figure 5. Concentrations of ^{90}Sr and ^{137}Cs in the Techa flood plain soil, Bq kg^{-1} dry weight. (*Name of locations see Fig. 2). [5].

3. Iset river studies

The contamination of the Techa river influences the ^{90}Sr and ^{137}Cs concentrations in water and sediments of the Iset river (Table 6). It appears that the Miass river (Figure 6) contributes to the contamination of the Iset river with ^{90}Sr and ^{137}Cs. The source of this pollution has not been identified in this study. An estimate of the annual transport of radionuclides in the Iset river, at Mekhonskoe, is 1.1 TBq ^{90}Sr and 0.04 TBq ^{137}Cs. This may be compared to the transport of radionuclides in the Techa river at Pershinskoe.

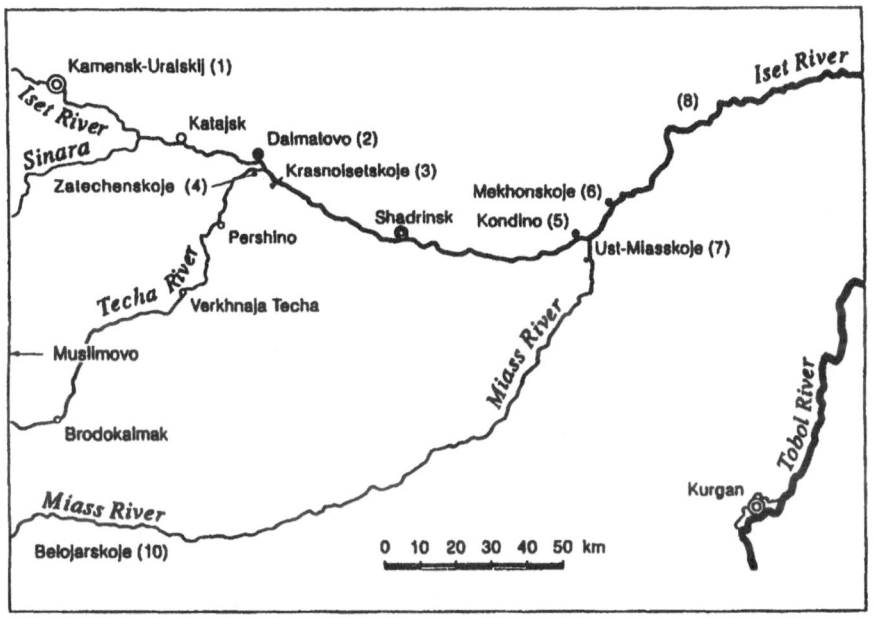

Figure 6. Iset river.

TABLE 6. Radionuclides in Sediment and Water of the Iset and Miass Rivers, in July 1992 [4].

LOCATION	Sediment		Water	
	^{90}Sr	^{137}Cs	^{90}Sr	^{137}Cs
	Bq kg^{-1}	Bq kg^{-1}	Bq m^{-3}	Bq m^{-3}
No 2 Dalmatovo	27	1.2	120	1.3
No 3 Krasnoisetskoje	14	3.1	910	13
No 5 Kondino	10	1.5	620	2.6
No 6 Mekhonskoje	38	9.1	430	15
No 7 Ust-Miasskoje	51	8.3	49	23
No 10 Belojarskoje	15	10.1	490	65

4. Tobol river studies

Sediment samples have been collected in the Tobol river at Kurgan (below the inflow of Techa-Iset river) and at Jalutorovsk before and after the Iset river has joined the Tobol river (Figure 7). Table 7 shows the results.

Table 7. ^{90}Sr and ^{137}Cs deposition densities measured in sediment columns along the Tobol river 1992-1993.

			(IPAE data)		
Site	River (Fig. 7)	Loc. no.	Sediment layer cm	^{90}Sr kBq m^{-2}	^{137}Cs kBq m^{-2}
Kurgan	Tobol	4	0-32	8.1	5.2
(Oct. 1992)	Tobol	5	0-29	3.5	8.6
	Iset	1	0-33	8.1	4.3
Jalutorovsk	Tobol	2	0-27	2.5	2.6; 0.5$^{a/}$
(Aug 1993)	Tobol	3	0-54	1.5	1.6; 0.3$^{a/}$

$^{a/}$ Risø data; loc 2: 0-40 cm; loc. 3: 0-32 cm;

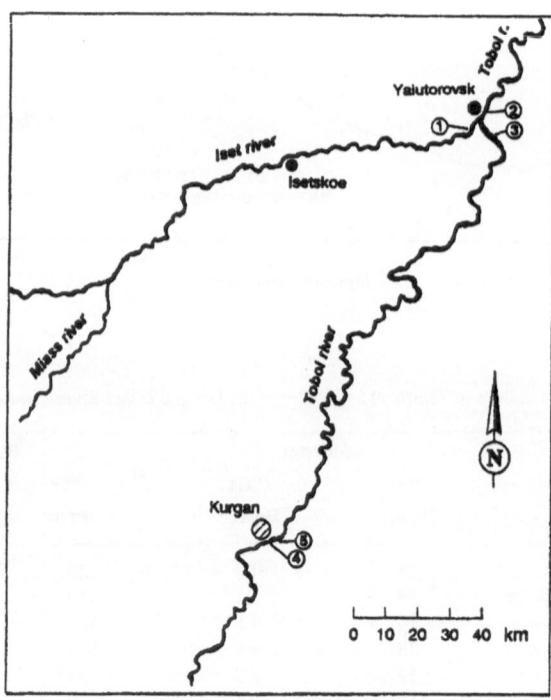

Figure 7. Tobol and Iset rivers.

The levels found at Kurgan are surprisingly high and indicate the possible presence of some unknown sources in the upper parts of the Tobol river. The samples collected at Jalutorovsk show that the Tobol sediments have higher levels after the confluence with the Iset river than before, because the Iset river is more contaminated than the Tobol river. The discrepancy between the IPAE and Risø data may be because the samples were not identical, however, it may also be due to analytical problems.

The vertical distribution of the activity in the Tobol sediments did not show any clear trend. In most cases the activity seemed to have been fairly homogenously distributed between the sediment layers. The mean concentrations (Bq kg^{-1}) were of the order of 10.

5. Irtysh river studies

For the time being results from only two sediment samples from Irtysh river are available. The samples were collected in 1993 9 km before the confluence of Irtysh with the Ob river (Figure 1). The thickness of the sediment layer analysed was 35 cm. The levels measured were 2.2 and 2.1 kBq ^{137}Cs m^{-2} and the mean concentration in the sediment columns was 6 Bq kg^{-1}. There was no evident pattern in the vertical distribution of the activity in the sediment columns.

6. Ob river studies

In order to get a first estimate of the contamination of the Ob river, sediments were collected in 1992 near Salekhard and Labitnangi (Figure 1), situated about 300 km from the outlet of the Ob into Obskaja Guba at the Kara Sea. As a control the samples from the Hanmej river, in the same region, were used. The Hanmej river has no connections to the Ob. Finally 2 sediment samples were collected in Ob 5 km before and 5 km after confluence with the Irtysh river Results are summarized in Table 8.

TABLE 8. ^{90}Sr and ^{137}Cs in Ob-river sediment columns collected in 1992-1994.

Location	Sediment layer cm	^{90}Sr kBq m^{-2}	^{137}Cs kBq m^{-2}	Laboratory
Surgut (before confluence with Irtysh)	0-20	1.6;0.81	1.14;0.94	IPAE
Gusing Jahr (before confluence with Irtysh)	0-25	0.5,0.84 4	1.2; 1.56	IPAE
Neulyovaya (before confluence with Irtysh)	0-25	1.72;1.78 8	2.6; 1.15	IPAE
5 km before confluence with Irtysh	0-35	-	2.2; 1.42	IPAE
5 km after confluence with Irtysh	0-30	-	1.54	IPAE
Lamskaya (after confluence with Irtysh)	0-20	2.54;1.53 3	0.85; 2.22	IPAE
Labitnangy (Fig. 1)	0-34	2.2	3.9	IPAE
Labitnangy	0-32	0.15	0.45	RISØ

Hanmej river sediments (0-26 cm) contained 1.84 kBq ^{90}Sr m^{-2} and 2.7 kBq ^{137}Cs m^{-2} (IPAE); another Hanmej sediment column (0-20 cm) was analysed by Risø, where 0.32 kBq ^{137}Cs m^{-2} were found.

It appears that the ^{137}Cs levels in the Ob sediments before and after confluence with the Irtysh do not change significantly. If the various sediment layers are considered it is difficult to see any systematic changes in ^{137}Cs concentrations with depth. In some cases the top 5 cm show higher concentrations than the deeper layers, in other cases peak concentrations are found at a depth of 15 cm. The mean concentration was of the order of 10 Bq ^{137}Cs kg^{-1}. An attempt to date the sediments from Labytnangi with the ^{210}Pb method was unsuccessful. The sediments were totally mixed, so it was impossible to determine the age of the various layers. Livingston et al [9] have succeeded to date the sediment layers from a station in the Ob estuary. They found that the maximum ^{137}Cs concentration was found in the layer of 1966. This suggests that the ^{137}Cs in this sample primarily has come from instant run-off of global fallout deposited in the Ob-river drainage area.

Although the deposition of ^{90}Sr and ^{137}Cs was 20-40% higher than in the Hanmej, this observation is not sufficient to prove that Ob sediments are contaminated by liquid discharges from the Techa or Tom (from Tomsk-7) rivers. Further investigations are necessary in order to obtain final evidence.

7. Conclusion

The studies carried out so far in the Techa, Iset, Tobol, Irtysh, Ob river system by the Joint International Study Group have not been able to identify any contamination from MAYAK or Tomsk-7 outside the Techa and Iset rivers. The studies have mainly been based on analysis of ^{137}Cs in river sediments collected to about 30 cm depth. It is difficult to find undisturbed sedimentation layers in the river system and dating of the sediment layers has until now not been possible in the river itself. Studies of so called sorlakes along the Ob river may overcome these difficulties, because the sedimentation regime in such systems is more calm than in the river itself.

8. References

1. Academy of Science.(1991) Conclusion of the Commission on the estimation of the ecological situation in the region of production association "Majak", organized by the direction of Presidium of Academy Science. n. 1140-501// Russia J. Radiobiology 31, (3): 436-452.
2. Nikipelov, B.V.; Nikiforov, A.S.; Kedrovsky, O.L.; Strakkov, M.V.; Drozhko, E.G. (1990) Practical rehabilitation of territories contaminated as a result of implementation of nuclear material production defence programme. English translation of Nov. 9, 1990 from Oak Ridge National Laboratory.
3. Hydrometeo (1973) Resources of surface water. Volume 11. Leningrad: Hydrometeo Publishing: 848 (in Russian).
4. Trapeznikov,A.V.; Pozolotina, V.N.; Chebotina, M. Ya.; Chukanov, V.N.; Trapeznikova, A.V.; Kulikov, N.V.; Nielsen, S.P. & Aarkrog, A. (1993) Radioactive contamination of the Techa river, the Urals. Health Physics, 65, 481-488.
5. Trapeznikov,A.V.; Aarkrog, A.; Pozolotina, V.N.; Nielsen, S.P.; Polikarpov, G.; Molchanova, I.; Karavaeva, E.; Yushkov, P.; Trapeznikova, A.V. & Kulikov, N.V. (1994) Radioactive Pollution of the Ob River System from Urals Nuclear Enterprise "MAJAK". J. Environ. Radioactivity, 25, 85-98.

6. Strand, P.; Rudjord, A.L.; Salbu, B.; Christensen, G.; Nikitin, A.I.; Chumichev, V.B.; Valentova, N.K.; Kryshev, I.I.; Namiatov, A.A., Cheliukanov, V.V.; Føyn, L.; Lind, B.; Iosjpe, M.; Sickel, M.; Østbye, G.; Selnæs, T.D.; Bjerk, T.O. (1994) Radioactive contamination at dumping sites for nuclear waste in the Kara Sea. Results from the Russian-Norwegian 1993 expedition to the Kara Sea. Joint Russian-Norwegian Expert Group for Investigation of Radioactive Contamination in the Northern Areas. ISBN 82-993079-3-7 Østerås: Norwegian Radiation Protection Authority.
7. MAYAK (1991) Short Characteristics of the Industrial Amalgamalion "MAYAK", Chelyabinsk-65, pp 36 (in Russian).
8. Materials of the Commission (1991) (Decree of the President of the USSR N RP-1283 on January 1991) Vol II. Ecological Characteristics of the Cheliabinsk District, pp 157 (in Russian).
9. Livingston, H.D. & Panteleyev, G.P. (1994) Personal communication.

OVERVIEW OF WATER QUALITY MANAGEMENT IN THE AREAS AFFECTED BY THE CHERNOBYL RADIOACTIVE CONTAMINATION

O. VOITSEKHOVITCH
Ukrainian Hydrometeorological Institute
Prospect Nauki, 37
CIS-252028 Kiev, Ukraine

1. Introduction

Numerous post-Chernobyl case studies describe the extensive radioactive contamination of large regions of the Ukraine, Belarus, Russia and parts of Western Europe, caused by the 1986 accident in Chernobyl Reactor 4. Most radioactive atmospheric fallout was deposited within the Dnieper river catchment area which adjoins the site of the Chernobyl Nuclear Power Plant (CNPP). This area and adjacent drainage basins form an extensive area from which contaminated water and sediments flow downstream through the Pripyat and Dnieper rivers across the Ukraine to the Black Sea (Figure 1). Erosion material and runoff from contaminated landscapes contribute to a further transport of radionuclides from the Chernobyl area to the greater Dnieper region. From the more than 20 millions inhabitants of the area, about 9 millions use river water as drinking water. Fishery products and products from agriculture, which uses water from the Dnieper for irrigation, are also linked with water from the Dnieper reservoirs. Therefore, the contamination is, even nine years after the accident, a hot topic, and discussions on water protection in the Chernobyl exclusion zone are continuing. The water protection and remediation efforts at the CNPP site have a dramatic history and are worth assessing for possible lessons.

2. Lessons for the future

An analysis of the actions taken to mitigate the effects of water contamination after the Chernobyl accident offers decision-makers a unique opportunity to optimize their approaches to surface- and ground-water protection. Most engineering measures inside the Chernobyl 30-km exclusion zone focused on preventing secondary contamination from entering the Pripyat river and the Kiev reservoir. Both surface and ground water may act as secondary sources. These countermeasures required huge financial and human resources for their implementation. This paper describes nine years of countermeasures on contaminated water bodies surrounding the 30-km exclusion zone. The results demonstrate that the effectiveness of mitigating measures depends not only on proper application of technology but also on a selection of projects which offer a possibility for significant risk reduction. Economic considerations require that environmental mitigation projects maximize risk reduction and cost effectiveness.

F. F. Luykx and M. J. Frissel (eds.), Radioecology and the Restoration of
Radioactive-Contaminated Sites, 203–216.
© 1996 *Kluwer Academic Publishers.*

204

Figure 1. The Dnieper River cascade system and its catchment area after the Chernobyl accident.

3. Evaluation of the radioactive contamination of the Dnieper cascade.

An overview of the pollution of the Dnieper and Pripyat river watersheds is given by Valculovsky et al. [1, 2]. The highly contaminated flood-plain area near Chernobyl is described by Voitsekhovitch et al. [3]. The drainage basins of the described areas form the main potential secondary contamination source of the Dnieper cascade system. Most important are ^{137}Cs and ^{90}Sr from runoff and erosion processes. Their transport via the Pripyat river and Dnieper cascade system to the Black Sea is described by Voitsekhovitch et al. [4,5].

3.1 DETAILED DESCRIPTION OF THE CONTAMINATION

To estimate the transport of radioactivity in the Pripyat and Dnieper rivers via liquid phase and suspended sediments, special measurements were carried out near the boundary of the Chernobyl site. Values were averaged over 10-day periods, months and years since the accident. Some of these data are presented in Figure 2. The highest radioactivity level in the water of the Pripyat river was observed during the initial period after the accident. For some time, its level exceeded the permissible sanitary levels in natural waters. From 1986 to 1994, the radionuclide concentration in the river decreased continuously, but, during each high-water period, it increased temporarily due to wash-off and erosion phenomena around the borders of the rivers. Therefore, the ratio ^{90}Sr to ^{137}Cs in water changed significantly with time. The same holds for the ratio between suspended and soluble radionuclides.

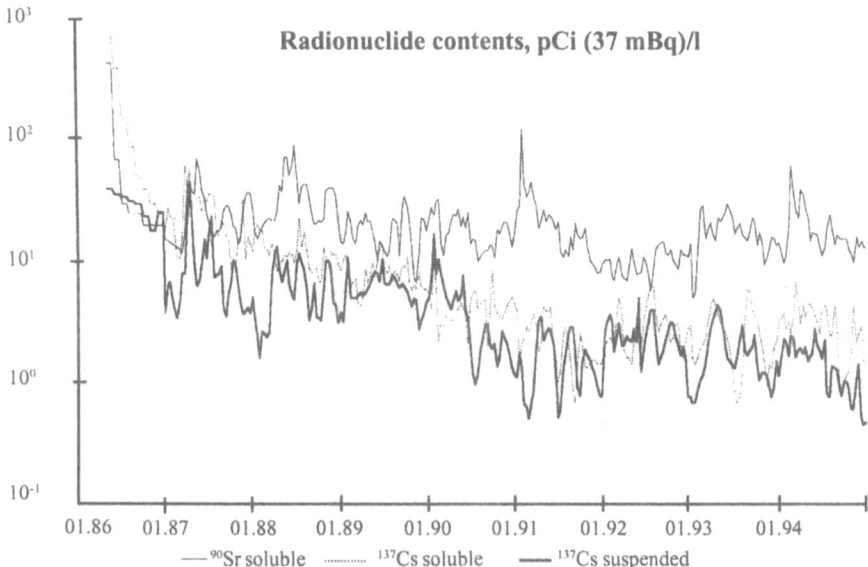

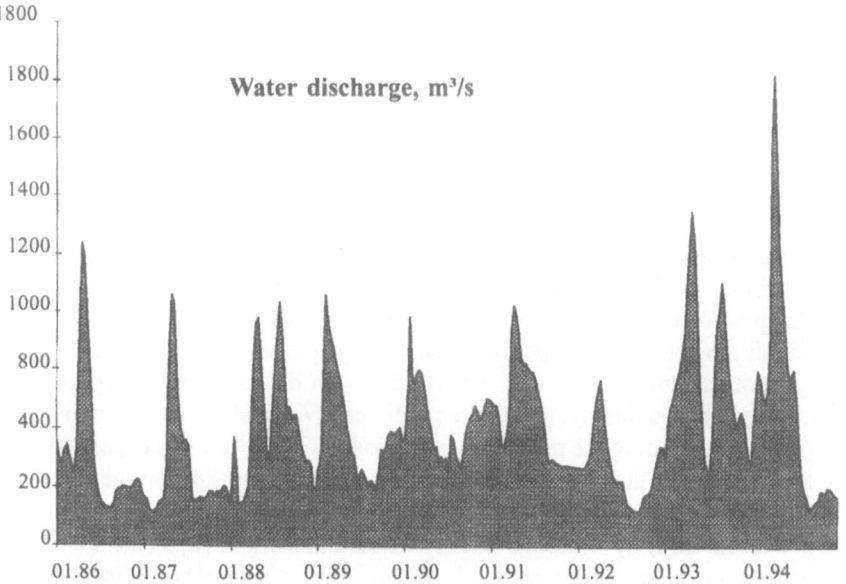

Figure 2. ^{137}Cs and ^{90}Sr concentrations in the Pripyat river averaged over 10 days.

TABLE 1. Average annual radionuclide inflow into the Kiev reservoir (GBq/y)

year	Pripyat river			Dnieper river		
	^{137}Cs solution	^{137}Cs suspended	^{90}Sr total	^{137}Cs solution	^{137}Cs suspended	^{90}Sr total
1986	56,380	9,700	27,600	27,300	3,600	10,700
1987	9,030	3,700	10,400	9,700	4,200	8,000
1988	5,900	3,600	18,700	6,700	2,600	5,200
1989	3,600	2,800	8,900	5,800	1,300	3,600
1990	2,300	1,600	10,100	3,700	1,400	3,700
1991	1,400	1,400	14,400	1,100	1,000	4,500
1992	900	800	4,100	800	500	670
1993	2000	1,500	14,800	600	260	930

Table 1 shows that the wash-off of ^{137}Cs decreases more with time than that of ^{90}Sr. In particular in 1992 and 1993, the ^{90}Sr wash-off into the Kiev reservoir would have been 1.5 times higher if the engineering measures on the flood plain zone near Chernobyl had not been taken (see below). A graphic presentation of the concentrations of ^{137}Cs in soluble and suspended forms is shown in Figure 2 for different seasons.

It appears that a large part of ^{137}Cs, as well as of some other radionuclides, is associated with suspended particles. The average size of these suspended particles varies from 0.02 mm in low-water periods to 0.25 mm during floods. Particles smaller than 0.05 mm represent 55-60% of the total suspended sediments. In contrast, bed sediments of the Pripyat river consist almost exclusively of sand with very little silt and almost no clay [6, 7].

The monthly average ^{137}Cs and ^{90}Sr concentrations in the inlet and outlet water flows of different reservoirs are shown in Figure 3. The values confirm that the reservoirs can be considered as temporary sinks for radionuclides. Table 3 shows also the important role of sedimentation as a self-purification process in water reservoirs. For instance, on the basis of the presented data, it is clear that only 1-2% of the total inlet of ^{137}Cs into the Kiev reservoir reaches the Dnieper river estuary. Most of the ^{137}Cs accumulated in the Dnieper bottom sediments and was firmly fixed on the solids by adsorption on the suspended particles and direct uptake by the sediments. An analysis of the data shown in Figure 3 shows that during the post-accident period, 98% of ^{137}Cs was deposited on the bottom layers of six of the reservoirs, with more than 50% deposited in the Kiev reservoir.

Results of a recent study by Voitsekhovitch, Kanivets and Sansone [6] on the radionuclide content in bottom sediments are shown in Figure 4. It appears that the most polluted bottom sediments of the Kiev reservoir are found in the upper area of the reservoir near the mouth of the Pripyat river and in the submerged river bed of the Dnieper river. The sedimentation of relatively clean material on top of the initially contaminated bottom sediments over the recent years buries most polluted layers. The thickness of the most contaminated layer of bottom sediments, dating from 1986, ranges from 10-30 cm near the upper part of the Kiev reservoir to 2-3 cm near the dam. The investigation shows that the Dnieper reservoir serves as, and will remain, the main receiver of suspended radioactive sediments.

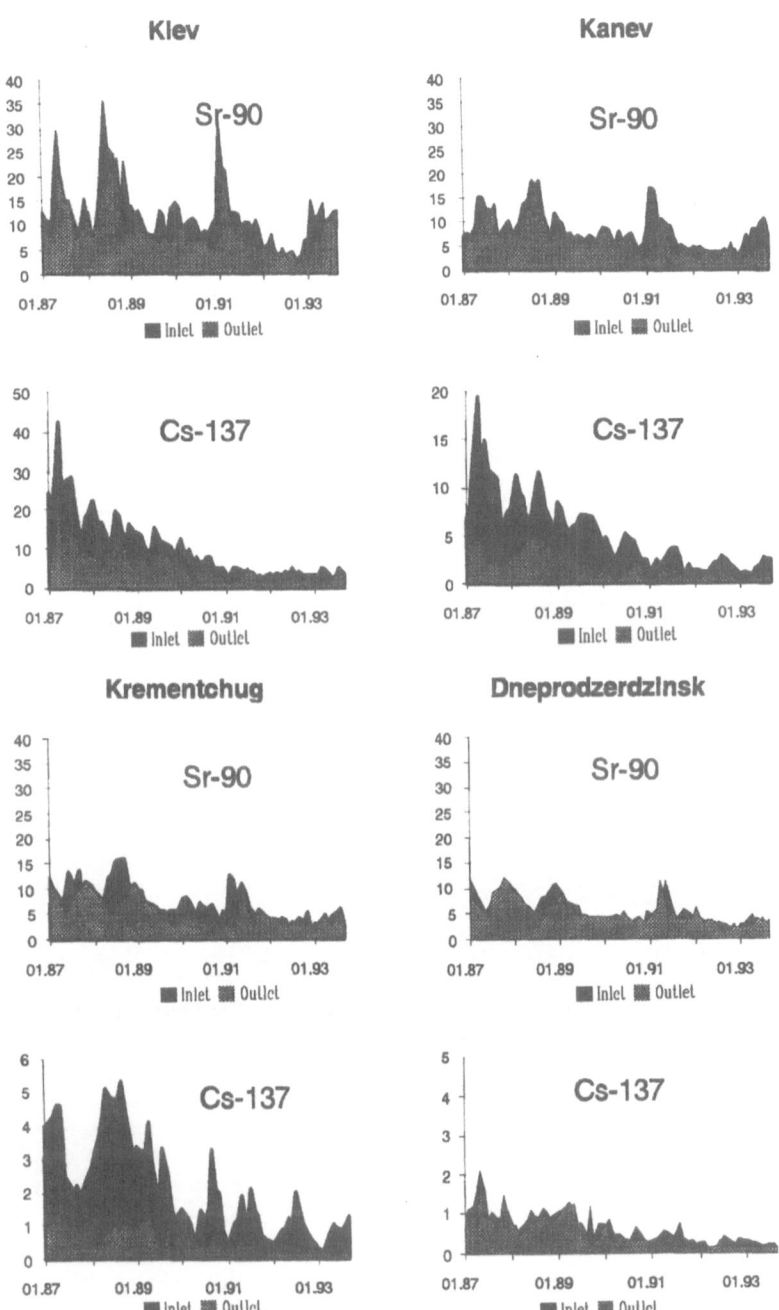

Figure 3. ^{137}Cs and ^{90}Sr concentrations in the inlet and outlet of the Dnieper reservoir. Monthly averages by the Ukrainian Hydrometeorological Institute. pCi/l (37 mBq/l).

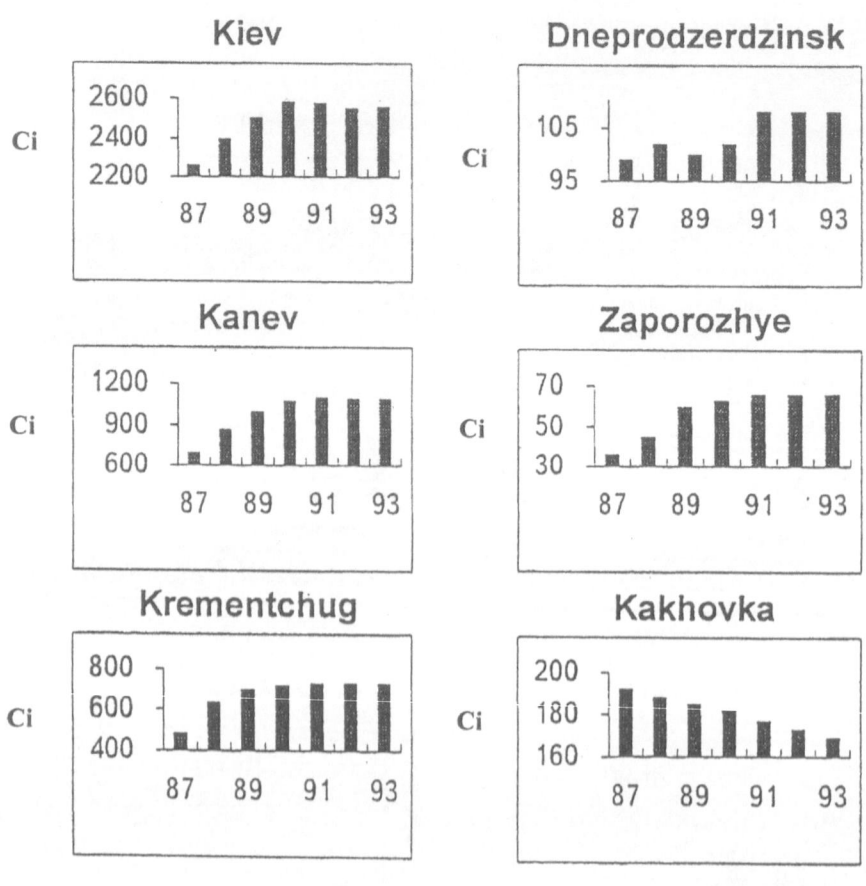

Figure 4. ^{137}Cs in bottom sediments of the different reservoirs of the Dnieper cascade.
(1Ci = 37 GBq)

The following can be concluded from the above studies:

- After the accident in Chernobyl, the most significant sources of surface water contamination of the Dnieper river, are surface runoff from the initially contaminated flood plains and catchment areas. Ground-water pollution contributes no more than 3-7% of the total transfer from the terrestrial environment.

- Since the accident in Chernobyl, there has never been a long period of very high water and flooding of the contaminated areas. The spring water flow in the Pripyat river did not exceed the normal flood conditions (600-2,000 m³/s). The maximum possible water flow can exceed 5,000 m³/s. Therefore, the contaminated area remains a potential hazard for the future.

- The present radionuclide concentration in the Dnieper river is not considered a factor of health risk. However, taking into account the above fact and the risks of existing chemical pollution, some measures for preventing and mitigating the risks have to be economically and socially approved.

- A strictly limited number of justified water-remedial actions should play a positive role in the future. The new plan on water remediation should take into account all findings of the post-Chernobyl water remedial experience.

3.2 EVALUATION OF DIFFERENT STAGES OF WATER REMEDIATION

Since the accident, both technical and administrative countermeasures have been taken to mitigate risks.
In this respect, three phases can be distinguished:

3.2.1 Emergency phase: (early post-accidental period, i.e., 2-3 months after the event)
Countermeasures during this period were based mainly on administrative decisions and mainly aimed to control the situation. These countermeasures included
1) attempts to regulate the flow of contaminated water through the Kiev reservoir by the dam operation system,
2) increased use of ground-water sources by municipalities,
3) avoiding the use of contaminated surface water resources where possible,
4) supplementary purification of drinking water in municipal water-treatment plants and
5) construction of supplementary ground-water supply wells.

Most of these measures have been implemented without cost-benefit analysis. The main reasons were the lack of time to carry out the research required and the emotional state of the affected society and decision-makers during the early period after the accident. Therefore, most measures to reduce the radiation risk to the public from water usage were very expensive and limited in success. Different examples of implemented methods are described in [3-5, 8, 9].

Decision-makers made many errors because of lack of adequate information. For example, due to the lack of experimental data and due to controversies between knowledgeable scientists and unexperienced decision-makers, the first assessment of the adsorption and desorption parameters for radionuclide liquid-solid interaction was wrong and radionuclide wash-off coefficients were much overestimated. As a result, many useless water protective actions were taken during the first months after the accident, as for instance, the addition of unprocessed zeolite to the river downstream from Chernobyl. Another example: early May 1986, surface gates were opened and bottom gates closed in the dams of the Kiev reservoir. It was thought that clean water was thus being let out of the reservoir so that the highly contaminated runoff water from the spring rains could be captured in the reservoir. In reality, during the first week after the release of the radioactivity, the vertical mixing of water was slow and, therefore, the lower water layers of the reservoir were much less contaminated than the upper layers which were contaminated directly by atmospheric fallout. A better approach to decrease the activity level in the Kiev reservoir immediately after the accident would have been to open the bottom dam gates and to close the surface gates. This would have reduced the levels of radioactivity in downstream drinking water in the first weeks after the accident, when the main exposure from drinking-water intake took place.

3.2.2 Early intermediate phase (Summer 1986 to 1988)
In the summer of 1986, protective dikes of several km were constructed along the Pripyat river to catch the contaminated urban runoff from the cities of Chernobyl and Pripyat. This action was not effective since runoff from the broad landscapes could not be readily controlled.

Several canal bed traps were dredged in the Pripyat river during the first summer after the accident to increase the crosssection of the river and, thus, reduce the water velocity in an attempt to increase sedimentation of suspended radioactive particles. However, subsequent studies [8, 10] showed that these traps were ineffective, because the suspended radioactive particles were much too small to settle in a large natural river such as the Pripyat river with huge water flows and turbulent flow conditions.

Attempts to isolate the CNPP cooling pond from the Pripyat river were a major task in the early years after the accident. A special drainage well system was built around the cooling pond to catch infiltrating radioactive water. Until now, the drainage system is not operational because of uncertainties on the consequences of its operation. Indead, recent research [9] has shown that pumping water from the wells back to the cooling pond may cause problems with the water balance and dissolved salts in the pond. Anyway, the construction and maintenance costs were and are very high.

Another action in this period was the construction of a slurry wall and a series of drainage wells to prevent sub-soil migration. Additional studies have shown that migration of radionuclides within the soil is much too slow to allow the drainage wells to be effective. Moreover the slurry, wall could not prevent the contamination of the surrounding ground water from expanding. Therefore the project was stopped.

During 1986 and early 1987, more than 100 special zeolite-containing dikes were completed to adsorb radionuclides from smaller rivers and streams. Subsequent studies on the effectiveness of these dikes indicated that only 5 to 10% of the entrained ^{90}Sr was adsorbed by the zeolite barriers. Moreover, the streams, which were dammed during this activity, contributed only a few percent to the total flow of the Pripyat and Dnieper drainage basins. So the whole project did not make much sense. In 1987, the construction of a new dam was terminated and it was decided to destroy most of the existing dams.

In 1987, an early mitigating measure, after the evacuation of the local population and the extinction of the reactor fire, was the construction of Sites of Temporary Radioactive-Waste Localization (STRWL) near Reactor 4. They were used to bury contaminated soils, vegetation, debris and even small buildings. Also wood from the highly contaminated Red Forest pines, killed by high radiation levels, were buried here. These measures were deemed necessary to protect the emergency response groups (known in Russia as liquidators) and power plant workers from high doses. The burial of the highly contaminated wood occurred in shallow trenches without any protection to prevent contamination of ground water. As a result, severe local ground-water contamination from these buried trees will be an important, long-term, problem for the environment.

3.2.3 Later intermediate phase (1988 to 1991)

A new phase of hydrologic remediation began after the summer flood of 1988. High water levels covered much of the contaminated flood plain and caused secondary ^{90}Sr contamination of the river. Surface hydrologic modelling shows that a realistic, more dangerous or worst-case, scenario, i.e., one that would cause the highest radionuclide concentration in rivers, would be a spring flood with a maximum water flow of 2,000 m^3/s. Such a flood has a probability of 25%/year. But, when in the computed results the wash-off processes were simulated for isolated radioactive sources on the flood plain, the ^{90}Sr concentration decreased 2-4 times.

Several approaches have been proposed to reduce the radionuclide concentration in the river and the potential effectiveness of each of them has been simulated. The construction of a dike around the contaminated area on the left (east) bank of the river has been chosen as the best protective option. This measure, supplemented by decontamination of the soil on the right bank (or by constructing dikes, too), together with an acceptable solution for the cooling pond's seepage water, will significantly diminish the ^{90}Sr concentration downstream. The flooding of the flood plain in January 1991 and in the summer of 1993 confirmed the correctness of the simulated results [4, 5, 11]. The construction of the dike was finished at the end of 1992. As a result of this action during the spring flood of 1994, more than $3.7 \cdot 10^{12}$ (100 Ci) ^{90}Sr were prevented from being washed from the flood plain of the Pripyat river into the Dnieper river.

3.2.4 Future actions

Gradually, it became clear that other countermeasures cannot be planned without the designation of a "General Strategic Scheme for Water protection". Most engineering works were completed in 1993 and, at present, the development of a scientific basis for a further "Strategy and Technology for Water protection" is in progress.

4. Present understanding of the problem

At present, there is still a huge amount of radioactive material concentrated in the Chernobyl area, for instance in the flood-plain areas (Figure 5), which have each spring a 40-50% probability of being flooded by the river. Also the soils of polder areas in the 30-km zone may annually be inundated by river and ground water during snow melt and rain-fall periods. These polders contain more than $3.7 . 10^{14}$ Bq (10,000 Ci) ^{90}Sr and about $7.4 . 10^{12}$ Bq (200 Ci) plutonium. Also huge amounts of radioactive waste are present in the shallow underground waste-disposal sites, where they are in contact with ground water flowing in the direction of the Pripyat river. Details are shown in Figure 5 [4-6].

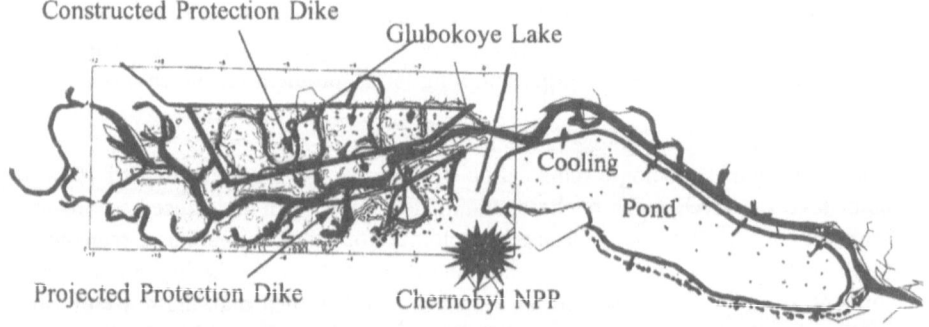

Constructed Protection Dike

Glubokoye Lake

Cooling Pond

Projected Protection Dike

Chernobyl NPP

Figure 5. The Pripyat river flood plain area near the Chernobyl nuclear power plant.

However, the technological possibilities to control the existing sources of radioactive contamination of such a large catchment area are very limited. Therefore, the main conclusion of the earlier studies is that countermeasures should be optimized, taking into account both averted human doses and economic costs.

4.1 DOSE CALCULATIONS

The collective effective dose to the public, integrated over 70 years, as a result of ^{90}Sr and ^{137}Cs intake via different water pathways, was calculated for some regions of the Ukraine, using water from the Dnieper cascade. Results are shown in Figure 6. The dose estimations were based on monitoring results and on predicted radionuclide contents in the Dnieper reservoirs up to the year 2056. The dotted curves on this figure present the annual collective dose. The structure of irrigated land and the tap water consumption are taken into account.

In 1986, the collective dose from water use in the Kiev region exceeded by 6-7 times the current dose level. An opposite effect was observed in the Crimea region, where the initial contamination level of the lower Dnieper was mainly caused by primary fallout. The separate contributions of different water pathways to the collective dose are shown in Figure 6. [12]. The studies by Berkovsky et al. [12] and Voitsekhovitch et al. [4, 5] show that, for different regions of the Ukraine, the structure of the dose pattern differs. These studies also show that the annual average individual effective internal doses for different regions of the Ukraine vary strongly. For instance, in 1993, the water pathway contributed 6% to the total radiation dose for Kiev citizens but 20-30% for people living in the south of the Ukraine .

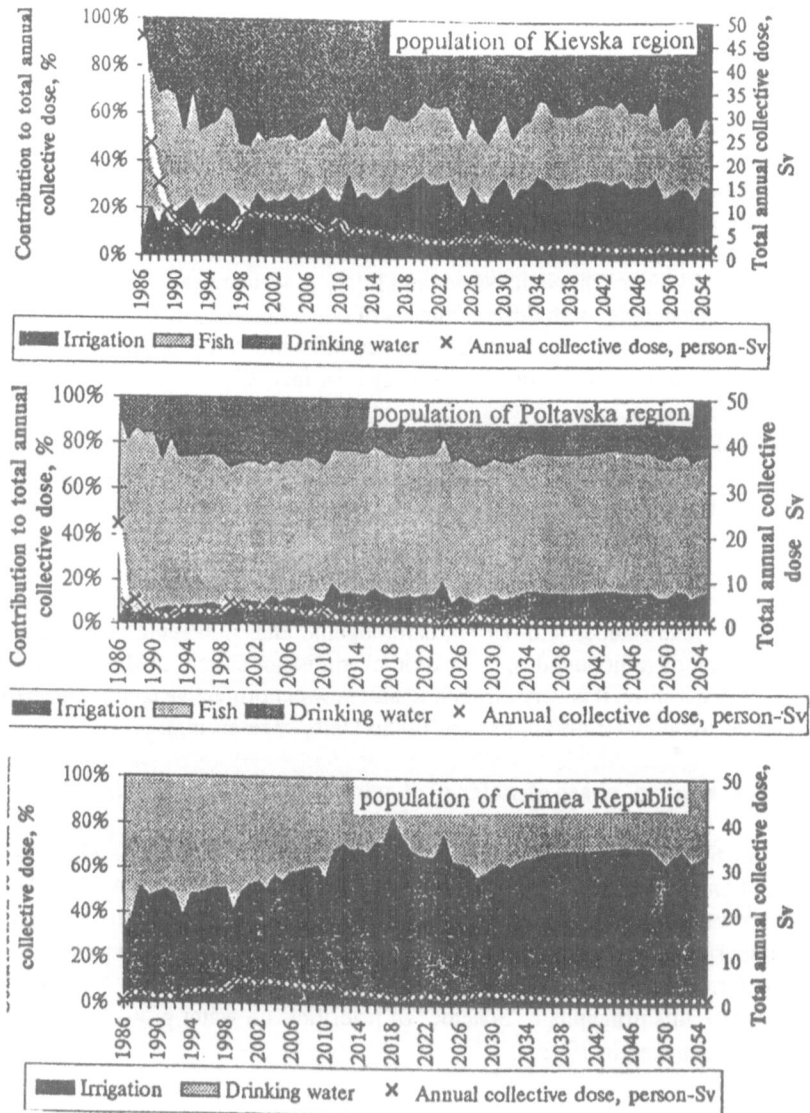

Figure 6. Computed retrospective (1986-1993) and simulated contribution of ^{137}Cs and ^{90}Sr via different water exposure pathways to the annual collective committed dose(manSv) for Kiev, Poltava (Krementchug reservoir) and Crimea (Kakhova reservoir).

5. Assessment of health risk from water usage

Using ICRP's nominal probability coefficient for stochastic effects of 7.3×10^{-2} per Sv [13], the number of stochastic effects due to the use of Dnieper water during 70 years was estimated. It results in about 200 cancer cases for 21 million people (70 years of exposure) and in about 60 cases over the period 1986-1992 [12]. A calculation of the dose to the total population shows that the individual human radiation risk from Dnieper water is not higher than 1×10^{-5}. In some critical groups, affected more by "water pathways" than average persons, the expected individual risk may be a factor 4-5 higher. In particular cases, the risk may even be higher. Implementation of the most effective countermeasures can reduce the calculated risks from water usage up to a maximum of 3-4 times. In fact, the radiation risk from water consumption is very low compared to other radiation risks. The stress component caused by the psychological reaction of the population, may dominate the purely physical component of the risk.

From a study by Voitsekhovitch, Nasvit and Los'y [14], it appears that more than 30% of the interviewed people, with different levels of education but without special knowledge on radiation protection, supposed that the actual health risk is higher from water consumption than from other exposure pathways. In fact, the average individual effective dose from natural radionuclides like ^{226}Rn, ^{222}Rn and ^{238}U in drinking water is two orders of magnitude higher than the dose from post-Chernobyl exposure.

Unfortunately, at present there is no clear method available to estimate the partial and total risks due to environmental contamination or from water use in particular. All Dnieper reservoirs are situated in industrial and agricultual areas with high non-radioactive pollution. Toxicological investigations show the presence in the water reservoirs of a number of other toxic substances with strong carcinogenic and mutagenic properties. In comparison to that of radionuclides, their toxic origin is not certain or their releases are not controlled. Therefore, it is very difficult to assess water protective countermeasures when many other toxic materials are present in the Dnieper reservoirs.

6. Current strategy phase of water protection in the Chernobyl site

In spite of the low exposure level of the population of the Ukraine via aquatic pathways, radionuclide transfer by river water will remain a sensitive factor. When all the sources of radioactive contamination of the water are known, it is preferable to derive an optimal set of water protective countermeasures rather than to be passive. However, any countermeasure should be decided according to the well known "ALARA" principle. Based on this principle, a "Scheme" of priorities on water protective countermeasures was established in 1993.

Scheme of countermeasures

The immediate goals of the planned "Scheme" of countermeasures are

    1) to mitigate a further significant expansion of the impact of the accident by controlling and minimizing radionuclide transport via water outside the Chernobyl area and

    2) to develop a radioactivity monitoring system for ground and surface water inside and beyond the Chernobyl area.

To provide safe water usage at any point of the Dnieper reservoir, it was proposed to use a main criterion for decision-makers to select optimal countermeasures, i.e., a water contamination level of 2 Bq/l ^{90}Sr in the Pripyat river downstream of the boundary of the Chernobyl exclusion zone.

Mitigating actions that could fulfil the above-mentioned tasks are the following.

1. To build dikes around the flood-plain areas with extremely high contamination levels on the left and right banks of the Pripyat river near the CNPP, to protect these areas from direct flooding and soil erosion.

2. To develop a project to solve the complex clean-up problem of the bottom sediments of the cooling pond of the CNPP.

3. To regulate the water level of the severely contaminated wetlands near Chernobyl, in agreement with the concepts adopted in Belarus, to protect peat bog areas against fire.

4. To provide an expanded soil water monitoring system in the Chernobyl exclusion zone and around the temporary waste disposal site (STRWL).

5. To provide reliable monitoring and control of transuranic material transport beyond the presently contaminated site.

6. To prevent expansion of radionuclide transport by ground water migration beyond their present localization in the waste-disposal site by the construction of technical and geochemical barriers around the temporary waste disposal site.

Any other proposal related to water remediation seems ineffective at this moment and its funding should, therefore, temporarily be suspended.

7. Acknowledgements

This study was mainly initiated and funded by the Ministry of Chernobyl Affairs of Ukraine. The author wishes to thank Dr. Kanivets, Dr. Nasvit and Dr. Berkovsky, who were directly involved in the analysis of the data discussed and also Dr. S. Kazakov and E. Panascevitch from SPA "Pripyat and Engineer",and O. Zvekov for his permanent attention and assistance in this study.

216

8. References

1. Vakulovsky S.M., Voitsekhovitch et. al. (1990). Radioactive Contamination of Water Bodies in the Area Affected by Releases from the Chernobyl Nuclear Power Plant Accident, *Environmental contamination following a major nuclear accident.* (Proc. Int. Symp. IAEA, 1989), IAEA, Vienna, 231-246 .

2. Vakulovsky S.M., Nikitin A.l. and Voitsekhovitch (1994). Cesium-137 and Strontium-90 Contamination of Water bodies in Areas Affected by Releases from Chernobyl Nuclear Power Plant Accident: An Overview, *J. Environ. Radioactiv.*, **23**, 103-122.

3. Voitsekhovitch O. V., Borsilov V.A. ,Konoplev A. V. (1991) Hydrological Aspects of Radionuclides migration in water bodies following the Chernobyl Accident, Proc. Seminar. *Comparative Assessment of Radionuclides released during three Major Nuclear Accident: Kyshtym, Windscale and Chernobyl.* Luxembourg, October 1990, Brussels, CEC, 527-548.

4. Voitsekhovitch O., Kanivets V., Laptev G., Biliy 1. Y. (1993) Hydrological Processes and Transport by Surface Water Pathways as Applied to Water Protection after the Chernobyl Accident, Proc. *Hydrological Considerations in Relation to Nuclear Power Plants.* Paris. September, 1992,85-105.

5. Voitsekhovitch O.V. Zhelesnyak M.I., Onishi Y. (1993). Chernobyl Nuclear Accident. Hydrological Analysis and Emergency Evaluation of Radionuclide Distribution in The Dnieper River, Ukraine. PNL. Report-9980. p96.

6. Voitsekhovitch O. V. Kanivets., Sansone (1994). Suspended ^{137}Cs Transport from the Rivers located in the Chernobyl area to the Kiev Reservoir, Progress Report (1993-1994) *Modelling and study of the mechanisms of the transfer of Radioactive material from the terrestrial ecosystem to and in water bodies around Chernobyl.* ANPA-DISP, Italy, 59-66.

7. Vishnevsky V., Voitsekhovitch. O.V. (1994) Suspended Sediment Transport in the Pripyat River System, Proc. Conference. *Dynamic and Thermic of rivers, reservoirs...* , **1.** (in Russian).

8. Zheleznyak M.I., Voitsekhovitch O.V. et al. (1992) Simulation of Effectiveness of Countermeasures Designed to Decrease Radionuclide Transport Rate in the Pripyat-Dnieper Aquatic Systems., Proc. Int. Sem., Cadarache, 1991,CEC, Intervention Levels and Countermeasures for Nuclear Accidents.

9. Waters R, Gibson D., Bugay D. Dzhepo S., Voitsekhovitch O. (1994) A review of Post-Accident Measures Affecting Transport and Isolation of Radionuclides Released from the Chernobyl Accident, Proc. *Environmental Contamination in Central / Eastern Europe.* Budapest, pp 19-24 (In press).

10. Voitsekhovitch O.V, Kanivets V.V., Shereshevsky A.I. (1988) The Effectiveness of bottom sediment traps created with aim to catch contaminated matter transported by suspended particles. Proc. of Ukr. Hydromet. Institute. USSR. **228**, 60-68.

11. Biliy I.Y., Voitsekhovitch O.V., Onishi Y., Graves R. (1994), Modelling of Sr-90 wash-off from the Pripyat flood plain by four-year flood, Proc. Isotopes in Water Resources Management, Vienna, 1994, IAEA, pp 9-10.

12. Berkovsky V., Ratia G, Nasvit O. (1994) Forming of internal doses to Ukrainian population in consequence of using of the Dnieper water. 39th Health Physics Society meeting, June, 24-28, San-Francisco, USA . 10 p.. (Submitted for publication to "Health Physics").

13. ICRP Publication 60 (1990) 1990 Recommendations of the International Commission on Radiological Protection, Annals of the ICRP **21**, Nr1-3.

14. Voitsekhovitch O., Nasvit.O, Los'y (1994). Present Concept on Water Remedial Activities on the Areas contaminated by the Chernobyl Accident, *Freshwater and Estuarine Radioecology*, Proc. Int. Symp., CEC, Lisbon, (submitted for publication).

THE APPLICATION OF CAESIUM SELECTIVE SORBENTS IN THE REMEDIATION AND RESTORATION OF RADIOACTIVE CONTAMINATED SITES

V. P. REMEZ
JU "Compomet Cantec"
8th March street 5, Suite 414, Ekaterinburg 620014, Russia

1. Introduction

Nuclear activities have resulted in the contamination of large areas with caesium. This manuscript describes sorbents which absorb caesium with a very high selectivity and can be used to:

1) decrease the uptake of caesium by animals;
2) concentrate caesium -and if specifications are changed- other radionuclides in liquids, which may range from milk to large water bodies, to facilitate the analyses;
3) remove radionuclides from liquid wastes.

The sorbents offer therefore various options for the remediation or restoration of contaminated environments and industrial areas.

2. Composition of the sorbents

The process which has been developed at the JU "Compomet Cantec" to produce FEZHEL-type composite sorbents is based on ferric-ferrocyanide. It uses granules of wood cellulose coated with layers of sorbing material [1]. These materials are both specific and highly selective to a variety of radionuclides, possess good recovery kinetics and are stable over a broad pH range. The manufacturing process enables materials to be synthesized according to various specifications of sorption capacity, sorption layer thickness, granule size, etc., depending on the task to be done.

Mixed ferrocyanides of transitional elements (Co, Cu, Fe) and alkali metals (K, Na) are well known as sorbents [2, 3]. The high selectivity for heavy alkali metals is a very interesting one, because it allows the selection of caesium radionuclides (^{134}Cs and ^{137}Cs) from aqueous systems with a complex composition. The production of an inorganic ion exchanger with good exploitation qualities, based on composite ferrocyanides as sorbents, is complicated. One solution is to fix the ferrocyanide sorbent on a bearer [4], another one to integrate the ferrocyanide group in the exchanging resin [5, 6]. Examples are

217

F. F. Luykx and M. J. Frissel (eds.), Radioecology and the Restoration of Radioactive-Contaminated Sites, 217–224.
© *1996 Kluwer Academic Publishers.*

sorbents based on mixed iron-potassium ferrocyanide on a wood cellulose bearer. They have the appearance of irregular granules, blue, tinged with light and dark nuances depending on the ferrocyanide concentration.

The sorption characteristics and exploitation possibilities of the system: 'cellulose bearer mixed iron-potassium ferrocyanide' were studied in detail in 1978-90. The study included the influence of the conditions on the manufacturing of sorbents, on the formation of hydrogenous and co-ordinative relations between bearer and inorganic phase, on the strength of the sorbents, on the specific surface and on the sorption capacity for various radionuclides.

2.1 SPECIFICATIONS

The specific surface of sorbents was determined by heat desorption of argon and equals 13-14 m²/g for FEZHEL with an initial surface of the wood cellulose bearer of 2-4 m²/g. The relation between pore diameter and specific surface area is shown in Figure 1.

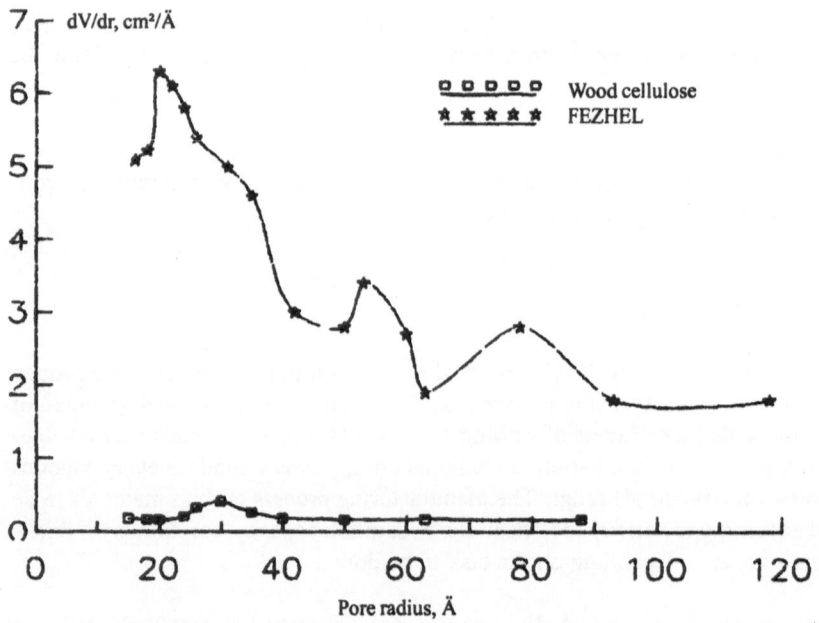

Figure 1. Relation between pore diameter and specific surface area, relative units.

FEZHEL and its varieties ANFEZH and BIFEZH are suitable for the removal of all types of cations [3]. The highest absorption coefficients were observed for Cs, Tb and Tl. The contamination caused by the accident in Chernobyl caused an enormous growth of the rise of such composite sorbents.

3. Applications

3.1 REMOVAL OF CAESIUM FROM MEAT AND MILK BY BIFEZH SORBENTS

The crucial question in the contaminated areas of Ukraine, Byelorussia and Russia is how to obtain healthy animal products such as milk and meat with a sufficiently low radioactivity level. In some 17 regions in Russia, and especially in the Bryansk region, 17% of the territory (6050 km².) is contaminated with radioactive deposits. One of the possibilities to settle this problem is to apply inorganic sorbents (biological iron ferrocyanide) as a veterinary additive to animal feed.

The product has been tested by the Institute for Biophysics of the Ministry of Health of Russia in the Bryansk region, and its use is permitted by the Main Veterinary Administration of the Ministry of Agriculture of Russia.

Figure 2 shows the effectiveness of the removal of Cs from milk as a function of its contamination level. The tests were carried out by the Institute for Biophysics. BIFEZH was administered daily during two weeks, along with the feed in amounts of 10 - 20 g/head for sheep and 30 - 60 g/head for cow. The Cs concentration in muscular tissue decreased by a factor 12 - 13, and by a factor 25 - 90 in internal organs. Figure 3 shows the content of caesium in milk as a function of the time of administration.

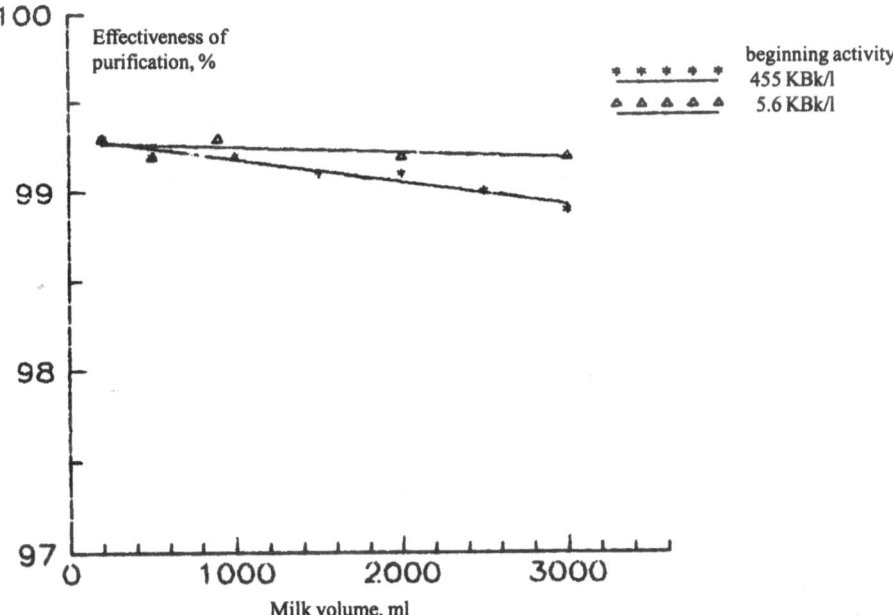

*Figure 2.*Effectiveness of the removal of Cs from milk as a function of its contamination level.

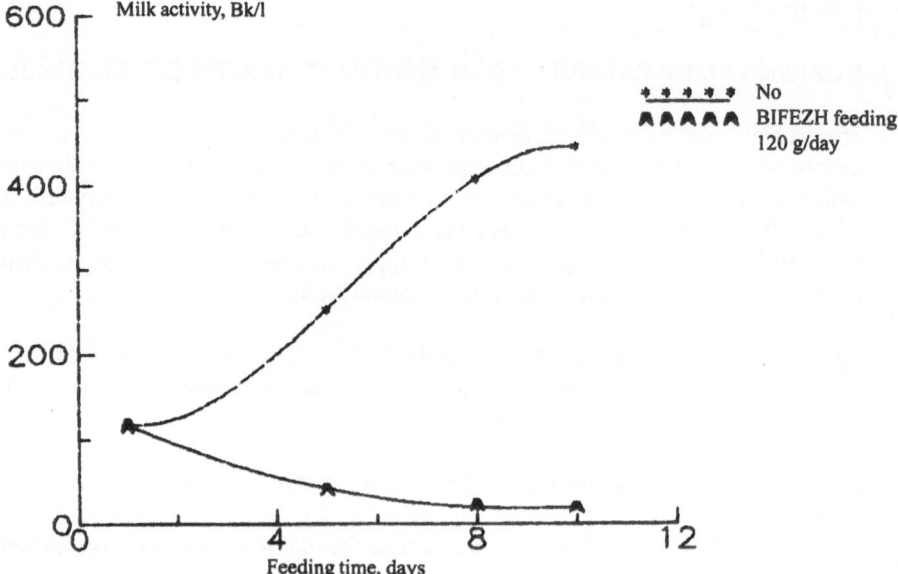

Figure 3. Concentration of caesium in milk as a function of the time of administration

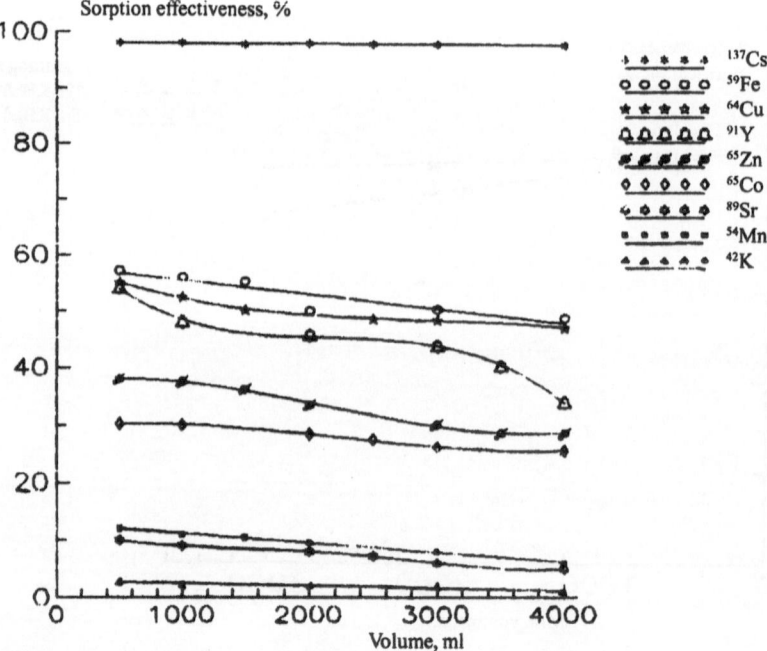

Figure 4. Effectiveness of the sorption of ^{137}Cs, ^{59}Fe, ^{54}Cu, ^{91}Y, ^{65}Zn, ^{65}Co, ^{89}Sr, ^{54}Mn and ^{42}K from marine water.

A test comparing BIFEZH to ferrocyanide boluses of German/Norwegian origin showed that BIFEZH is 2 - 3 times more efficient than the boluses. Additional advantages are that BIFEZH is cheaper and requires no special veterinary training.
Early 1993, 15 - 25 tons of BIFEZH have been used in the Bryansk region In four thousand cows, the Cs level in milk was reduced by a factor of 3 - 8.

3.2 THE USE OF ANFEZH TO CONCENTRATE RADIONUCLIDES FROM AQUE-OUS SYSTEMS TO FACILITATE ANALYSES

The analysis of small quantities of radionuclides in large water bodies is often difficult because a pre-concentration is required. The cellulose-based inorganic sorbent ANFEZH has been developed as a pre-concentration sorbent. It is based on ferrocyanide, sulfide and oxyhydrate cellulose/inorganic sorbents. It can be used to concentrate caesium in fresh and marine water, aqueous solutions and milk. It is easy to use in any application and requires no special training to be properly handled.
The application involves the loading of one layer at a time into the sorption column and the circulation of the liquid through the column at the rate of 150 ml/m². min, its cross-section being perpendicular to the water flow.
In an example, a column with a sorbent volume of 150 ml was loaded with ferrocyanide during an hour. Thereafter, a sample volume of 1 m³ was circulated through the column during 4.5 h. The column was equipped with a NaI gamma-spectrometer to follow the accumulation of Cs in the column. The technique proved especially efficient compared to the traditional pre-concentration via coprecipitation with ammonium phosphomolybdate (APM).
The sorbent is easy to compact. This property, in conjunction with its other advantages, enabled to sample any required volume.
By removal of the Cs from the column and re-use of the column the sample volume can be reused dozens of times, although the total efficiency may decrease to 90-92 %.

3.2.1 Application in natural waters
Rapid analyses of natural waters were carried out in the 30-km zone around Chernobyl using ANFEZH sorbents. Other areas which were analysed in this way include the rivers Yenissey (Krasnuyarsk-26) and Techa, the Beloyarka lake (which receives cooling water from the Beloyarskoe nuclear power plant) and the North-Western part of the Pacific Ocean.

The effectiveness of the sorption of ^{137}Cs, ^{59}Fe, ^{54}Cu, ^{91}Y, ^{65}Zn, ^{65}Co, ^{89}Sr, ^{54}Mn and ^{42}K from marine water to which radioactive tracers were added is shown in Figure 4. Water sample were led through the column with sorbent at a speed of 100 ml/h during 2 hours. The distribution of ^{137}Cs determined with ANFEZH in samples from the Pacific Ocean up to a depth of 2,000 m is presented in Figure 5.

222

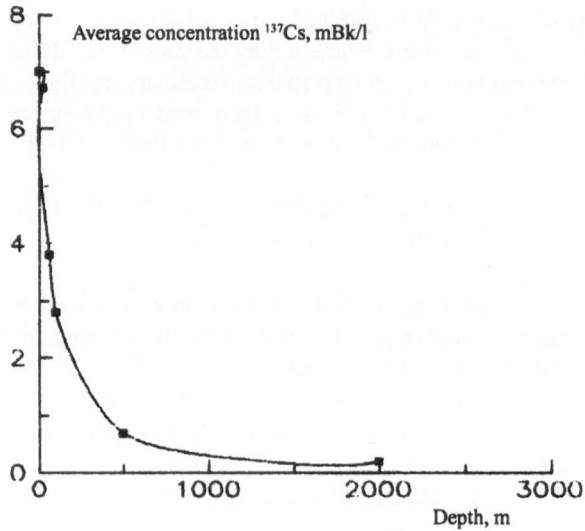

Figure 5. Distribution of caesium-137 in samples from the Pacific Ocean up to a depth of 2000 m.

3.2.2 Application to the analyses of milk

Investigations of the extraction of caesium from milk with ANFEZH were carried out in the Radiological Laboratory of the 'Gomel diary plant for processing milk from areas contaminated with Chernobyl release'. Caesium was shown to be virtually removed from milk. The resulting concentrate had a specific activity tens of times higher than the activity of the original milk. This allows both a saving of analysis time and a more accurate measurement of Cs.

At the Radiological Laboratory of the Dnieper water supply station a comparative test between domestic and imported ion-exchange resins was carried out. It was shown that ANFEZH reached a sorption percentage of up to 95%, while the concentration rate was 2-3 times faster than with resins and the sorbent's exchange capacity was 10-15 times higher. In addition, ANFEZH is simpler and easier to handle and requires no lengthy preparation prior to use.

3.2.3 Comparison of techniques

TABLE 1. Comparative characteristics of some techniques for concentrating caesium

Technique of sample concentration	Time per analysis,h	Cs yield, %	Analysis costs, $ per analysis
APM*	72	85-90	400-600
ANFEZH	1-2	95-98	20-35
Ion exchange resins	5-6	95-98	200-300

* APM: Cs analyses via coprecipitation with ammonium phosphomolybdate (APM). Salts of molybdenum or caesium are added to the sample to be analyzed, the sample is treated with acid, Cs is coprecipitated with ammonium phosphomolybdate (APM)[8].

Table 1 shows that the ANFEZH method is much cheaper, more rapid and more efficient than the APM and resins methods. Furthermore, it permits an easier organisation of mass sampling of water.

The ANFEZH has been approved by the Regulatory Section of the Interagency Commission on Environmental Radiation Control of the Russian State Committee on Hydrometeorology. The State Interdepartment Commission on Radioactive Control recommends the method for fast analyses as well as for general use. The method is employed in the Chernobyl R&P Merger "Prypyat", at the Dnieper water supply station in Kiev, as well as in many specialized institutes in the Ukraine, Siberia, Kamchatka and the Urals.

3.3 THE USE OF FEZHEL FOR THE TREATMENT OF LIQUID RADIOACTIVE WASTE FROM NUCLEAR POWER PLANTS

Nuclear power plants generate liquid radioactive wastes (LRW) of different chemical compositions and activity levels. LRW are stored liquid-waste storage tanks after being concentrated in evaporating units. These liquids have a complex radiochemical composition, a high salt content and include soaps and oils. The separation of long-lived radionuclides is difficult and, consequently, large sums of money are required to store LRW. A problem is that in the evaporating units radioactive effluents of various origins and compositions are often mixed. The removal of the main radionuclides of these effluents with selective sorbents lowers the cost of LRW storage considerably.

During 1986-1991, FEZHEL sorbents were tested for the purification of LRW in the nuclear power stations in Kalinin, Rovno, Zaporozhye, Balakovo and at the Kurchatov Institute.

The radioactivity of solutions of various compositions was reduced by a factor of 3-4 for caesium and by a factor of 2 for cobalt and manganese. The sorbent, used for acid reclamation of desalting units, allows to treat at least 10,000 liters of solution per liter of sorbent, at least 1000 l/l for trapped waste water and 500 l/l for fuel detention basins, with an overall decontamination factor of about 2000 (Table 2). The sorbents are used without regeneration; after their service life, they are cemented or bituminized.

TABLE 2. Treatment of LAW from nuclear power stations

Type of solution	Regenerates of desalting units	Regenerates of special water	Trapped waste water	Fuel detention basin water
Filtration speed m/h	16	23	8	2.4
Column volume flowed	3,000	700	3,000	330
^{134}Cs, Bq/l	78 - 6	-	4,440 - 16	144,300 - 80
Original filtrate	< 10		< 10	< 10
^{137}Cs, Bq/l	81 - 7	274 - 9	4,810 - 20	92,500 - 72
Original filtrate	< 10	16	< 10	< 10
^{60}Co, Bq/l	67 - 6	266 - 9	814 - 14	-
Original filtrate	18 - 3	< 10	26	
^{54}Mn, Bq/l	144 - 7	-	4,070 - 18	30,710 - 56
Original filtrate	< 10		56 - 6	< 10

The Department for Sanitary Supervision of Russia keeps in some areas of Russia all three varieties FEZHEL, ANFEZH and BIFEZH in stock as a precaution for possible emergencies.

4. References

[1]   Patent WO 93/12876
[2]   Bara V., Caleyka R., Tympl M. et al. Application of the solgel method for the preparation of some inorganic ion exchangers in spherical form. (1975) J. Of Radioanalytical Chem. vol.24, N2, p. 353 - 359.
[3]   Tanaev I.V. (1971) Chemie ferrocyanide. 320 p.
[4]   Amflett Ch. (1966) Inorganic ionites. 188 p.
[5]   Europatent (1983) N 217143 BOIj 39/02
[6]   Patent USA (1964) N 4448711 BOIj 27/24,
[7]   Demidov V.V., Remez V.P. et al. In russian
[8]   Standard Methods for the Examination of water and waste water 18th edition, (1992) USA p7-15

THE EFFECT OF IONIZING RADIATION ON COMMUNITIES OF ORGANISMS

V. POZOLOTINA
Institute of Plant and Animal Ecology
Urals Division, Russian Academy of Sciences
ulitsa 8 Marta, 202
CIS-620219 Yekatarinburg, Russia

1. Introduction

Radiobiology has accumulated a huge amount of data on the effects of ionizing radiation on cells and whole organisms.

These data are important for understanding the response of populations and ecosystems to radiation, but are insufficient for a full explanation of the actual situation. The properties of populations and ecosystems are not just the simple sum of the properties of the organisms forming them. They have their own laws of functioning, which determine the ultimate effect.

The effect of radiation on communities of organims was studied along three main lines.

1. Observation in zones with increased levels of natural radionuclide contents (such zones are in Brazil, Czechia, the State of Kerala in India, the Colorado plateau in the USA, a number of regions in Russia, etc.) or in zones polluted by artificial radioisotopes (nuclear weapon test sites, industrial nuclear sites or zones of accidents, the gravest of which were Kyshtym in 1957, Windscale in 1957, Chernobyl in 1986 [1, 2].

2. Large-scale experiments of exposure of natural communities to powerful reference sources, conducted in the tropical woods of Puerto Rico and in the broad-leaved forests of the state of Wisconsin, in woody species of the Mediterranean type in France and in mixed coniferous-leaved woods in Georgia and in Central Russia [3].

   For large-scale experiments and field studies, forecasts were made based on data of the radiosensitivity of organisms incorporated into ecosystems. The results showed good agreement for the wood stratum only. For the second stratum, the actual effects differed from the predictions by a factor of 5-7. The difference was even greater for the herb stratum and animals living under the forest canopy, with vectors of changes pointing to different directions, some of the species being suppressed at considerably lower doses than in the experiments and others exhibiting increased resistance [2].

3. Small-scale experiments, based on the isolation of the different components under controlled conditions, which provide a better understanding of phenomena

225

F. F. Luykx and M. J. Frissel (eds.), Radioecology and the Restoration of
Radioactive-Contaminated Sites, 225–234.
© 1996 *Kluwer Academic Publishers. Printed in the Netherlands.*

occurring in ecosystems. In nature, one can observe an integrated answer to the entire range of effects. Only simple models enable to study the effect of individual factors against the totally controlled background of other factors. This scientific line, i.e., experimental radiation biogeocenology, was developed by N.V. Timofeef-Ressovsky in the Southern Urals [4].

2. Main problems in the investigation of radiation effects on populations and ecosystems

2.1. INTRASPECIFIC VARIATION OF THE RADIOSENSITIVITY OF ORGANISMS

Integrated indicators for the radiosensitivity of species are not effective for forecasting the consequences of exposure to radiation. The ecological approach to this problem requires that all types of variation in populations be taken into account, in accordance with their response to radiation. Below we consider briefly the most important types of variations.

2.1.1 Individual variation of radiosensitivity
Individual variation mirrors the genotypic variety of members of a population. Each genotype can manifest itself phenotypically in a different way, depending on conditions. Thus, individual variation is the manifestation of a particular genotype in actual living conditions. All natural populations contain organisms which are highly resistant to radiation and others which are very sensitive to it. The most numerous group occupies an intermediate position. Let us illustrate with a model what happens to a population upon exposure to radiation.

We constructed model populations from half-sib seeds of the birch Betula pendula Roth by exposing them to different doses of radiation before sowing. Plants of the same age were grown in a uniform agrosetting within a sufficient edaphic space to take care of absence of competition, which enabled to reveal the variation fully.

The structure of the population and changes in it are best characterized by variational curves for the distributions of main morphological features. Thus, for the birch in the control, the distribution of the feature height of the shoot is close to Gaussian (Figure la). In a sample where seeds were exposed to 200 Gy, the variational curve is asymmetric and excessive, while the range of variation is very narrow. At a lower dose, the sample exhibits a differentiation, which is evidenced by a variational curve with two peaks.

In the next stage, differentiation in the shoot height is observed in both samples of experimental plants (Figure lb). Thus, the model populations contain, even though they may be genetically homogeneous, groups of plants differing in their response to radiation by greater or lower intensity of recovery processes. The results clearly show that estimation of radiation effects on the population level by average values is insufficient, and frequently totally inadequate, since average plants constitute less than 20 % of the samples.

In the next qualitatively new stage after exposure, the morphological structure of the population is restored (Figure lc). The variational curves approach a normal distribution and do not change during subsequent years of observation if no other factors add their effect [5].

To gain an understanding of these phenomena, we analysed the growth curves of hundreds of plants in control and experimental groups. As a result, we revealed four types of growth. The first type features a uniform increment from year to year, which means that part of the exposed plants had a powerful antiradiation protection from the start. This type of growth prevailed in the controls as well. The second type is characterized by intensive growth during the initial period. The third demonstrated a smaller increment in the first year than in subsequent years. The fourth type exhibited virtually no growth. The last group was eliminated almost completely, the heaviest extinction being observed in the 200 Gy variant. The plants of the second type appeared to be sensitive to other environmental factors; thus, after a particularly severe winter, some control and experimental plants lost the apical buds, which slowed down their rate of growth. The third type of growth is characteristic of plants damaged by radiation but possessing intensive systems of recovery on molecular-genetic and ontogenetic levels.

Thus, after exposure to radiation, the morphological structure of populations undergoes rehabilitation (Figure 1), basically at the expense of the most radiosensitive plants and as a result of recovery processes in cells and organisms. The result is a new sample which has eliminated radiosensitive organisms. According to N.V. Timofeef-Ressovsky, a change in the genotypic structure of the population, the emergence of a new variational curve with a mode different from the control, may be classified as an elementary adaptive phenomenon [6]. The number of individuals in radiation-resistant groups determines the further fate of the population.

The problem of individual variations in the radiosensitivity of organisms has important applied implications for man. It can be said a priori that, in a group of practically healthy people of the same age and sex, there are some who are more sensitive to radiation, working with radionuclides is counterindicated for them. Such persons should be the first to be evacuated from radioactive contaminated areas. Express methods for estimating radiosensitivity are being developed at the present time.

2.1.2 Age variation of radiosensitivity

The reaction of organisms to radiation varies throughout ontogenesis. When making predictions, it is necessary to take into account which stage of development the maximum dose rate falls in. Taking plants as an example [7] (Figure 2), the most sensitive stages are obviously gametogenesis, zygote and young seedlings. The stages of full maturity of seeds and vegetative growth are relatively resistant to radiation.

As in animals, the gametogenesis and the juvenile age exhibit the greatest sensitivity. Sensitivity grows towards older age as diseases pile up and the recovery systems become incapable of coping with an additional load. Invertebrates which have a complicated life cycle feature a different pattern of age variation. No matter at what stage exposure has taken place, it will manifest itself in the phenotype [8].

228

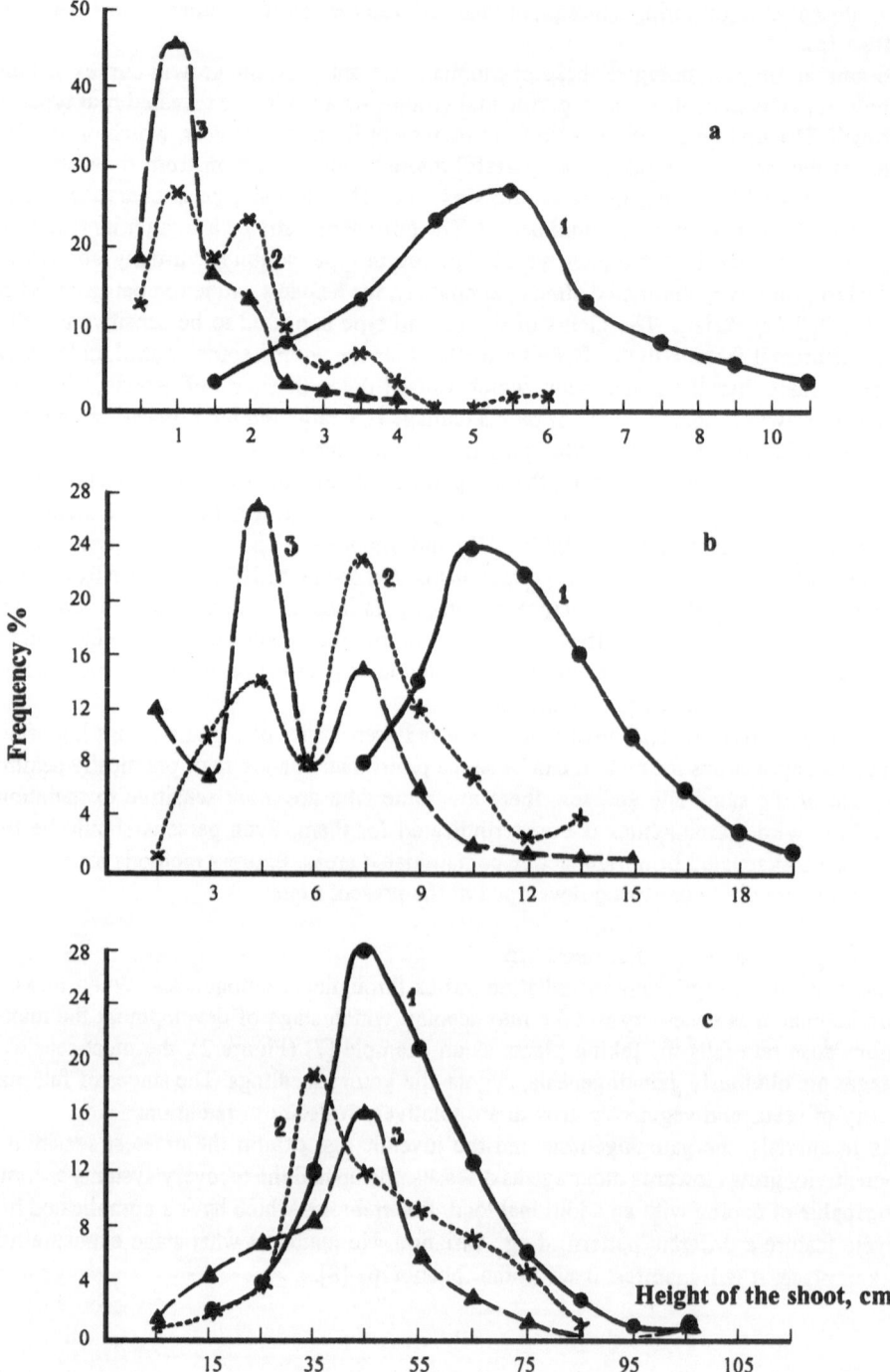

Figure 1. Variation of the height of the shoot of birch in the first (a), the second (b) and the fifth (c) year
after exposure to: 0 Gy (1), 150 Gy (2), 200 Gy(3)

Stages   - ← Levels of radiorestistance → +

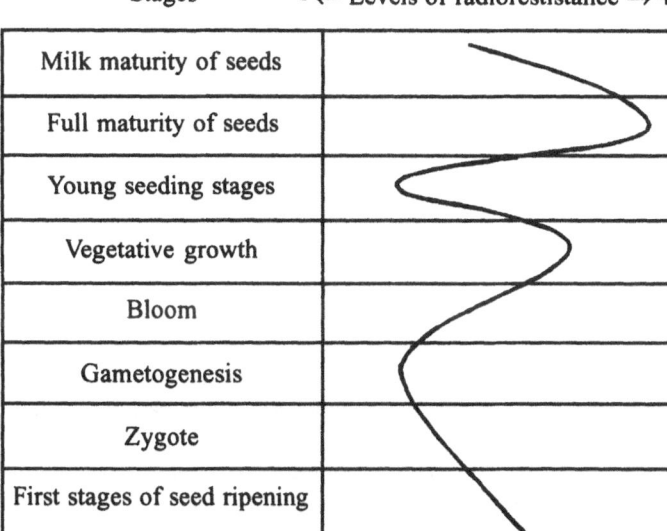

Stages	
Milk maturity of seeds	
Full maturity of seeds	
Young seeding stages	
Vegetative growth	
Bloom	
Gametogenesis	
Zygote	
First stages of seed ripening	

Figure 2. Age variation of radiosensitivity of plants

2.1.3 Sex variations in radiosensitivity

The essence of this type of variation is clear by definition. It is not observed in some plants and animals species and is marked in others. In particular, for human populations, the dose which doubles the rate of spontaneous mutation is 0.46 Gy for men and 1.25 Gy for women [1].

2.1.4 Ecological variations in radiosensitivity

This type of variation reflects the modifying effect of various abiogenic environmental factors such as temperature, humidity, illumination, mineral composition of soil and diet on the response to radiation. There are numerous experimental data and field studies throwing light on this problem. The complexity is that these factors occur in a great variety of combinations and each specific combination can orient the vector of influence in a different direction, even in an opposite one [2].

2.1.5 Geographical variation of radiosensitivity

This type of variation is similar to the previous one. Specific combinations of abiogenic environmental factors in this case are determined by the geographical location. Different geographical zones have specific conditions.

2.1.6 Temporal variation of radiosensitivity

Weather conditions in a particular area vary from year to year. Combinations of temperature and precipitation, frost or thaw, etc. create each year specific conditions for living organisms and, in particular, for radiation effects. One-year observations are not sufficient. Variation limits should be set off against the fluctuations of meteorological conditions in a given area.

The last three types of variation (ecological, geographical and temporal) are essentially similar and can be united in the following important radioecological category.

2.2. MODIFICATION OF RADIOSENSITIVITY BY ABIOGENIC ENVIRONMENTAL FACTORS

The complexity of the problem lies in the fact that combinations of factors are diverse, whereas the final effect is not their simple sum. As a rule, various factors interact or influence one another [3].

All organisms have developed protective reactions towards adverse conditions in the course of evolution. Thus, trees shed their leaves; animals grow light-adapted thick fur; a lot of them lapse into anabiosis. But sometimes the swing of physical factors is so large that the adaptive potential of the community fails. In these conditions, even a small addition of the irradiation factor would reduce the chances of survival for the population. This was demonstrated by Platt in his works [3], where the effect of radiation under conditions of drought turned out to be more effective than a simple sum of the effects of these factors.

Spamow et al. [3] showed that, in adverse climatic conditions, doses as low as 0.1-5 R/day limited the growth rate of pine.

Does this mean that ionizing radiation is a factor to which we may apply all the main ecological principles? Namely:

- organisms resident in optimal conditions can successfully resist the effect of radiation,
- harmful effects can be compensated by optimizing the conditions,
- the total effect of a group of adverse factors is not equal to the sum of the effects of each of them taken separately.

A huge body of factual evidence confirms these principles, specifically, the effects of pseudo-radiostimulation [9]. However, interpretation of results requires a profound analysis in each particular case. Take, for instance, the study conducted by Famelis [10] on two populations of molluscs, Lymnaea stagnalis, one of which resided in the favourable conditions of a river and the other in extremely adverse conditions (the stream dried up in summer and froze to the bottom in winter). The population subjected to periodic drying and freezing was three times more resistant to radiation than the river population, which lived in better conditions. A genetic analysis showed that, in adverse conditions, selection favours homozygotes, i.e., only the most viable and best adapted organisms possessing an optimal genotype survive in this case. General nonspecific resistance brings about better resistance to radiation.

2.3 BIOTIC INFLUENCE ON THE RESISTANCE OF COMMUNITIES TO RADIATION

Diverse biological relations permeate every community of organisms and make it a uniform system with qualitatively new properties. These relations may be based on competition or symbiosis as well as incorporation into a food chain; the relations may be direct or indirect.

Each species in an ecosystem occupies a certain ecological niche and is in a condition of dynamic equilibrium with the other components of the community. Suppression or loss of the most radiosensitive species from the cenosis may bring about events which would result in the restructuring of the entire ecosystem. Thus, damage to producers is undesirable for consumers; this drives them to search for new sources of food, which is fatal for these species or which may bring about a critical reduction in the number of the consumers themselves.

Auerbachs experiments [11] illustrate how strong and indirect the influence of biotic factors may be on radiation effects. A study where the object was a community of irradiated arthropods showed that, at a dose of 150 Gy (usually damaging), the number of animals increased by a factor of 2-2.5. This effect, however, did not have any bearing on radiostimulation. In fact, the relations between the arthropods and their predators-ticks had been broken. The ticks turned out to be more radiosensitive and the pressure of selection reduced, which caused the increase in numbers.

2.4 INHOMOGENEITY OF THE DOSE RATE WITHIN AN ECOSYSTEM

Sources of radiation are distributed in an extremely uneven way in natural communities. It is, therefore, difficult to establish the exact radiation dose for different species. The redistribution of radionuclides as a result of migration and accumulation in individual elements of the ecosystem increases the inhomogeneity of the dose even further [12].

When solving the dosimetry problem, it is necessary to take into account the effect of shielding, which acts as a protection for many species or for some populations. In plants, many of the vital organs are covered with a layer of soil. Tree trunks may present a good shield, too.

Animals living in deep holes have an additional chance to survive. Rats on the Marshall Islands thus survived after nuclear tests there. Difference in dose to which separate groups of organisms are exposed results in the fact that, even within one population, we may observe a whole range of radio-biological effects - from lethal to stimulating.

We have defined the main problems associated with the effects of radiation on communities of organisms. Now, let us draw generalizations and have a look at them from another point of view. All changes in an irradiated population or ecosystem may be divided into two groups: primary and secondary.

Primary changes are radiation effects proper. Secondary changes are disturbances in irradiated cenoses caused by changes in the interspecific biocenotic relations with abiogenic factors taken into account. These changes build up gradually and may result in a restructuring of the cenosis. In other cases, the recovery potential of the community permits it to survive in new conditions. Although the basic distinction between the primary and secondary changes is obvious, in practice it is difficult to separate them from each other; events follow in a continuous succession.

What effects does radiation cause?

1. lethal outcome;
2. reduced fertility;
3. suppression of growth, reduction in the biomass growth rate;
4. impairment of resistance to diseases, vermin and adverse physical environmental factors;
5. slowed down or accelerated completion of development phases, effect of radiostimulation;
6. changes in behaviour.

Let us consider these effects in detail.

a. Lethal outcome is the most characteristic type of restructuring whereby the most radiosensitive species perish in the ecosystem. A lot of observations of this type have been carried out in former test sites [3]. As a rule, radiation eliminates all woody and part of the herbaceous plants from the community. The canopy of the cenosis is opened, light-requiring species receive an advantage and eventually become dominant. In several years, the succession reverses its course. Stool shoots appear, seedlings emerge from seeds preserved in the soil. In 7-10 years, woody plants regain dominance, but the new community is qualitatively different from the previous one. Not all radiosensitive species regain the initial place; some of them are not restored at all [13].

b. Reduced fertility plays a key role in the fate of populations. Whereas damage to the blood system is the most important factor at the organism level, the key role at the population level belongs to the reproductive system. An individual organism with a damaged reproductive function can exist for a long time. For the population, life or death of one individual is not significant; but the level of reproduction and the quality of the progeny is very important.

Sterilizing doses are, as a rule, lower than lethal ones by an order of magnitude. The population, therefore, can decrease at doses of 5-10% of the semi-lethal dose. At the First World Conference on Peaceful Uses of Nuclear Energy in Geneva, Knipling et al. [14] presented the following calculations. If toxic chemicals kill 95% of the population of the fruit fly Ceratitis capitata, the remaining 5% will completely restore the population in eight generations. But if 95% of the flies are sterilized, the population will perish in eight generations. This method has found practical applications.

Intracellular recovery processes play a different role at the population level. Contributing to the survival of damaged organisms, they increase the mutation burden on the population, which has an adverse influence on subsequent generations.

c. Suppression of growth and development is observed frequently, especially in woody plants. A reduction in the biomass growth rate points to the damage to cambium. Such trees are susceptible to breaking under wind, and the greatest part of them is broken within 2 years.

A similar phenomenon is observed in the contaminated areas of the Chernobyl nuclear power station, although no severe storms have been noted in this region. The meristematic tissues of buds are damaged, too. Shoots appearing from resting buds commonly exhibit radiomorphoses: short internodes, ugly leaves or needles [15].

d. Reduced resistance to diseases, vermin and adverse abiotic factors is associated, as a rule, with the suppression of growth and development of organisms. Internal processes of destruction are aggravated by external effects, both biotic and physical.

e. Slow-down or acceleration in the passage through the main phases of development. The same dose can accelerate the rate of development of some organisms and slow it down in others. For this phenomenon, secondary changes caused by the destruction of biotic relations are of special importance. It is difficult to predict the remote consequences of these effects. Thus, for instance, a delayed or early mass appearance of pollinating insects can leave a large part of flowers unpollinated and the number of seeds will sharply decrease. However, we have observed a different result in our department. Exposure of plants to gamma radiation caused a delay in their growth and development, but these plants surpassed the control of the key indicator yield. As it was, the ripening of seeds of the control coincided with a mass appearance of sedating beetles. The delay in flowering in the irradiated plants was a salvation.

An interesting study was carried out by Willard [16] on birds. He collected 100 nests; half of the nestlings were exposed to radiation and replaced into the nests. Laboratory experiments suggested damage at a dose of 25 Gy; in reality, it was observed at a dose of 5 Gy. In fact, the nestlings featured the same behaviour as the controls. Being unfledged and weak, they tried to fly out from the nest and died. Also, parents stopped feeding those which failed to learn to fly in time.

f. Changes in behaviour patterns
The previous example has brought us to this vague and contradictory question. In animals living in a colony, disturbances of behaviour are especially marked. Thus, with ants living under chronic exposure to radiation from a powerful source, the working ants moved their regular paths deeper into the soil to take advantage of its shielding effect and practically stopped appearing on the irradiated sides of the tree stubs and trunks.

Bees demonstrated a reverse effect. Irradiation of bees at a dose of 50 Gy in laboratory caused just a 30 % reduction in life expectancy of the workers. In nature, the same dose completely eliminated a whole colony in 14 days due to the disturbance of the behaviour patterns [8]. Thus, populations and ecosystems exist as a single entity as long as biotic relations are not broken; these relations are very sensitive to radiation effects and they are hard to predict.

Radioadaptation is a phenomenon of essential importance which is specific of populations and ecosystems. It means the development of increased resistance to radiation by organisms living in conditions of long-term chronic exposure. Being limited by the scope

234

of this report, we would like to note just that radioadaptation is likely to be based on the selection of the most radioresistant organisms, as well as physiological restructuring of the metabolism within the limits of the genotype [9].

4. Conclusion

Three main components may be distinguished in the complex bundle of changes occurring in populations and ecosystems after exposure to radiation:
1. primary radiation damage;
2. processes enhancing suppression;
3. recovery processes.

Each of these components can be modified by the totality of environmental conditions. The mechanisms of all the processes operating at the level of populations and ecosystems have their own specific character. We have discussed damaging processes in some detail. As to the repair potential, it depends on the recovery capacity of individuals in populations as well as on the properties of species incorporated in ecosystems, and their general flexibility or stability.

5. References

1.UNSCEAR (1977, 1982, 1988, 1993) *Reports on ionizing radiation to the General Assembly. United Nations,* New York.
2. (1988) *Influence of ionizing radiation on ecosystems. Moscow:* "Science", 240.
3. (1968) *Questions of radioecology.* Moscow, Atompubl.,332.
4.Timofeev-Ressovsky N.V. (1962) *Some problems of radiation biogeocenology.* Report for defence of a thesis of Dr. biol.science. 34.
5.Pozolotina, V.N., Kulikov, N.V. (1988) *Ecology,* 1, 28-33.
6.Timofeev-Ressovsky, N.V., Voronzov, N.N., Jablokov, A.V. (1969) Short essay of theory of evolution. Moscow, *Science,* 1969. 407
7.Batigin, N.F., Savin, V.N. (1966) Use of ionizing radiation in plant growing. Leningrad, *Kolos,* 162.
8.Krivolutsky, D.A. (1983) Radioecology of community of terrestrial animals. Moscow. *Energyatom publ.* 87.
9.Kulikov, N.V., Molchanova, I.V. (1981) *Continental radioecology.* Plenum press, Nauka publisher, Moscow, 176.
10.Famelis, S.A., *Ecology.* (1975) 2, 84-86.
11.Auerbach, S.I. et al. (1958) *Proceedings 2nd International Conference Peaceful Uses Atomic Energy.* Geneva, 18, 494.
12.Tikhomirov, F.A. (1972) *Influence of ionizing radiation on ecological systems,* Moscow, 174.
13.Woodwell, G.M., Oosting, G.K. (1965), *Radiation Bot.,* 5 No.3.
14.Knipling E.F.et al (1955) *Proceedings 1st International Conference Peaceful Uses Atomic Energy. Geneva,* 12, 114.
15.Kozubov G.M., Taskaev A.I.(1994) Radiobiology and Radioecology Investigation of Tree Plants (in the region of the Chernobyl NPS accident). S.Petersburg, *Science,* 256.
16.Willard W.K.(1963) *Radioecology.* New York, Reinhold Publ. Cop. 345.

EFFECTS OF IONIZING RADIATION ON ORGANISMS OF TERRESTRIAL ECOSYSTEMS IN THE EAST URALS RADIOACTIVE TRACK TERRITORY

D.A. SPIRIN
'MAYAK' Experimental Research Station (ONIS)
CIS-454060 Cheliabinsk-60, Russia

1. Introduction

The present work is dedicated to investigating the effects of radioactive contamination on natural organisms in the East Urals Radioactive Track (EURT) territory, which resulted from the nuclear accident at the Mayak Production Association in 1957. Radiation effects on terrestrial natural organisms at different levels of their biological organization, excluding genetic effects, were investigated. Furthermore, data have been obtained from experiments carried out under natural conditions in the EURT territory (exposure of forest and herbaceous ecosystems to a point gamma source and other ones) to compensate for the absence, in a number of cases, of the necessary information on direct observations of radiation effects on organisms in the contaminated territory (e.g. peculiarities of origin and distribution of doses in different ecosystem components, variety of organism species, etc.).

The total area of the EURT territory is about 23,000 km^2 and represents part of the Southern Urals wooded steppe with a relatively plane relief. Approximately 50% of this area is covered by an ecosystem of birch trees and, to a lesser extent, by birch-pine forests. Herbaceous vegetation in which communities of herb grass and blue and oat grass meadows are predominant occupies 41% of the EURT territory. Fauna of the EURT territory is typical of the wooded steppe. Terrestrial and soil invertebrates, insectivorous birds and birds of prey are numerous. Rodents (mice, hares), hoofed animals (roe deer, elk) and small predators are the predominant mammals [1].

The radioactive product mixture released during the accident contained mainly short-lived radionuclides (^{89}Sr, [144]Ce, [95]Zr, [95]Nb (representing over 90% of the total activity), which determined the exposure levels of the natural objects during the first years after the accident. Subsequently, the long-lived [90]Sr (2.7% of the total activity) represented the main radiation hazard of long-term exposure of the organisms. [90]Sr was therefore adopted as a benchmark radionuclide: radioactive contamination was assessed on the basis the ^{90}Sr concentration.

F. F. Luykx and M. J. Frissel (eds.), Radioecology and the Restoration of
Radioactive-Contaminated Sites, 235–244.
© 1996 *Kluwer Academic Publishers.*

2. Sources of exposure of natural organisms in the East Urals Radioactive Trail

The development of radiation effects in plants and animals living in the contaminated territory was determined by the characteristics of the absorbed dose delivered to the main ecosystem components over space and time.

Three main factors can be distinguished [2].

In the first place, beta exposure constituted up to 15 % of the exposure energy in the fallout mixture. The dose distribution in different ecosystem components is closely related to the radionuclide distribution in these components, since beta energy is absorbed in biological tissues over a depth of several centimetres.

Secondly, radionuclides with half-lives less than 1 year (^{95}Zr, ^{106}Ru, ^{144}Ce and their daughter products) contributed to a large extent to the activity of the fallout mixture. Only 1 to 1.5 years after the accident, the long-lived radionuclides, mainly ^{90}Sr and to a lesser extent ^{137}Cs (0.04% of the total activity), became important. As a result, the dose accumulation occurred in two stages: an initial period of acute exposure (the first 1-1.5 years) and a period of delayed or chronic exposure. More than 80 % of the total absorbed dose arose in the initial period.

Thirdly, the radioactive contamination and a significant part of the acute exposure period occurred during the physiological rest of plants and many animal species, when processes in organisms are slowed down. Therefore, the degree of significance of somatic radiation effects was determined mainly by the integral dose accumulated in tissues and organs, and was to a smaller extent dependent on dose rate and dynamics of the exposure regime. These effects began to be revealed in the spring of 1958, when the physiological activity of the organisms resumed and continued for some years.

A summary of exposure data of organisms in the acute period is given in Table 1.

3. Radiation effects in terrestrial plants in the East Urals territory

3.1. WOODY PLANTS

Changes in rate of growth and development, deviations in morphological and physiological indices, loss of crop capacity, loss of reproductive capacity, withering of organs and their death are the most traditional assessment criteria of radiation effects in plants. In this work, the lethal dose (LD 100), inducing death of most of the individuals in a population of the examined species, is taken as a benchmark dose for convenience of consideration of the main radiation effects at different levels of biological organization. Pine trees (Pinus sylvestris) were found the most vulnerable vegetation component in the EURT territory. Radiation effects on pine were already revealed in the spring of 1958 and were expressed in complete or partial absence of sprout formation in the crown and drying up of needles depending on the dose received [3].

The complete death of pine trees in areas of the EURT territory occupied by birch and pine trees occurred at doses, accumulated in tree crowns (needles), of 40 Gy on average (for oppressed trees in the population, LD100 is 30 Gy; for prevailing ones, up to

TABLE 1. Absorbed dose*(μGy) to plants and animals in the EURT
territory during the acute exposure period
(autumn 1957- winter 1957-58)

Organism	per unit ⁹⁰Sr contamination (1 Bq/m²)	at maximum contamination (3.7x10⁺⁷-1.5x10⁺⁸ Bq/m²)
Woody plants		
pine		
needles	3 - 5	100 - 800
buds	2 - 3	50 - 400
seeds	0.5 - 1	20 - 200
birch		
buds	0.5 - 2	20 - 200
seeds	0.3 - 0.7	10 - 100
Herbaceous plants		
resumption buds	0 - 10	0 - 2,000
seeds on the soil	2 - 10	70 - 2,000
Soil invertebrates		
in forest litter	0.5 - 5	20 - 800
in soil at 1 cm depth	0.3	10 - 40
Birds		
hibernating insectivorous and graniverous	2 - 3	50 - 400
hibernating birds of prey (gastro-intestinal tract)	1	30 - 100
Mammals		
large herbivorous (gastro-internal tract)	3	100 - 400
mouselike rodents (whole body).	3 - 0.5	10-100
predatory (gastro-intestinal tract)	1	30-100

*) Dose assessments were made by means of calculations; in a number of
cases the results were confirmed in special experiments by modelling

50 Gy). The total area of complete pine death in the EURT territory was 20 km².

At doses of about half the LD 100 (15-25 Gy), a wide spectrum of radiation effects was observed within the range of separate trees: reduction of sprout growth by 50% and more, underdevelopment of needles, dying off of part of the needles in the crown (up to 70 %) and inhibition of radial increase of the wood.

Doses below 15 Gy induced decreasing growth processes (up to 25-30%).

They induced various physiological disturbances such as considerable fluctuation of photosynthesis intensity, transpiration and chlorophyll content. Moreover, blossoming and seed formation of pine were considerably reduced at these doses.

Finally, a certain stimulation of the growth process of grown-up trees and increasing length of sprouts and needles (up to 20%) were observed at doses below 1 Gy .

Thus, for the radioactive contamination situation in the EURT territory, three dose ranges

238

can be distinguished according to the significance of the induced radiation effect on pine (Table 2):
- range of lethal doses: 30-50 Gy,
- range of sublethal doses: 15-30 Gy and
- range of radiation effects of middle and low significance: 1-15 Gy.

TABLE 2. Radiation effects in pine trees in the EURT territory

Absorbed dose, in crown (Gy)	Effects	Extent of injury
30 - 50	Death of trees	lethal
15 - 20	Growth reduction of sprouts and needles by a factor of 2 Drying up of needles up to 80 % of the mass in crown Reduction of wood increase by a factor of 2 Reduction of viability of pollen and seeds by a factor of 2 or more Reduction of net primary productivity by a factor of 2 or more	sublethal
1-15	Growth reduction of sprouts and needles, up to 30% Drying up of needle mass up to 30% Disturbance of physiological processes Reduction of viability of pollen and seeds	middle and low
below 1 Gy	Stimulation of growth and development of separate trees	absence of visible injuries

Birch trees were found to be considerably more resistant to radioactive contamination. They were all dead only in areas of the EURT where buds received doses over 200 Gy (Table 3). On the average, according to common accepted criteria the radioresistivity of birch is 5 to 6 times higher than that of pine and it reaches a factor of 10 for separate indices (for example, growth of sprouts).

TABLE 3. Radiation effects in birch trees in the EURT territory

Absorbed dose in buds, Gy	Effects	Extent of injury
over 200	Death of trees	lethal
100 - 150	Reduction of sprout growth by a factor of 2 and more drying up of crown up to 50 % Inferiority of seeds	middle
40 - 60	Negligible reduction of sprout growth ($\leq$ 25 %) Reduction of seed Drying up of crown up to 10 %	low

However, for the EURT contamination conditions, besides the known differences in radiosensitivity of pine and birch trees, coniferous species received higher doses to critical pine organs due to the longer stay of radionuclides in the crowns.

An experimental assessment of the dose-effect relationship for different levels of contamination of the forest community has been carried out in exposing pine-birch forest to a mobile gamma point source (1.2 PBq ^{137}Cs) by 'MAYAK' Experimental Research Station.

This assessment reflects the reactions of organisms to exposure in a really acute regime, i.e., when the duration of the exposure is extremely short (several days) as compared with the EURT accident exposure time .Two exposures were carried out: one in autumn, i.e., in the period of physiological rest and one in spring, i.e., in the period of physiological activity.

The consequences of such an exposure for pine are given in Table 4; it allows to compare them with effects observed in the EURT territory.

TABLE 4. Radiation effects in pine trees in the 1st year after acute gamma-exposure in spring [4,5]

Absorbed dose in crown, Gy	Effects
20	Reduction of photosynthesis intensity by a factor of 2
4	Reduction of sprout growth by a factor of 2
2.4	Reduction of needle growth by a factor of 2
16.	Reduction of primary productivity by a factor of 2
30 - 35	Death of trees

These tests show that acute exposure according to a number of criteria has a greater injuring effect than the radioactive contamination in the EURT territory. In these experiments, differences in seasonal radiosensitivity of woody plants were determined. Radiosensitivity in the period of active growth (in spring) is higher than that in the period of physiological rest (in autumn and winter), on average by a factor 1.5 to 2. This is also true for herbaceous plants.

3.2. HERBACEOUS PLANTS

Herbaceous vegetation of both forest and meadow ecosystems of the EURT was subjected to a very wide dose range (from several Gy to hundreds of Gy) inducing, however, identical effects in different plant species. There are three reasons for this.

Firstly, the large variety of herbaceous plant species growing in the EURT territory (434 species) show a large spectrum of radiosensitivity (up to a factor of 100).

A second reason is the ability of some species to reproduce vegetatively.

A third reason is the location of resumption buds relative to the soil surface: above the soil surface, on the soil surface or in the soil.

On the basis of these three factors, herbaceous plants of the EURT were conditionally divided into two groups: highly vulnerable species (high radiosensitivity, buds of resumption above the soil surface or on the soil surface) and weakly vulnerable species (low radiosensitivity, buds of resumption in the soil). Only 2% of the total number of herbaceous plant species growing in the EURT territory belong to the category of highly vulnerable species. Therefore, the group of weakly vulnerable species determined the general picture of radioactive contamination effects on herbaceous plants in the EURT.

In areas where the absorbed doses to buds of resumption reached on average about 30 Gy, highly vulnerable species died after the period of acute exposure. The LD100 for weakly vulnerable species was more than 60 times higher, i.e., 2,000 Gy (Table 5). However, one should take into account that the total dose was received during the period autumn 1957 - winter 1958, when the above-ground plant parts were absent and the plants were in physiological rest. The importance of this factor is confirmed by the results obtained by acute exposure of herbaceous vegetation of forest and meadow ecosystems with the gamma source in the spring and summer period, when the above-ground plant parts were already formed and intensive growth was taking place. In these experiments, radiation effects in the group of weakly vulnerable species occurred at doses 1.5 to 2 times lower (Table 6). For the group of highly vulnerable species, the difference in lethal doses was much smaller and in a number of cases, the values coincided [6].

The biological radiosensitivity of herbaceous plants exceeds that of woody plants by a factor of 10 to 100, both on the basis of the radiation effects observed in herbaceous and woody plants of the EURT and during gamma exposure.

Interesting, the ratio of doses inducing stimulation effects in herbaceous and woody plants corresponds to the ratio of doses inducing injury effects in the same plants.

Death of vulnerable species occurred during the first three to four years after the accident. The vacant places were colonized by species capable of vegetative reproduction through rhizomes protected from beta exposure by a soil layer, by species transported with the wind from great distances, and from species uncontaminated areas. Due to these processes, which occurred sufficiently quickly, lethal effects had significantly less visible consequences for the communities of herbaceous plants than for communities of woody plants.

TABLE 5. Radiation effects in herbaceous plants in the EURT territory

Absorbed dose in buds, Gy		Effects
Highly vulnerable species	Weakly vulnerable species	
30	2,000	Death of plants
25	600	Stopping of growth and seed formation
15	350	Reduction of growth by 50 %, infertility of seed
5	100	Reduction of growth by 10-25%

TABLE 6. Radiation effects in herbaceous plants (weakly vulnerable species) from acute gamma-exposure in spring and summer

Absorbed dose in overground plant parts Gy	Effects
1,000	Plant death
420	Stopping of growth and seed formation
220	Reduction of growth by 50 %, infertility of seeds
80	Reduction of growth by 10 - 25 %
25 - 40	Stimulation of growth

4. Radiation effects on terrestrial ecosystem fauna in the East Urals radioactive track.

4.1. INVERTEBRATES

Among the invertebrates in the EURT territory species which occur in the litter of forest ecosystems, in turf or in the soil surface layer of meadow systems turn out to be the most sensitive ones to radioactive contamination. These species are especially vulnerable when in the embrionic stage.

Insects, spiders, worms, nematodes and others are amongst the most abundant in this group.

Observation of the number of the unpaired silkworms showed that the yield of caterpillar eggs of this species decreased during the first years after the accident by more than a factor of 2 in the areas where doses to eggs in the acute period reached 30-70 Gy. The number of flies-tachynids, the larvae of which parasitize silkworm caterpillars, had decreased too [3]. Pupae of these flies hibernate in forest litter.

The number of the most vulnerable invertebrates decreased on average by more than a factor of 2 in the area where doses exceeded 6 Gy. The largest decrease -up to 100 times- was observed in dew worms and in myriapods [7]. Similar results were obtained in experiments on forest ecosystem exposure to a gamma source.

The results are shown in Table 7.

TABLE 7. Radiation effects on invertebrates in the EURT territory and by gamma exposure of the forest ecosystem

Objects	Absorbed dose, Gy		Effects
	in EURT	by gamma exposure	
insects			
unpaired silkworm (eggs-caterpillars)	over 30		reduction of number by a factor of 2
fly-tachynida			reduction of number by 50-100%
mesofauna			
dew worms and myroapods	6 - 10	> 1	reduction of number by a factor of 10 to 100
spiders		>1	reduction of number by a factor 2 to 5
microfauna			
festaceous ticks and other soil ticks		250	reduction of number by a factor of 2 to 5

Radiation effects were practically not observed in flying insects (capable of quick settling elsewhere) or in insects having cover. Oppressing exposure effects on ants were not revealed after long-term observation, though they are on the surface of forest litter for most of their life [7].

4.2. BIRDS

Effects of radioactive contamination on birds in the acute exposure period were negligible. Only in the spring of 1958, the number of bird nests decreased in those areas where dose levels in tree cones reached 200 Gy and more. By 1959, however, the frequency of occurrence of the most distributed bird species in the EURT territory was identical, irrespective of the radioactive contamination density.

Much later, a reduction of the hatching percentage and a growth delay of chicks by 20% in nest volume at doses of 0.01 Gy/d was observed.

A retrospective assessment of potential doses to hibernating birds in the acute period has shown that granivorous and insectivorous birds had been subjected to extremely high doses; maximum doses to these birds ranged from 50 to 400 Gy (Table 8), what allows to suppose that at least part of these birds, living in areas with maximum contamination density, died (assuming an LD50 for birds of about 10 Gy).

Radioactive contamination of migratory birds, constituting the majority of those dwelling in the EURT territory, began only in the spring of 1958 , when the dose rate in tree cones had already decreased by a factor of 10.

Accordingly, the dose absorbed by the migratory birds in 1958 and 1959 did not exceed 1 to 2 Gy, which is significantly lower than the lethal dose. Radiation effects were thus considerably lower than for hibernating birds.

TABLE 8. Radiation effects in birds in the EURT territory

Absorbed dose, Gy	Effects
> 10 whole body dose (in "acute" period)	Partial death of hibernating graniverous and insectivorous birds *
>200 (in "acute" period)	Reduction of reproductive ability up to 2 times
0,01 per day (in nest)	Increasing mortality of embryos and delay of chick growth by 20 %

*) hypothetical data

4.3. MAMMALS

Mouse-type rodents received the highest whole-body dose among the mammals of the EURT. Systematic observation of their population in the EURT territory began only in 1962; therefore, there are no experimental data on radioactive contamination effects on these animals in the acute period. However, by comparison of the values of calculated doses in this period (Table 1) with the experimentally obtained LD50 for mice (on average 9 Gy), one can suppose that mouse-type rodents died in this period (Table 9).

As a result of the investigations on biological effects in mice populations in the subsequent period, it was shown that, in the first 15 years, during which doses decreased from 0.1 Gy/d to 0.001 Gy/d, mortality increase and reduction of fertility and life span were observed in mice.

Mortality increased by a factor of 2 to 10 times at doses of 0.01 Gy/d and life span reduced by a factor of 1.5 to 2, whereas fertility decreased by a factor of 1.25. Such primary radiation effects have led to a change of the population structure and to a weakening of the protection mechanism. In particular, the number of bloodsucking exoparasites, thriving on mouse-like rodents, increased by more than 5 times in the first years of the 15-year period.

At the end of this period, after 30 generations, their population was for all indices identical to the control, except for an increased radioresistance [8]. This was shown in experimental acute gamma exposure of animals with doses ranging from 4 to 11 Gy. The LD50 was 6.6-9.5 Gy for mice from the population dwelling in the EURT, and only

5.6 Gy for the animals from the control group.

More than 15 years after the accident, radiation effects on mouse-type rodents were recognized only at the fine structure level. In particular, deviations in the morphological structure of peripheral blood and marrow were observed at doses of 0.18-1.8 Gy/y. Radioactive contamination of large herbivorous and predatory mammals in the acute period could also have led to their death if they had been permanently dwelling in those areas of the EURT where contamination densities were 3.7 mBq ^{90}Sr/m^2 (dose to gastrointestinal tract: above 100 Gy). However, taking into account the migratory behaviour of these animals, losses were most probably negligible and quickly compensated by animal inflow from adjacent territories.

TABLE 9. Radiation effects on mousetype rodents in the EURT territory

Absorbed dose, Gy	Effects
> 9 in "acute" period *	Death of animals
> 0.01 per day	Increased mortality by a factor of 2-5
	Reduction of fertility by a factor of 1.25
0.18 - 1.8 per year	Reduction of life duration by a factor of 1.5 -2
	Changes of structure of peripheral blood and marrow, symptoms of leucopenia

*) hypothetical data.

5. References

1. Spirin D.A., Smirnov E.G. et al. (1990) The effect of radioactive contamination on natural systems. *Priroda (Nature)*, **5**, 58-63.
2. Tikhomirov F.A. and Romanov G.N. (1993) Exposure doses of organisms under conditions of radioactive forest contamination, *Ecological consequences of the radioactive contamination in the South Urals*, Moscow, "Nauka", 13-21.
3. Tikhomirov F.A. and Karaban' R.T. (1993) Radiation damage of the forest under conditions of the radioactive contamination, *Ecological consequences of the radioactive contamination in the South Urals*, Moscow, "Nauka", 85-95.
4. Spirin D.A., Mishenkov N.N., Karaban' R.T. et al. (1981) Effects of acute gamma-irradiation on assimilative apparatus of a birch-pine stand, *Lesovedeniye*, **4**, 75-82.
5. Spirin D.A., Aleksakhin R.M., Karaban' R.T. et al. (1985) Influence of acute gamma-irradiation on productivity of pine-birch forest, *Radiobiologiya*, **25**, N 1, 125-128.
6. Spirin D.A., Michenkov N.N., Karaban' R.T. et al. (1985) Postirradiation effects in grass cover of irradiated forest, *Radiobiologiya*, **25**, N 1, 122-125.
7. Krivolutsky D.A., Usachev V.L., Arhireeva A.I. et al. (1993) Changes of the structure of animal population (terrestrial and soil invertebrates) under the influence of Sr-90 area contamination. *Ecological consequences of the radioactive contamination on the South Urals*, Moscow, "Nauka", 241-250.
8. Il'enko A.I. and Krayivko T.N. (1993) Ecological consequences of radioactive contamination for populations of small mammals - strontiephorous, *Ecological consequences of the radioactive contamination in the South Urals*, Moscow, "Nauka", 171-180.

PHYSIOLOGICAL ADAPTATION OF SMALL MAMMALS TO RADIOACTIVE POLLUTION

B.V. TESTOV
Natural Sciences Institute
Perm State University
ulitsa Genkelya, 4
CIS-614005 Perm, Russia

1. Introduction

Life has been subjected to the influence of ionizing radiation since its beginning. Radiation levels were certainly much higher during earlier geological periods. This appears clearly from geochronological investigations where the equilibrium of long-lived radionuclides, such as ^{238}U, ^{232}Th, ^{40}K and ^{87}Rb, with their stable decay products is determined [1] and from which one can conclude that the radioactivity of the radioactive parents was much higher than now. Even today, radiation from natural sources accounts for 70-80% of all exposure to man and even more to animals living in non-contaminated areas, with doses varying over a considerable range depending on the geological underground and the altitude. This explains why small increases in exposure to ionizing radiation are not expected to be harmful to the evolution of living organisms. Our study aimed to illustrate how small mammals adapt to a changing radiation environment.

2. Populations of small mammals in radioactive areas

Comparative investigations of the populations of small mammals were carried out in territories with high natural (Ural and Uhta regions) and high artificial radiation [East Ural Radioactive Trail (EURT) and Chernobyl region] (Figure 1).
In these areas situated in different climatic zones, the following mammals were investigated: Clethrionomys rutilus, Clethrionomys glarcolus, Microtus occonomus, Microtus arvalis, Apodemus agrarius, Apodemus sylvaticus, Apodemus flavicollis and Mus musculus.
In our experiments plots with equal ecological characteristics, but different radiation levels, were selected in each of the areas to be investigated. We used the plots with the lowest radioactive contamination as controls (Figure 2).

In the areas with high natural radioactivity, the dose to animals (voles) is caused by gamma irradiation from radionuclides present in the soil. Alpha and beta emitters are incorporated by inhalation and by food consumption; these nuclides accumulate in dif-

245

F. F. Luykx and M. J. Frissel (eds.), Radioecology and the Restoration of
Radioactive-Contaminated Sites, 245–253.
© 1996 *Kluwer Academic Publishers.*

246

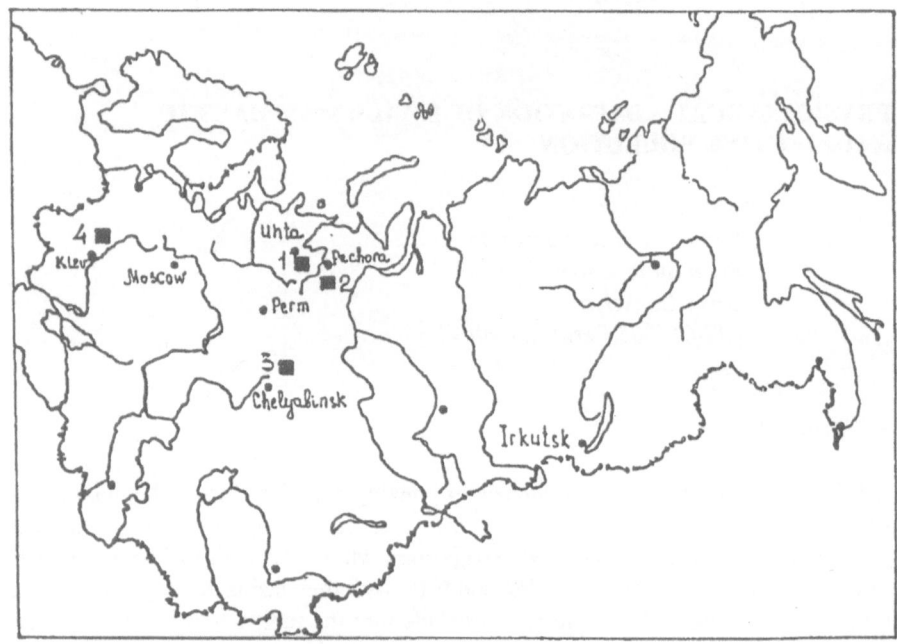

Figure 1. Investigated areas: 1- Uhta; 2 - Ural; 3 - EURT; 4 - Chernobyl.

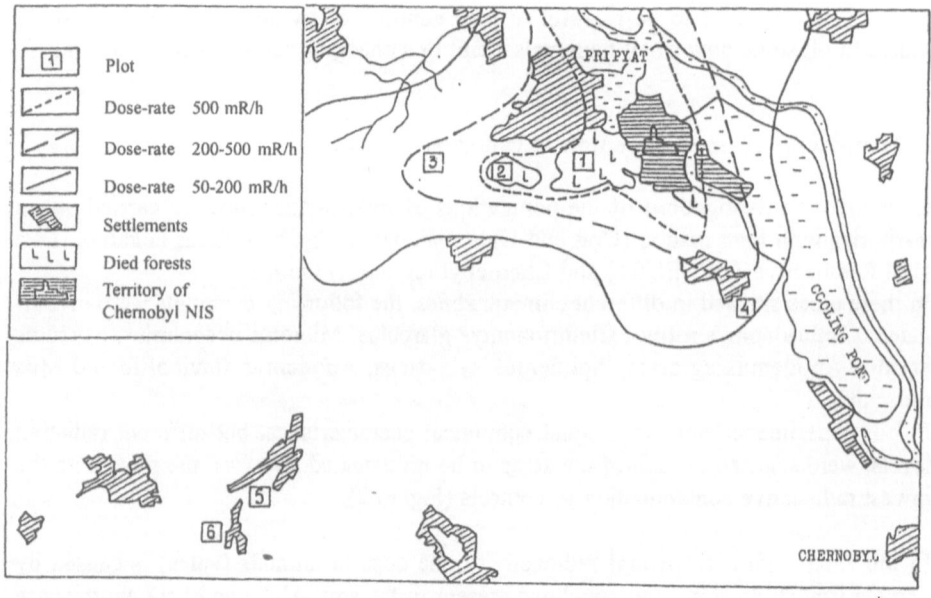

Figure 2. Investigated plots in the Chernobyl area:
1-3 - plots with high radiopollution;
4- plots with average radiopollution; 5,6,- control plots.

ferent organs and tissues of animals. Analysis of radionuclide distribution in the organisms revealed that the dose to the lungs is mainly caused by radon and its short-living decay products; the dose to bone tissues is mainly due to radium isotopes and the highest dose to soft tissues came from ^{210}Po (Figure 3).

In areas with high artificial radioactivity, the dose to animals (voles) is mainly caused by ^{90}Sr (EURT) or ^{137}Cs (Chernobyl region). To compare the irradiation levels of different plots, the effective dose, as defined by the International Commission on Radiological Protection (ICRP), was used [3]. The effective doses were averaged for all plots (Table 1). This allowed us to compare the irradiation conditions of animals.

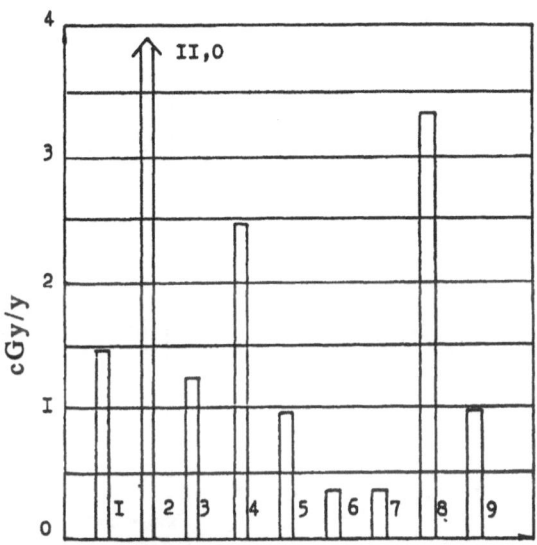

Figure 3. Irradiation doses to organs and tissues of voles:
1.) blood, 2.) Lungs, 3.) liver, 4.) kidneys, 5.) spleen, 6.) testicles,
7.) muscles, 8.) skeleton, 9.) brain.

TABLE 1. Average effective doses to animals in the investigated areas.

| AREA | mSv/year | |
	Radioactive plots	Control plots
Urals	16	1.2
Uhta	56-158	307
EURT	65-6,500	32
Chernobyl (1 year after accident)	880-86,640	16.2-48.6

The dynamics of the number and age range of small mammals in both the radioactive and control plots follows a normal cyclic development. Coincidence of the phases of the cycle (except during the first year after the Chernobyl disaster) provides evidence of a lack of disturbance of population homeostasis. Under conditions of high exposure, which was observed in the first months after the Chernobyl accident, most of the population died (Figure 2). But subsequently, the number of animals recovered rapidly and, since 1988, does not depend any more on contamination conditions (Figure 4). From 1988 on, the age ranges of the animals' population in the radioactive and control plots of the Uhta, Ural and Chernobyl areas were similar and were defined only by the cyclic development.

The reproduction of animals in the radiocontaminated areas is characterized by a common process. At the time of minimum population and while the number of animals is increasing, intensive reproduction of the animals which had not yet attained adulthood was observed in all plots. But at the time of maximum population and decreasing popu-

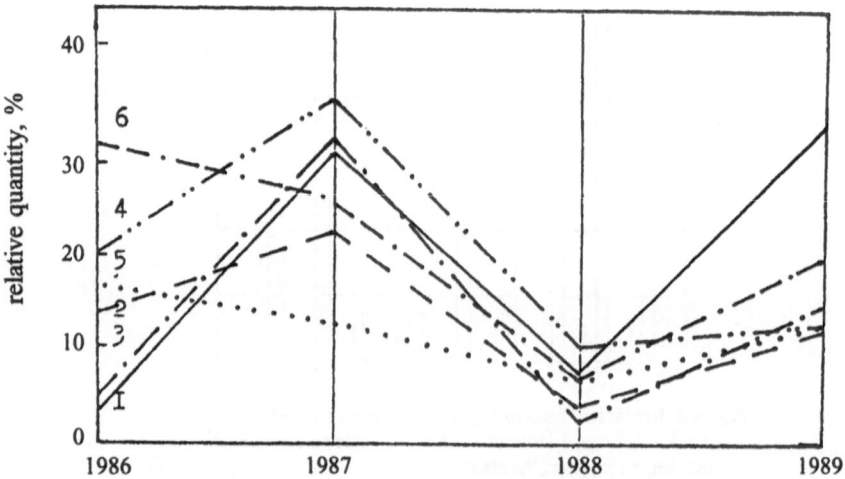

Figure 4. Population dynamics of populations of small mammals at the radio-polluted areas after the Chernobyl accident. Values of absorbed doses of gamma-irradiation in August 1987: 1- 228 cGy; 2,3 - 26-28; 4 - 3 cGy; 5,6 - 0,02 cGy.

lation, mostly adult animals were reproducing. A distinctive characteristic of the animals' reproduction in the radioactive plots was an increase in death rate of embryos. This was approximately twice the rate observed in control plots and was independent of the contamination level. Simultaneously, we observed an increase of the potential fertility of these animals. Since the increase in mature ovules led to an increase of born animals, but death of embryos led to a decrease, the number of young animals per reproducing female was not different in the radioactive and control plots.

The most effective indicator regulating the growth rate of the small-mammal populations is the rate of juvenile growth, which is especially high when the population is at a minimum. Differences between animals from radioactive and control plots were not observed.

Differences in morphometric indicators may identify the direction of the natural selection of the organims due to ecological conditions. It is known that population homeostasis is caused by physiological alterations affecting almost all parameters of life. In this case, morphological alterations appear, but do not exceed the usual limits. The interval between observed alterations characterizes the reaction rate of the population. A comparative analysis of the morphological parameters of the animals showed a greater deviation in animals from radioactive plots than in those from control plots. But these alterations did not exceed the limits characteristic of a normal reaction of the population.

As regards hematological indicators, the cells in the peripheral blood and marrow are rather sensitive and easy diagnostic indicators for high exposure. Therefore, hematological indicators are widely used for radioecological investigations. Our research showed that these indicators strongly varied according to the phase of the population cycle. Furthermore, these alterations (erythrocyte and karyocyte values) are comparable with the changes observed after exposure to high doses. Such signs as depression and recovery of haemopoesis, alterations of blood elements and appearance of a greater number of chromosome aberrations in the animals from the radioactive plots were observed [4,5]. However, the lack of clearly expressed specific alterations, testing the radiation nature of the observed deviations, allows to attribute these signs to variations of parameters.

Thus, the follow-up of natural populations of small mammals exposed to long-term radioactive contamination over many years did not reveal essential deviations from the controls. This may be seen as evidence of the adaptation of natural animal populations to long-term exposure [6].

3. Reactions of warm-blooded animals to irradiation

Experimental investigations showed that irradiation of small mammals to high doses causes hyperthermia, as shown by an increase of the rectal temperature. This phenomenon was observed in a wide scope of mammals (from mouse to man) [7,8,9,10]. However, a clear explanation of this effect has not been obtained. Interaction of ionizing irradiation with matter liberates thermal energy due to the conversion of kinetic energy of irradiation into heat. However, this effect is only important at very high doses. Even at lethal doses (10 Gy), radiation energy can only increase the temperature of an organism by 0.003 °C.

Our experiments showed that animals exposed to a radiation dose rate of 3 mGy/hour, whereby they absorbed 7.2×10^{-7} cal/g.hour of radiation energy, emit at the same time about 1 cal/g.hour of thermal energy. This is possible only by using metabolic energy.

A mechanism which allows metabolic heat production may be the separation of respiration and oxidative phosphorylation. The separation of respiration and phosphorylation during irradiation is described by numerous Russian and foreign authors [11,12,13]. During the separation, partial or complete cessation of ATP synthesis occurs and free energy is released as additional heat. This process, often used by organisms to protect themselves against supercooling [14], may lead to excess heat production under conditions of radioactive irradiation at normal temperature. To protect proteins and ferments against heat inactivation, the organism therefore emits heat into the environment. The emitted heat during irradiation may be sufficient for a temperature increase of 5 ˚C. This was observed by measuring the liver and spleen temperatures of dogs after radon injection [7].

Radiation-induced thermogenesis is probably a factor, not considered previously, which causes a high radiosensitivity of living organisms. Irradiation, as shown by numerous measurements and calculations, fulfils the role of a trigger, which initiates mammals' energy production. It results in a high radiosensitivity of these organisms [6]. But the mechanism of separation of respiration and phosphorylation is probably not the only process causing radioactive thermogenesis.

4. Protection of small mammals from irradiation

From the above, it may be concluded that a rapid release of surplus heat may protect organisms against irradiation. By artificially increasing heat emission by decreasing the temperature in an irradiation chamber, by wetting the animals' fells before irradiation or by increasing the room ventilation, we obtained a 60-70% increase of animals survival during irradiation with lethal doses [6]. Such a protection method is used by the organisms to protect them- selves against the effects of continuous irradiation. Experimentally, it was shown that short irradiations of animals increase their temperature during each period, with subsequent return to normal temperature (Figure 5). But prolonged and continuous irradiation leads to temperature normalization by decrease of oxygen consumption (Figure 6). A decrease of oxygen consumption by 8-10% compared to the controls [15] was observed for small mammals living permanently in the radioactive areas. It may be a way to decrease the surplus of heat production. A decrease of oxygen consumption arises as a natural reaction to radiation-induced hyperthermia. Therefore a shift of the temperature optimum to lower temperatures is observed for animals living in radioactive-contaminated areas. This was confirmed by investigating the average temperature of animals and also by studying the duration of swimming in warm and cold water of animals from radioactive and control plots [6].

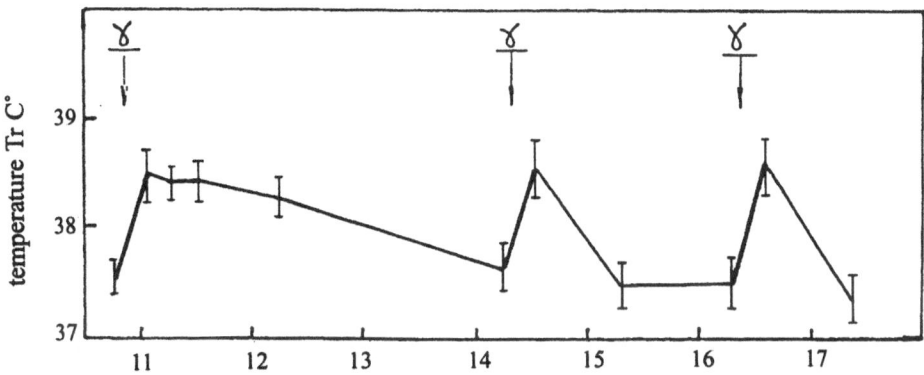

Figure 5. Rectal temperatures following three fractions of 25 mR gamma rays to white mice
(dose rate: 100 mR/h)
Irradiation time: 15 min.

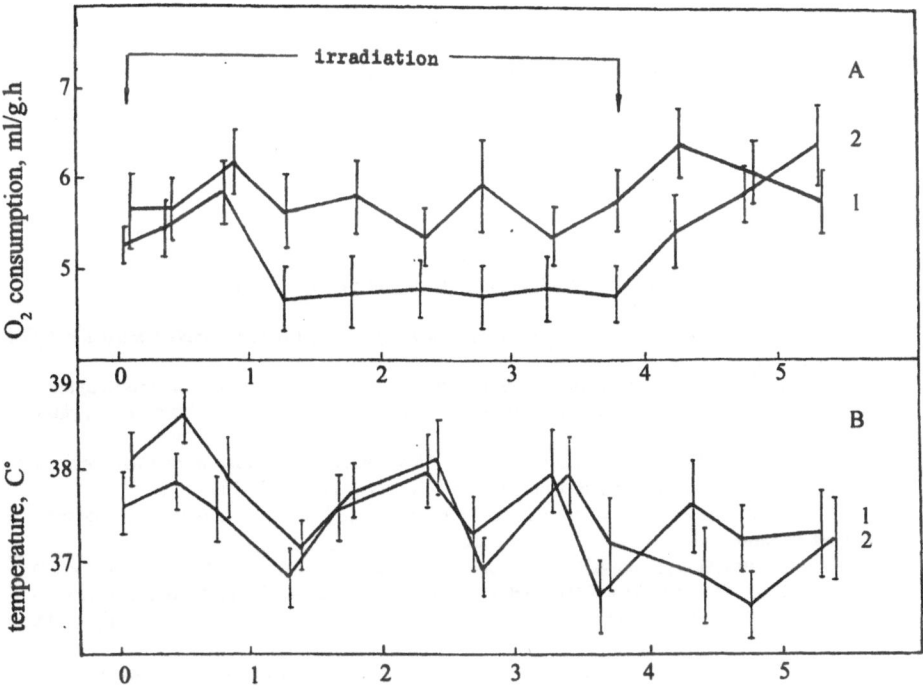

Figure 6. Oxygen consumption (A) and rectal temperature variation (B) during prolonged irradiation of white mice with ^{137}Cs at a dose rate of 300 mr/h: 1- irradiated group, 2 - control group.

5. Physiological adaptation processes of small mammals

At present, many scientists admit the existence of adaptation of living organisms to radiation. But most investigators ascribe adaptive alterations to enzymatic reactions, protein changes and the synthesis of substances characterized by anti-oxidative properties.

If we consider that in the case of chronic irradiation organisms may decrease their metabolic activity and that hypoxia reduces the negative effect of radiation, a postulate for adaptive structural alterations is not needed any more.

The prolonged evolution during considerable fluctuations of the temperature allows a wide range of adaptation mechanisms which permit warm-blooded animals to compensate a surplus heat production without decreasing their metabolic activity. These possible adaptations include the choice of locations with different levels of humidity, the change of occupation times in sheds, the change of reproduction time and the use of colder times of the day for activities. Therefore, it is necessary to pay more attention to a comparative analysis of the behaviour of animals during radioecological investigations.

It is also necessary to consider that the physiological adaptation of animals to radiation is possible only under continuous exposure. During periodic irradiation (some hours per day), adaptation is impossible due to a constant readaptation to the changes of the radiation regime. Under these conditions, even small irradiation doses may have negative effects.

6. References

1. Atomic Energetics (brief encyclopedia) (1958) *M.: Big Soviet Encyclopedia.*
2. Kusin, A.M. (1991) *Natural radiation background and its significance for the biosphere of the earth,* M. Nauka, 177.
3. ICRP Publication 32 (1981), *Limits of inhalation of Radon Daughters by Workers, Annals of the ICRP, 9, Nr 1,* Pergamon Press.
4. Maslova, K.l., Materii, L.D. (1974), *Morphological alterations in peripheral blood and spleen of the voles during their existence in a sphere with high radioactivity,* Problems of the earth biosystems, Syktyvkar, 74-85.
5. Materii, L,D. *(1989) Morphological violations in the blood system of the voles in the Chernobyl region,* Chernobyl-88. Abst. of All-Union Conference, Chernobyl, 3. Part 2, 142-164.
6. Testov, B.V.(1993) *Influence of radioactive pollution on micromammalia populations.* Abst. of doct. dissertation. Ekaterinburg, 36.
7. Pegel, V.A., Dokshina, G.A., Khilo, Z.V. *(1970) About radon influence on heat formation and protein exchange in organisms,* Transactions of SRI of biology and biophysics, Tomsk University, Tomsk, l, 125-131.
8. Veninga, T.S. (1971) *Implications of bioamines in the X-ray temperature response of cats and rabbits,* Radiate. Res.,48, 358-367.
9. Farland, W.I, Willis, J.A. (1974) *Cerebral temperature changes in the monkey (macaca mulatta) after 2500 rad ionizing radiation,* Scientific report R4-7, Armed. Radiobiology Research, Bethesda, Maryland.
10. Kandasamy, S.B., Hunt, W.A., Nickley, G.A. *(1988) Implication of prostaglandine and histamine H1 and H2 receptors in radiation induced temperature responses of rats//Radiat.* Res., 42-53.
11. Neifakh, S.A. (1959) *Oxidative phosphorilation and formation of animals' heat,* IX Congress of the All-Union Society of Physiologists, Biochemists, Pharmacologists, M.I. AN SSSR, 193.
12. Okada, Sh.(1974) *Radiation biochemistry of cell,* M.: Mir, 407.

13. Recer, E. *(1979) Bioenergetic mechanisms new vision*, M.: Mir, 216.
14. Skulachev, V.P,(1963) *Energy accumulation in cell*, M.: Mir, 440.
15. Ilienko, A.l., Krapivko T.P.*(1989) Animals' ecology in radiobiogeosystems*. M.: Nauka, 224.

RADIATION EFFECTS ON FORESTS IN THE CHERNOBYL AREA

P.I. YUSHKOV
Biophysical station of the Institute of Plant and Animal Ecology
Urals Division, Russian Academy of Sciences
ulitsa 8 Marta, 202
CIS-620219 Yekatarinburg, Russia

1. Introduction

1.1 THE SITUATION OF THE FORESTS NEAR CHERNOBYL

The area round Chernobyl was as early as 1960 used for radioecological studies on forests. Therefore, in 1986 there existed a fair knowledge of the ecological conditions of the forests in the 30 km zone. It was already noticed that the forests formed an effective natural filter capable of absorbing dust and aerosols due to the large area of their foliage. It was shown that from aerial radioactive contamination, forests capture up to 40-90 % of the radioactivity in their crown [1] [2] [3].

The Chernobyl power plant is located in the Eastern Polesye on the river Pripyat. The site around the plant is occupied by coniferous and deciduous broad-leaved forests. Conifers constitute 80 % within the 30 km zone. At the time of the 1986 accident the trees were 30-40 years old. Many of the woods contained a considerable fraction of birch and aspen. Lime, oak, maple and forest nut are occasional. The river banks are colonized by several species of willow and black alder. At about 5 km north-west of the Chernobyl plant there were 10 year old plots of pine, spruce and birch, and young plots of some other species of trees and shrubs.

After the accident experimental planting of juvenile spruce took place [4].

1.2 OBSERVED DAMAGE

External indications of radioactive damage to the forest around the plant appeared in pine as early as a few weeks after the explosion. Higher radiosensitivity of conifers as compared to deciduous trees was already experimentally established [5, 6].

Kozubov and Taskaev [4] distinguished four zones of radioactive damage to pine within the 30-km area:

1) Zone of lethal damage,
2) Zone of sublethal damage,
3) Zone of medium damage and
4) Zone of weak effects.

Data on dose rates and estimated total absorbed doses are given by Yuskov [6a] who shows zones of different damage. The damaged zones are characterised by a number of

255

F. F. Luykx and M. J. Frissel (eds.), Radioecology and the Restoration of
Radioactive-Contaminated Sites, 255–267.
© 1996 *Kluwer Academic Publishers.*

features [3, 4, 7], which are described below.

    1) Zone of lethal damage. The above ground organs of pine died; the needles developed a brick-red colour; the crowns of the deciduous woody plants were strongly damaged. The pine trees perished in late 1986 - early 1987.

    2) Zone of sublethal damage. Around 20-40 % of mature trees and a greater fraction of 7 to 10 year old pines died. In the remaining trees up to 90% of the crowns were damaged; 90-95 % of young shoots were necrotized. Many of the trees had yellow needles. Signs of radiomorphoses appeared in the needles, shoots and leaves of the younger plants of oak, maple and other species. Growth was largely suppressed.

    3) Zone of medium damage. Some trees died. Many of the shoots were partially necrotized. In the lower part of the crown, the needles were damaged. Part of the lateral dormant buds were initiated. The needles and shoots demonstrated a change in shape and size. There were signs of genetic disturbance. Reproductive capability was completely suppressed.

    4) Zone of weak effects. The pine trees showed weak damage. In 1987, the reproductive capacity was suppressed; the number of infertile seeds increased by 10-15 % and the frequency of chromosome aberrations increased by a factor 2-2.5.

Table 1 shows that data on absorbed doses from different sources are in good agreement. Differences are greater in the zone of weak effects, probably because estimates were obtained from different sampling sites.

TABLE 1. Characteristics of radioactive burden on forests within the
30-km zone around the Chernobyl Nuclear Plant
a) 01-10-1987 [4], b) summer 1987 [3]

Degree of damage to trees	Calculated absorbed dose, Gy		
	1 m above ground(a)*	needles leaves(b)	apical meristem(b)
lethal		500	100
lethal	100-130	100	25
sublethal	10-25	20-100	15-25
medium	5- 8	20-50	5-10
weak	0.7-1.2	10-20	3- 5

* by gamma radiation

2. Type of radiation exposure

Ionising radiation causes radiobiological damage and provokes repair processes in organisms at all levels, from molecular to organism structures. The relationship between these processes is determined by the dose, type of exposure (acute, chronic or intermittent), physiological condition of the organism and environmental factors. Radiobiological effects represent the resultant of radioactive damage and post-accident repair processes. The trees of the Chernobyl accident zone suffered from external gamma, beta and alpha radiation as well as from internal irradiation due to the incorporation of radionuclides. The initial distribution of radionuclides in the forests had a mosaic pattern. Thus, the forests displayed a "fringe effect", whereby fringe trees surrounding openings in the forests were contaminated more than trees deeper in the woods [6]. Various parts in the trunk and individual parts in the crown of the same tree were exposed to different doses, which determined the mosaic distribution of damaged metameres in the trees.

Such a pattern of damage distribution is caused by beta emitters which accumulate in the crown of trees. In the first year after the accident [4], the beta radiation dose rate to the crown was a factor of 2-3 to 5-10 higher than the gamma radiation dose rate, but the contribution of both types of radiation to the total absorbed dose was comparable.

3. Plant characteristics and radiation sensitivity

3.1. EFFECT OF GROWTH STAGE

In their ontogenesis from seed formation to natural death, trees pass several stages characterized by specific morphological features and physiological properties, including resistance to unfavourable environmental conditions. Dormant seeds show maximum radioresistance. They show decreased radiation resistance for some time after swelling. For instance, exposure of air-dry birch seeds to gamma radiation at a dose of 10 Gy did not affect the number of viable seedlings; but when exposed to the same dose 6 and 48 hours after wetting, seeds either did not emerge or were only viable for 32 % of the reference [8]. The age of the tree also affects resistance to radioactivity: nearly all 7-10 year-old plants died and only 20-40 % of the adult trees survived.

Experiments show that the relatively high dose rates in 1987 (200-250 mR/h and more) [8a] could not have impeded the natural recovery of conifers and other tree species. To a certain extent, this is in agreement with the results of an experiment in which seedlings were gamma-irradiated (0.5 and 5 R/day) for 140 days upon seeding. These doses did not affect plant morphogenesis, accumulation of dry matter, the outflow of C-14 assimilants from the needles and their use for biosynthesis in needles and roots [9, 10]. The success of natural recovery in the woods within the 30-km zone in the post-accident period depended on seed properties and the edaphic conditions of the territory.

3.2. EFFECT OF ANNUAL CYCLES

After the juvenile stage, perennial woody plants are characterized by annual development cycles, in which several periods can be distinguished. For the temperate zone these are:

- a period of shoot growth (from early April till late May);
- a period of latent growth (from early June till late August);
- a period of deep rest (from early September till late October);
- a period of forced rest (from early November till early April) [11].

The annual development cycles of pine are characterized by a specific distribution of assimilants formed in the needles among the organs and parts of the tree. Thus, during the period of growth, young shoots use the assimilants that have accumulated in the needles in previous years; they go partly to the trunk and roots. In the trunk they are used for the formation of wood and phloem. During the period of deep dormancy, assimilates are mainly stored in the trunk and in the above ground and underground organs [12a,b, 13].

The Chernobyl accident occurred in the middle of the growing season, when the main forest-forming species (pine, birch and spruce) were proliferating, specifically the apical meristems of the early shoots, needles, leaves, and reproductive organs which were formed in the post-accident year 1987.

In accordance with the statement of Bergonie and Tribondeau [14] about the highest damage by irradiation to young, non-differentiated cells with intensive metabolism, irradiation at high doses (90-120 Gy and higher) and high dose rates mainly damages the developing shoots of pine.

During the period of latent growth, vegetative and generative buds are formed, the cambium is activated and the trunk begins to thicken. In 1986, this latent period coincided with a period of considerably reduced irradiation as compared with the previous period. The effects of radiation in this period occurred in the early stages of organogenesis, which predetermined the appearance in 1987 of radiomorphoses of various organs. Some of these are:

a) buds with necrotised apex which developed only the covering flakes of the bud;
b) distorted bud flakes which assumed a lancet shape;
c) multi-bud (15-20 and up to 30 buds on one shoot) and multi-apexness;
d) shots with shortened growth;
e) gigantism of leaves and needles, change in form of needles in spruce (from crooked to hook-like);
f) deviation in giongenesis;
g) shifts in rhythm of growth, disturbance of spatial orientation of shoots;
h) increase in the growth of lateral shoots as a result of the removal of apical domination with the death of top buds [4].

Kozubov and Taskaev [4] suggest that a number of radiomorphoses (proliferation of cell flakes, formation of elongated shoots instead of needle pairs, overgrowth of the covering flakes on shortened shoots, etc.) provide evidence of the de-blocking of "atavistic" genes at the initial stage of organogenesis.

The next two periods in the annual cycle of development, the periods of deep and forced rest, are characterized by a reduced metabolic activity, suspension of growth processes and post-exposure recovery processes. During half a year, the effects of irradiation accumulate and can manifest themselves at the beginning of a new development cycle, just like after an acute irradiation dose [3].

The above data, as well as the results from investigations in the Southern Urals forests [15, 16, 17] provide evidence of the considerable importance of the physiological condition of the tree during the period of dose delivery.

The hypothesis of radiation damage dynamics and post-irradiation recovery in pine is illustrated by the data in Table 2. It can be seen that, in 1986 and 1987, the growth of shoots in length and thickness declined; in 1987, the quantity of needles decreased and the shoot weight, the average needle length and the mass of 100 needles sharply increased as compared with 1986 . From 1988 onwards, recovery processes in the damaged trees became predominant. Specifically, the needles of 1987 and 1988 were twice as long as the needles of the pre-accident year 1985. From 1989 onwards, the growth processes in the crown in the sublethal zone of damage were stabilized on the whole.

TABLE 2. Biometric indices of vegetative shoots of 40-50 years old pine for an absorbed dose of 20-25 Gy Epicormic shoot of the secondary from the middle part of crown [4]

year of shoot formation	mean shoot length (cm)	mass of shoot (g)	weight entire shoot (g)	number of needles of entire shoot	mean needle length (mm)	mass of 100 needles (g)
1985					56.5±2.5	4.5
1986	3.5±0.4	1.9±1.9				
1987	6.0±0.5	9.3±1.9	6.6	48.9	102.9±0.1	13.5
1988	16.1±1.7	31.7±5.8	26.3	187.0	104.2±5.5	13.5
1989	16.8±0.7	24.1±1.4	20.9	224.2	96.4±2.8	0.5
1990	17.8±1.4	24.6±2.0	21.0	253.6	78.9±2.5	8.3
1991	20.0±1.1	26.8±2.1	20.1	264.1	75.1±2.7	7.9
1992	14.7±2.2	21.5±4.1	10.0	133.8	69.8±3.1	7.2

The stimulating effects observed in trees in 1987 and 1988 were probably caused by an important increase of the nutrition of the live parts of the crown coming from the intact root system. Uniform experimental removal of 10-20% of the crown of pines 43 and 76 years old led to an elongation of the needles by 3-6% [18]. The author believes that this is associated with a change in the ratio of assimilating organs and roots.

Various disturbances were caused by ionizing radiation in assimilating organs: needles and leaves [19, 20, 4]. Due to the damage to internal structures, mass death of pine needles occurred at absorbed doses of 8-12 Gy, and at 3.5-5 Gy and higher in spruce needles. In trees with intensive recovery only young needles remained in the autumn of 1987. For pine and spruce the changes in needle shape and the structure and size of resin passages are characteristic.

TABLE 3. Number of pine plantlets in pinetum (May 1990)[21]

Exposure dose rate mR/h	General protective coating of soil,(%)	Turfclady (cm)	Number of plantlets (plants per m²)		
			min.	max.	average
low degree of injury (safety of stock > 80%)					
6	80	20		0	3.2
8	90	10	44	2	2.8
8	30	96 16	4.3		
middle degree of injury (safety of stock 50-80%)					
15	60	10	6	0	1.1
16	80	20	3	0	0.8
strong degree of injury (safety of stock < 50%)					
20	60	30	4	0	0.08
22	30	30	2	0	0.05
total stock destruction (lethal dose)					
24	80	80	1	0	0.2

Damaged photosynthesizing cells of needles show enlarged vacuoles, plastides and mitochondria with a damaged membrane system and damaged grains in the chloroplast [4, 19]. In 1987, part of the leaves of birch in the zone of sublethal damage had yellow peripheral strips due to a lack of chlorophyll [22]. The observed increase in the starch content of needles subjected to strong irradiation [4] is associated with the retarded transport of assimilates from the needles to the axial organs.

Earlier, in the 35- and 70-day seedlings of pine, which were chronically irradiated in a gamma field at a dose rate of 50 R/day, an increased starch content of the needles and a retarded inflow of ^{14}C assimilates from the needles to the roots [10] was observed. This was explained by cessation of apical growth and retardation of the growth of the roots.

"Hot" particles deposited on the leaves and needles caused local damage [4]. However, the total damaged area compared to the area of the indicated organs is small, and the author assumes that they do not substantial affect the functions of these organs. Probably, "hot" particles form one of the sources of radioactive contamination of the axial organs of trees.

4. Effects on wood production

Ionizing radiation causes an effect on the activity of the lateral meristem-cambium, which ensures the thickeningof the trunk and branches.

According to Kozubov and Taskaev [4], in 50-year old pines, which received an absorbed dose of 70-80 Gy (up to 01-05- 87), the growth of the radius of the tree trunk decreased in 1986 to 50% as compared with 1985, and to 75% in 1988. In 1987-1988, a reduction by 34-58 % of the radial increase of the trunk as compared with the growth in 1985 occurred at doses of 5-60 Gy. In subsequent years, radial increases of pine trunks were in these areas nearly the same as in 1985. The observed reduction in wood production of pine trees in the post-accident period is primarily because in the accident year the greater part of young, most productive, needles died or had a smaller size. The death of buds and young shoots in the majority of pines in 1986 in the zones of sublethal and medium damage led to disturbances in the hormonal regulation of growth processes in the trunk.

Similar effects were observed in pine trees in experiments of artificial defoliation. The removal of part of the needles led to the appearance of lateral buds, increase of potential photosynthesis in the remaining needles, inflow of newly formed assimilates and stored substances from needles and roots into the trunks to stimulate the formation of young shoots [23].

In the more radiosensitive species (spruce), where the absorbed dose was only 1.5-3.0 Gy in 1986, a reduction in radial increase of the trunk by a factor of 2-2.5 as compared with the growth in the previous year, was observed. In the subsequent two years the growth of the upper part of the trees and roots was reduced, but in 1990 and 1991 it increased again. In the deciduous species (birch, aspen, alder), a radial decrease of the trunk was also observed; however, an association with the absorbed dose was less clear. In birch, the increase in radius of the trunk at absorbed doses of 40-60 and 70-150 Gy in 1986 was lower than in 1985 [4]. In all areas with high absorbed doses the radial increase of the trunks of birch in 1987 and 1988 was higher than in 1986.

A decrease in radial increment of the trunk of aspen and alder was observed in 1986 in areas with absorbed doses of 120-150 and 40-60 Gy, respectively. In the subsequent two years, a considerable increase in radial increment of the trunk was observed. This growth was facilitated by a mass death of young shoots, the absence of generative organs and use of stored substances and current photosynthesis.

Comparison of data on the effect of ionizing radiation on the increments of length and radius of trunks shows that the cambial meristem is more radioresistant than the apical one, which agrees with other studies [16].

5. Effects on sowing qualities

Strong damage was caused by ionizing radiation on the generative sphere of woody plants in the zones of lethal,sublethal and medium damage. Thus, absorbed doses of 4-6 Gy and more in 1987 during the first period after the accident, caused death in nearly all male and female generative shoots of pine. In 1986, doses of 5-8 Gy to pine caused 65 % of empty seeds, and doses over 10-12 Gy nearly 100 %.

The viability of pollen in pine in areas with a dose of 0.7-1.2 Gy and 5-8 Gy, was in 1987 and 1988 respectively 1.3-1.6 times lower than for pine in the control lot. In 1989 pollen and 5 to 75 % of buds of the first and second development were necrotised in those areas where the absorbed dose was 0.7-8 Gy.

The damaging effect of ionizing radiation on female gametophytes in pine in 1986 had a negative effect on the quality of seeds as well (Table 4). With the increase of the absorbed dose from 0.5-1 Gy to 8-10 Gy one has observed a reduction in seed mass, in germination quality as well as an increase of the number of empty seeds. In the subsequent two years, such effects did not appear.

TABLE 4. Sowing qualities of pine at different absorbed doses [24]

absorbed dose, Gy	mass of 1000 seeds, g	germination energy, %	germination rate, %	number of empty seeds, %
		1986		
control	7.09±0.38	89.60	90.05	18.00
0.5-1.0	6.34±0.32	75.82	70.00	19.00
3.0-4.0	2.94±0.18	33.20	33.60	71.33
8.0-10.0	1.99±0.17	3.0	3.0	85.20
		1987		
control	6.09±1.24			10.00
0.8-1.2	6.89±0.21			14.86
0.9-1.3	6.41±0.24			9.40
1.7-2.3	7.07±0.59			13.14
		1988		
control	5.68±10.23	75.13	75.51	14.9
1.0-1.4	6.90±10.24	75.32	82.95	10.17
2.1-2.9	6.68± 0.23	87.94	89.31	6.52
4.4-5.8	6.42± 0.32	79.45	82.18	5.36

A study of the sowing qualities of the seeds of another widespread woody species within the 30-km zone, birch, showed that they do not depend on the dose received [25]. At an absorbed dose of 10-12 Gy, part of the female and male catkins of birch had split apical parts [25, 4].

In seeds collected in 1987 from a zone of sublethal damage, with a gamma dose rate of 20-49 mR/h, the germination rate was 70.5%, which exceeded by a factor of 1.5 that of the seeds from the areas where the gamma dose rate was 0.3-0.6 and 3.0 mR/h. The mass of 1000 seeds was 1.2-1.4 higher in the first area than in the latter two [25]. The germination rate of birch seeds of the 1987 crop remained practically unchanged for the first three years of storage at a temperature of 3 to 5 ^{0}C, while by the end of the forth year of storage, it reduced to 53-68 % of the initial level. The rate of ageing of the seeds of individual trees did not correlate with the ^{134}Cs + ^{137}Cs content of the seeds, which varied from 20.1 to 279 Bq/g of dry mass. Evidently, the rate of ageing of these lots of birch seeds at the given concentration levels was determined by the genetic and morpho-physiological properties rather than by the irradiation doses.

Thus, newly formed birch seeds in 1987 exposed to an external gamma dose rate of 0.3-40 mR/h, show higher sowing qualities. There were not negative effects from internal irradiation. These seeds caused the natural reproduction of the forest.

The latter is supported by results of investigations carried out in 1990 in the accident zone. It appeared that the preservation of pine reduces with the increase of exposure rate from 0 to 20 mR/h (gamma dose rate as of 1 May 1990). The success of natural recovery in areas of increased level of radiation depends on the growing conditions, the degree of turfing of the soils and the presence of seeds.

In the zones of lethal and sublethal damage in the first and second period after the accident, all pine seedlings, as well as the seedlings and plantlets of deciduous species died (Table 4). In the dead pine forests, at a gamma dose up to 60 mR/h in 1990, we observed a change from pine to deciduous species (birch, aspen, willow) with a predominance of birch [21]. The forestry and radiobiological characteristics of birch permit it to occupy new areas and areas away from other forest-trees. In case of death of the trunk, a birch can produce shoots. The author assumes that the mentioned renewal of birch occurred mainly with seeds from the 1987 reproduction.

6. Natural regeneration

Results of the authors studies and literature data indicate that the process of natural regeneration of forest ecosystems, which were damaged by ionizing radiation, proceeds basically with the same mechanisms as for ecosystems damaged by factors of a non-radioactive nature.

The above data show a wide range of injuries to both individual woody plants and their communities. Repair of the intracellular structures of the woody plants harmed by chronic ionizing irradiation at dose rates exceeding substantially the natural background takes place in all periods after a nuclear accident. However, the efficiency of these processes and their importance for life activity of trees varies with the reduction in dose rate. Even in the first period after the accident the importance of the post-irradiation recovery of genetic and other systems of cells differed within individual trees. This is caused by different levels of radiation damage of various organs within the crown.

If the fraction of the lethally damaged parts of the branch exceeds a certain critical value, the complete branch dies. As dose rates decrease with time and, with it, the level of newly induced damage, the biological recovery of intracellular structures increases. The death of trees in the second and subsequent periods of development of the radiological situation is determined by the degree of damage to correlatively balanced relation-ships between various organs and parts of the trees, through the hormonal relation, through the metabolism of organic and mineral substances and unbalance between the above-ground part of tree and root systems due to radioactive damage to the crown.

In a radiation damaged stand, a complex of processes develops which act sometimes opposite to each other. E.g. the suppression of fruit production creates a reserve of sub-stances for the formation of the trunk part. Defoliation and needlefall in strongly irradi-ated trees leads to an increase in the illumination of the plant, which has an effect on the daily variation of the temperature of the tree and the surface layer of soil. The changes in the temperature entail a change in intensity of photosynthesis, in intensity of transpira-tion, and thus in the rate of growth processes.

Moreover, greater illumination leads to increased growth of herbaceous plants, the den-sity of which has a negative effect on the natural seed reproduction of the woody plants. Many of the phenomena observed in pine trees in the zones of sublethal and medium damage were observed in earlier experiments with the removal of needles of various ages from the whorls of various years. Removal of one- or two-year-old needles led to the death of the buds on the shoots of the lower storeys and for a number of trees also of the apical buds.

Part of the needle shoots of the 2nd and 3rd storeys formed dormant buds with a pair of bright-green needles of increased size. In defoliated trees a 1.3 to 1.8 higher intensity of photosynthesis was observed than in intact trees . However, the lack of assimilates reduced the growth of the shoots in length down to 70-80 % as compared with the increments of the previous years [27].

7. Influence of industrial pollution

Apart from radiation damage, a specific feature of the Chernobyl and Kyshtym nuclear accidents is that there are also industrial areas with high levels of chemical contamina-tion.

Through the release of industrial gases, vast amounts of material deposited on forests and other ecosystems, particularly sulphur and nitrogen dioxides, fluorine, hydrogen chloride and ozone [28]. Data on the effect of smoke on the radio-stability of trees are practically lacking, which does not permit to predict the influence of contamination of forests suffering from intensive smoke and industrial emissions.

Laboratory tests and pot experiments with birch and pine seeds from control plants and from plants which had been subjected for several decades to releases of a number of acidic gases (SO_2, SO_3, NO_x, HF) showed that over a wide range of gamma doses, there was no substantial difference in radio-resistance for between groups of seeds [29, 30]. This is possibly caused by the elimination of damaged microspores, oospheresy and zyzgotes and the underdevelopment of seeds. The pollution of tree stands by gases leads to a sharp reduction of cones in pine.

The author presents some unpublished data from a series of experiments on the com-bined effect of nitrogen dioxide (NO_2) and gammaradiation on young birch plants. In this experiment four groups of 14-day old seedlings of birch, in the phase of seed leaves, were exposed to gamma doses of 150 and 250 Gy at a dose rate of 0.30 Gy/s and subsequently fumigated for one hour by nitrogen dioxide at concentrations of 0.2 and 0.6 mg/l (Table 5). It turned out that gamma irradiation only did not affect the survival rate of the seedlings by the 30th day, while fumigation of the seeds caused a sharply increasing death rate of the seedlings with the increase of NO_2 concentration . Exposure of the seedlings to a concentration of 0.2 mg/l NO_2 and to gamma radiation did not reveal any interaction between these factors, while, at a concentration of 0.6 mg/l and irradiation at a dose of 250 Gy synergic effects were observed. Synergism was even more pronounced for the combination of gamma radiation and fumigation of young birch plants as regards the retarded appearance of leaves. In longer lasting experiments (2-3 years) with birch seedlings (1-2 years) and in areas of the pine-birch forest the author noted an increase of lethal effects of acute gamma radiation as a function of time after irradiation [30, 15]. These results indicate a high probability of heavy damage to tree stands severely polluted with acidic gasses.

TABLE 5. Effect of gamma radiation and nitrogen oxide
on 14-day old plants of birch

Index	dose, Gy	Concentration of NO_2 , mg/l		
		0	0.2	0.6
% of seedlings lost at 30 days age	0	2.8±0.5	10.8±3.2	76.1±2.7
	150	3.2±0.8	9.5±3.8	84.0±2.1
	250	3.1±0.5	12.1±2.9	92.5±0.9
number of 30 day seedlings with 1-2 leaks, % of control	0	100	87±2.2	33±4.3
	150	90±2.5	72±1.7	16±1.4
	250	83±3.1	67±2.9	10±0.8

8. Conclusion

The radioecologists at the present time have data of experimental irradiations conducted in forest areas and in forests which have suffered from radiologic consequences of nuclear accidents. The greatest part of the radiobiological effects observed in both groups of studies proved to be similar. They confirm a higher radio-stability of herbaceous species as compared with woody plants. The available differences in dose levels causing similar effects in woody plants irradiated in experiments and during the accidents are caused by specific features of the received dose. As a rule, under experimental irradiation of forest areas differences in the doses of individual trees and their separate parts are determined by the distance from the source of radiation of known activity. Exposure of a tree in the event of a nuclear accident results from external irradiation from passing clouds and from radioactive substances deposited on this tree and neighbouring trees as well as on the soil. This causes considerable variation in radioactive exposure of the various parts of the crown and trunk. This, as well as the absence of dosimetry in forest plants in the first hours after the accident, does not allow to give a precise estimate of dose rates on plants. In experimental conditions, one can obtain sufficiently correct data on the dependence of radiobiological effects on dose and dose rate. However, these data are insufficient for predicting the conditions of forest ecosystems in the areas of nuclear accidents. Only the combination of data obtained in experiments and under radioactive exposure in forest ecosystems can give a more reliable forecast of consequences for forests. It should be noted that for predicting the consequences of radioactive contamination of forests under severe chemical industrial contamination there are at present insufficient data on the combined effect of ionizing radiation and harmful industrial gasses.

9. References

1. Tikhomirov, F.A., Aleksakhin, R.M., Fedorov, Ye.A. (1972) Migration of radionuclides in forests and the effect of ionising radiation on forest plants, *Peaceful uses of Atomic Energy*, **11**, 675-688.
2 Aleksakhin, R.M., Naryshkin. M.A. (1977) Migration of radionuclides in forest biogenocenoses. Moscow, *Nauka* . 144.
3. Tikhomirov, F.A., Shcheglov, A.I. (1990) The radioecological consequences of the Kyshtym and Chernobyl radiation accidents for forests ecosystems. *Proceedings of Seminar on Comparative Assessment on the Environmental Impact of Radionuclides Released During Three Major Nuclear Accidents; Kyshtym, Windscale, Chernobyl*, Luxembourg. 1-5 October, 1990, Report EUR 1991, **1**, 867-877.
4. Kozubov, G.M., Taskaev, A.I. (1994) Radiobiological and radioecological studies of woody plants, S. Petersb, *Nauka*, 256.
5. Sparrow, A.H., Woodwell, G.M. (1962) Prediction on the sensitivity of plants to chronic gamma radiation *Radiat. Bot.* **2**, No. 1, 37-53.
6. Kozubov, G.M., Taskaev, A.I., Kozlov, V.A., Patov, A.I. (1993) Dynamics of growth of woody and basic forest-forming species in repair period. *Radioecological Studies in 30-km accident zone of Chernobyl Power Plant. Syktyvkar:* 89-104 (Trudy Komi Scientific Centre, Ural Division of Russia Academy of Sciences. No. 127).
6a. Yuskov, P.I. This issue.
7. Taskaev, A.I. (1990) Forest and radioecological characteristic of experimental areas. *Radioactive impact on coniferous forests in the zone of Chernobyl accident. Syktyvkar* 16-29.
8. Yushkov, P.I., Chueva, T.A. (1989) Radioresistance of dormant and germinating seeds of common birch, *Radioresistance and post-exposure recovery of plants. Sverdlovsk*, 65-71.
8.a Kozubov, G.M., Taskaev, A.I., Fedotov, I.S., Arkhipov, N.P., Davydchuk, V.S., Abaturov, Yu.D. (1991) Schematic map of radioactive damage to forests in the zone of the Chernobyl Nuclear Power Plant (Scale 1:100000) with descriptive note, *Syktyvkar.*
9. Yushkov, P.I., Rakov, V.S., Karavayeva, Ye.N., Mironova, N.V. (1970) Effect of chronic gamma-radiation on the formation and growth of seedlings of pine , The effect of ionising radiation on hydrobionts and terrestrial plants. *Transactions of the Institute of Plant and Animal Ecology*, the Urals Branch of the USSR Acad. Sci. Sverdlovsk, Nr 74, 57-64.

10. Yushkov, P.I. (1970) Effect of chronic gamma radiation on distribution of tracer assimilates in the seedlings of pine, The effect of ionising radiation on hydrobionts and terrestrial plants. *Transaction of the Institute of Plant and Animal Ecology*, the Urals Branch of the USSR Acad. Sci. Sverdlovsk, Nr 74, 46-56.

11. Sergeyev, L.I., Sergeyeva, K.A. (1967) Structural metabolic mechanisms of adaptation of woody plants to unfavourable environmental factors, *Seasonal Structural Metabolic Rhythms and Adaptation of Woody Plants. Ufa*, 1-36.

12a. Yushkov, P.I. (1965) On carbon feeding of growing branches of young pines, Physiology and ecology of woody plants, *Transactions of the Institute of Biology*, the Urals Branch of the USSR Acad. Sci. Sverdlovsk, Nr 43, 25-32.

12b. Yushkov, P.I. (1965) Distribution of products of photosynthesis in pine, Physiology and Ecology of Woody Plants, *Transactions of the Institute of Biology* Nr 32, 17-23.

13. Yushkov, V.I., Yushkov, P.I. (1979) Some regularities in radial transfer of C-14-assimilates in radial transition of woody plants, *Population biogeocenological studies in the mountainous park-confer forests of the Middle Urals*, Sverdlovsk, 105-122.

14. Bergonie, Tribondeau. (1906) Interpretation de quelques resultats de la radiothérapie et essais de la fixation d'une technique rationelle. *C.R. Acad. Se. Paris*. **143**, 383-419.

15. Karaban, R.T., Mishenkov, N.N., Prister, B.S., Aleksakhin, R.M., Tikhomirov, D.A., Fedorov, E.A., Romanov, G.N. (1979) Effect of acute gamma radiation on forest biocenosis, *Problems of Forest Radioecology*, Moscow, 53-67.

16. Spirin, D.A., Aleksakhin, R.M., Karaban, R.T., Mischenkov, K.N. (1985) Radiation and post-exposure exchanges in forest biogeoconosis under acute gamma-radiation; effect of acute gamma-radiation on the productivity of pine- birch forest, *Radioecology*, **28**, 125-128.

17. Krivolutsky, D.A., Tikhomirov, F.A., Fedorov, E.A., Pokarzhevsky, A.D., Taskaev, A.I. (1988) The action of ionising radiation on biogeocenosis, *Nauka*, 240.

18. Kravchenko, G.A. (1972) Regularities in pine growth, *Forest Industry*, Moscow, 168.

19. Goltsova, N., Abaturov, Y., Abaturov, A., Mclankholin, P., Girbasova, A. and Rostova, N. (1991) Chernobyl accident: effects on the shoot structure of pinus sylvestris. - *Ann. Bot. Fennici*, **28**, 1-13.

20. Ladanova, N.V. (1992) Ultrastructural organization of needles. of spruce by radiation effects, *Radiobiology*, **32**, 640-645.

21. Pautov, Yu.A., Ilchukov, S.V. (1993) Condition of stand and natural reproduction in pine forest exposed to radioactivity, *Radioecological Studies in the 30 km Zone of Chernobyl Power Plant. Syktyvkar*: 118-132.

22. Yushkov, P.I., Chueva, T.A., Karavayeva, Ye.N., Molchanova, I.V., Kulikov, N.V. (1993) Content of Cs-134, Cs-137 and Sr-90 in the crown of birch trees in the first years after the accident in the Chernobyl Power plant, *Ecology*, Nr 1. 86-91.

23. Yushkov, V.I. (1991) Distribution of photo-assimilates in some woody species, *Ecological features and recovery dynamics of dark coniferous forests in the Middle Urals.Sverdlovsk*, 42-58.

24. Artemov, V.A., Kozubov, G.M., Ostapenko, Ye.K. (1990) Reproductive processes, *Radiation effect on coniferous forests in the region of the Chernobyl power plant*. 90-126.

25. Yuskov, P.I., Chueva, T.A., Kulikov, N.V. (1993) Effect of increased background of ionising radiation on common birch in the accident zone of Chernobyl, *Ecologiya*, Nr 5, 40-45.

26. Makhneva, O.V., Yushkov, P.I. (1991) Biological properties of birch seeds formed under increased radioactive background, *Ecology and introduction of plants in the Urals*, Collection of papers, Ural Scientific Centre, Academy of Sciences of the USSR, Sverdlovsk: 280.

27a. Yushkov, V.I. (1976) The effect of artificial defoliation on the growth and intensity of photosynthesis in pine, *Ecological-physiological studies of Coniferous Woody Species in the Urals*, Sverdlovsk, 14-23.

27b. Yushkov, V.I. (1976) Some specific features in the distribution of assimilated carbon-14 in pine, *Ecological-physiological studies of Coniferous Woody Species in the Urals*, Sverdlovsk, 41-53.

28. Nikolayevsky, V.S. (1979) Biological bases of gas stability of plants, Novosibirsk, *Nauka*, 280.

29. Yushkov, P.I., Yushkova, O.P. (1980) Radioresistance of birch seeds from stands exposed and unexposed to industrial gases, *Radiostability of plant seeds and its variability*. Sverdlovsk, 46-49.

30. Yushkov, P.I. (1991) Some questions of forest radioecology, Radiation mutagenesis and its role in evolution and selection, Moscow, *Nauka*, 142-151.

RADIATION EFFECTS IN THE CHERNOBYL AND KYSHTYM AQUATIC ECOSYSTEMS

G.G. POLIKARPOV AND V.G. TSYTSUGINA
A.O. Kovalevsky Institute of Biology of the Southern Seas
Prospekt Nakhimova 2, CIS 335011 Sevastopol, Crimea, Ukraine

1. Introduction

This paper evaluates studies on the effects of radioactive pollution on hydrobionts in natural conditions as published in the former USSR and the CIS.
Presently, the results of many studies on the consequences of radioactive pollution of the aquatic environment on hydrobionts are available. For the Kyshtym accident, these are found in references [1-13, 16-26, 28-30] and for the Chernobyl accident in [27, 31-73, 77]. However, the results of these investigations are often contradictory. The purpose of this paper is to summarize and to analyse the available data as well as to consider probable reasons for discrepancies between the results obtained.

2. Classification of investigations

The investigations are divided into four groups depending on the completeness and the character of the dosimetric data.

2.1 THE FIRST GROUP

The first group consists of publications reporting absorbed dose rates or doses, which are based on radiometric measurements [7, 10, 17, 23-25, 28, 36, 38, 39, 54, 55, 59, 71] or measurements by thermoluminescence dosimeters [8, 9]. The last two studies present a comparison of measured and calculated dose rates, from which it appears that measured figures were larger than the calculated ones.
The same group includes studies [37, 73] which consider, besides calculated dose rates, data on the pollution of biotopes with heavy metals and chlororganic compounds.
Doses resulting from the Kyshtym accident are mainly based upon radiometric measurements of ^{90}Sr- and ^{137}Cs, whith ^{90}Sr being the main contributing term. In the case of the Chernobyl accident, the absorbed dose rates were calculated either only from results of gamma-spectrometry analyses [37, 38, 54, 55, 73] or derived from ^{90}Sr data [71]. Some publications [36, 39, 59] do not indicate the base for their dose-rate calculations. In 10 out of 18 studies, deleterious effects were registered over a wide range of dose rates (3μGy/d up to 0.1-0.2 Gy/d). No damage was observed below dose rates of 0.004 to 0.0008 Gy/d in the other 8 studies.

269

F. F. Luykx and M. J. Frissel (eds.), Radioecology and the Restoration of
Radioactive-Contaminated Sites, 269–277.
© 1996 *Kluwer Academic Publishers.*

2.2 THE SECOND GROUP

The second group contains studies in which the exposure dose was explicitly controlled [52, 63, 67]. Individual effects were observed in organisms at dose rates below 3-28 µSv/h.

2.3 THE THIRD GROUP

The third group considers studies based on radiometric data only [12, 13, 18-22, 29, 30, 41, 48-51, 53, 62, 64-66, 68-70]. Deleterious effects are described in 15 out of 22 studies. It is stressed that expressing radiation effects as a function of radionuclide concentrations is insufficient; without a correct estimation of the absorbed doses, it is not correct to interpret radiation effects [74, 76, 80, 81].

2.4 THE FOURTH GROUP

The fourth group covers studies which contain neither radiometric nor dosimetric data, but only an indication of the location of the site studied within a contaminated area which was surveyed [11, 33-35, 40, 42-47, 56-58, 60, 61]. Disturbances were registered in 10 out of 16 studies.

3. Effects in different species

3.1 GENERAL EFFECTS IN THE FOUR GROUPS

It is remarkable that in groups 1, 3 and 4, the number of studies in which radiation effects were observed exceeds the number of studies in which no effects were registered. The majority of the effects was discovered in populations of benthic invertebrates (92% of the studies) and amphibia (86% of the studies) compared to other hydrobionts (hydrophytes: 80% and fishes: 47% of the studies) (Figure 1). Among the investigations, supported by dosimetric data, (Figure 2) the largest radiation effects were also observed in invertebrates in all studies [37-39, 71, 73], but for fish only in 6 out of 14 studies [7-10, 39, 71]. Radiation impact was observed in 2 studies with hydrophytes [36, 71].

The relatively high effects on invertebrates may be influenced by benthic biocenosis, because the major part of the radionuclides are deposited in bottom sediments. Another explanation is that 62% of the studies on invertebrates and 71% of the studies on amphibia use the most sensitive criteria: the genetic ones. For fish and hydrophytes these values are much lower (35% and 20%, respectively).

3.2 STIMULATION

Some studies [36, 71] demonstrate stimulation of the development of separate species of hydrobionts in combination with deleterious effects for dose rates below

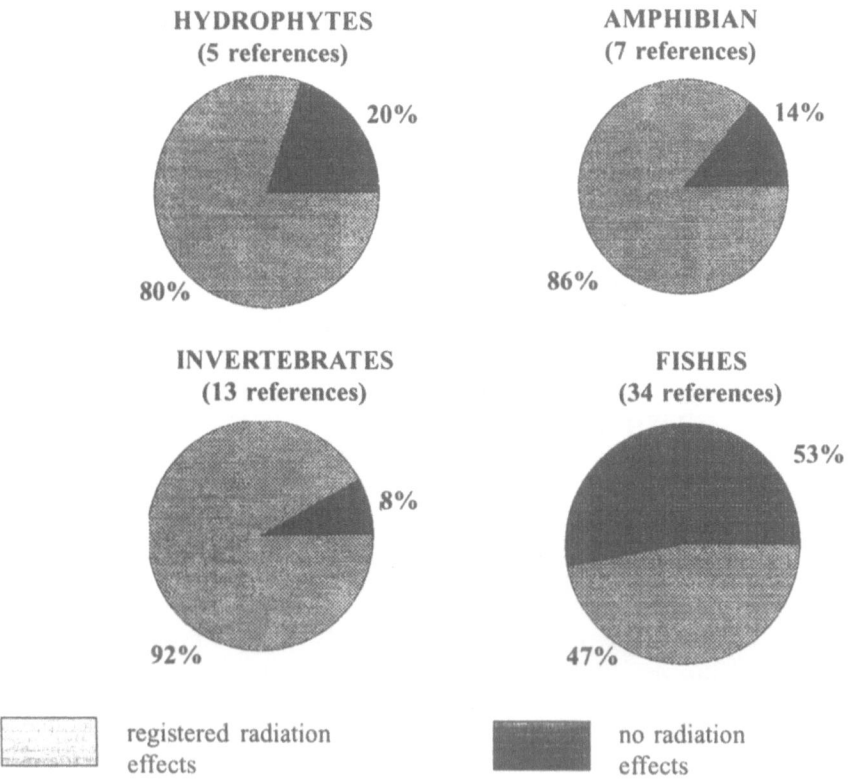

Figure 1. Percentage of references which describe radiation effects in hydrobionts in the Chernobyl and Kysthym zones. The references [39] and [71] were taken into account 2 and 3 times, respectively, since they describe effects in populations of different taxonomic groups.

0.008 - 0.000003 Gy/d. Yet, such an effect should not be considered as a beneficial reaction of biocenosis,since it can to entail disorder of the homeostasis, change of dominating species [36] and reduction of biological diversity.

3.3 STUDIES ON SELECTED SITES

An analysis of doses observed in the territory of the East Urals Radioactive Track is shown in Table 1. One of the authors proposed in 1977 a classification of chronic effects of ionizing radiation [74]. According to this classification, effects are mainly in zones of ecological masking (Figure 3). This means that radiation induced effects are dominated by other ecological factors (Table 2). The authors investigated populations of oligohaetes Stylaria lacustris in water bodies of the 10-km zone of the Chernobyl nuclear power plant and in a water body adjacent to the 30-km zone [37, 73]. It appeared that the doses absorbed by worms in these water bodies are in the range of the ecological masking zone (Table 2).

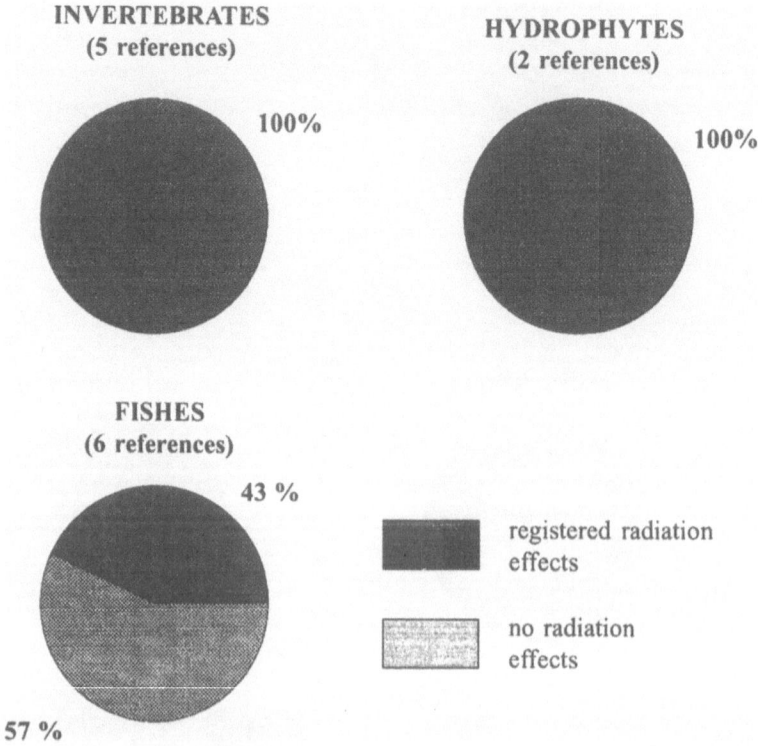

Figure 2. Radiation effects discovered in investigations in which data were supplied on dose rates.

Listed according to decreasing absorbed doses by worms, the studied water bodies are : Kopachi Village area > Creek of the Pripyat River > Strakholesye Village area (outside the 30-km zone). Another order is obtained if the mean level of chromosome mutagenesis in populations is considered:

Creek of the Pripyat > Kopachi > Strakholesye. Obviously, the intensity of chromosome mutagenesis in oligohaetes populations is determined not only by radioactive pollution, but also by other ecological factors, in particular by chemical pollution. Indeed, bottom sediments of the Pripyat river creek proved to be more polluted by pesticides than the other two water bodies [73].

TABLE 1. Overview of publications with and without registered effects as a function of the dose rate on hydrobionts in natural systems.

Gy/h	Effects of ionizing radiation in natural systems			
	Registered		Absent	
	Hydrobionts	References	Hydrobionts	References
10^{-3}	Invertebrates	[39]		
	Invertebrates and fish	[71]		
10^{-4}	Hydrophytes, invertebrates and fish	[71]	Fish	[17, 28]
	Fish	[7-10]		
10^{-5}	Invertebrates	[37, 38, 73]	Fish	[23-25, 54-55]
	Hydrophytes, invertebrates and fish	[71]		
10^{-6}	Invertebrates	[73]		
	Hydrophytes and invertebrates	[71]		
10^{-7}	Hydrophytes,	[36]		
	Hydrophytes and invertebrates >	[71]		

TABLE 2. A model for the zones of chronic effects of ionizing radiation on natural populations, [based on 74-76].

Zones		Dose rates		Effects
		Gy/y	10^{-6} Gy/y	
I.	Well being zone	<0.001-0.005	<0.11-0.6	No effects
II	Physiological masking zone	0.005-0.05	0.6-6	No or masked effects
III	Ecological masking zone			
	a) terrestrial animals	0.05-0.4	6-46	Masked effects
	b) hydrobionts and terrestrial vegetation	0.05-4	6-480	Mutation doubling
IV	Ecological manifesting zone:			
	a) dramatic	(>>0.4) >>4	(>>46) >>480	Genetic, reproduction, growth and other changes, morbidity
	b)catastrophic	>100	>120	Death of populations, ecosystem structure changes

The damage to cells in worms from the first water body was larger as well (8.86 aberrations per 100 cells, compared to 7.33 in the second one). It confirms that a comparison of populations from a polluted and a reference water body on the basis of arithmetic means is not correct. Processes as synergism and antagonism may influence the effects.

Analysis of cell distributions of chromosome aberrations may also indicate significant damage by chemical pollution to natural populations of hydrobionts. The authors observed [73] Poisson distributions of chromosome aberrations in cells of hydrobionts after ionizing irradiation. However, the distribution tends more to a geometric one under the influence of chemical mutagents. In case of a combined action of ionizing radiation and chemical mutagents, and a higher efficiency of the first factor, the distribution is closer to the Poisson distribution. The distribution approaches a geometric one when the effects of radiation and chemical factors are cmparable. In nature, when radiation can be excluded, the geometric distribution dominates.

An analysis of the distribution of chromosome aberrations in cells of hydrobionts from water bodies of the 10-km zone showed that they tend more to geometric ones. Obviously, aberrations of chromosomes in worms and gammaridae were induced by combined radioactive and chemical pollution; the latter contributes significantly to the damage of heredity structures.

It is also necessary to take into account the termination of human activities in the zone influenced by the Chernobyl accident, for example in the power plant's cooling pond [77]. From published studies, it may be concluded that it is still difficult to obtain correct dose relations. Absorbed doses are presented only in 18 out of 59 publications. In reference [73], chromosome mutagenesis was investigated in the populations of Stylaria lacustris in two biotopes with different levels of radioactive pollution. Levels of chemical pollution were also significantly different in these two biotopes. Since induced effects in areas with low dose rates are masked and are modified by natural and human factors, knowledge of these factors, particularly of chemical pollution, is very important.

The difficulties described above are also reported with many similarities in studies in Northern America [78-79] and Western Europe [76].

4. Conclusion

It must be concluded that, for a correct assessment study in a complex radioecological system, other pollutants have to be considered. The effect of the latter ones may mask radioecological effects. Much work has to be done. For instance, in the Karachay lake, important chemical pollutants (such as heavy metals, chlororganic and related compounds and xenobiotics) are insufficiently studied. Such investigations have to combine an ecosystem approach (assessment of species composition and diversity, domination, productivity) with a more detailed research of separate populations.

5. Acknowledgement

This work was partly supported by the Task Force on "Effects of Ionizing Radiations on Populations in Natural Conditions" of the International Union of Radioecology.

6. References

1. Nikipelov B.V., Romanov G.N., Buldakov L.A. et al. (1991). Atomnaya Energiya, 67 (2). 74-80. (in Russian)
2. Results of Studies and Experience on Elimination of Consequences of Accidental Contamination of Territory by Products of Uranium Fission. (1990). Ed. A.I. Burnazyan. Energoatomizdat, Moscow. 145p. (in Russian)
3. Romanov G.N., Voronov A.C. (1990). Priroda, No. 5, 50-52. (in Russian)
4. Romanov G.N., Spirin D.A., Alexakhin R.M. (1990). Priroda, No 5, 53-58. (in Russian)
5. Spirin D.A., Smirnov E.G.,Suvorova .I., Tikhomirov F.A.(1990). Priroda. No 5. 58-63. (in Russian)
6. Conclusion of the Commission on assessment of ecological situation in the region of activity of the industrial unit Mayak of the Ministry of Atomic Energy and Industry of USSR N 1140-501 on 12.06.90 .(1991). Radiobiologiya. 31 (3). 436-452. (in Russian)
7. Veronika E.A. Peshkov S.P., Shekhanova I.A.(1974). Trudy VNIPO. 100. 74-79. (in Russian)
8. Shekhanova I.A., Peshkov S.P., Muntyan S.P., Ermokhin V.N.(1978). Trudy VNIRO. 134. 105-121. (in Russian)
9. Peshkov S.P., Shekhanova I.A., Romanov G.N. et al. (1978). Trudy IERZh [Proc. of Inst. of Ecol. of Plants & Animals], Urals Sci.Center of USSR. Sverdlovsk. Issue 110, 47-55. (in Russian)
10. Voronina E.A., Shekhanova I.A., Peshkov S.P., Muntyan S.P. (1978). Trudy VNIRO, 134, 122-130. (in Russian)
11. Milishnikov A.N., Novikova T.A. (1977). Inform. Bull. of Sci. Council on Radiobiology, Issue 20, 110-111. (in russian)
12. Altukhov, Milishnikov A.N., Novikova. (1977). In: Radioecology of Animals. Nauka, Moscow. 70-71. (in Russian)
13. Shevchenko V.A., Pechkurenkov V.L., Fedorov E.A. et al. (1986). Voprosy ikhtiologii. 26 (3), 494-503. (in Russian)
14. Luchnik N.V. (1957). Biofizika, 2 (1). 86-93. (in Russian)
15. Tsytsugina V.G., Risik N.S., Lazorenko G.E. (1973). Natural and Artificial Radionuclides in Life of Hydrobionts. Naukova Dumka, Kiev. 152 p. (in Russian)
16. Zavitaeva T.A., Sevankaev A.V., Palyga G.F. (1986). Radiobiologia, 26 (1), 711-713. (in Russian)
17. Pitkyanen G.B., Shvedov V.L., Safronova N.G. (1878). Trudy IERZh, Issue 110, 56-60. (in Russian)
18. Ermokhin V.Y., Muntyan S.P. (1977). In: Radioecology of Animals.Nauka, Moscow. 76-77. (in Russian)
19. Muntyan S.P. (1977). Ibid. 81-82. (in Russian)
20. Muntyan S.P. (1977). Ibid. 82-83. (in Russian)
21. Muntyan S.P. (1978). Voprosy ikhtiologii, 18 (6), 1153-1157. (in Russian)
22. Muntuan S.P. (1993). In the book: Ecological Consequences of Radioactive Contamination on the South Urals. Nauka. Moscow. 187-191. (in Russian)
23. Pechkurenkov V.L., Pokrovskaya G.L. (1978). Voprosy ikhtiologii, 18 (6). 1118-1127. (in Russian)
24. Shevchenko V.A., Abramov V.L., Pechkurenkov V.L. (1992). Radiation Genetics of Natural Populations. Genetic Consequences of the Kyshtym Accident. Nauka, Moscow. 219p. (in Russian)
25. Shevchenko V.A., Abramov V.L., Pechkurenkov V.L. (1993). In the book: Ecological Consequences of Radioactive Contamination on the South Urals. Nauka, Moscow. 258-303. (in Russian)
26. Zaitsev Y.A., Pitkyanen G.B. (1970). In the book: Radioactive Isotopes in the External Environment and an Organism. Atomizdat, Moscow. 103-105. (in Russian)
27. Tsytsugina V.G. (1991). IInd All-Union Conf. on fishery toxicology, devoted to 100-anniversary of the problem of water quality in Russia, Nov.1991. Abstracts. Vol. 2. Sankt-Peterburg. p.246. (in Russian)
28. Fetisov A.N., Peshkov S.P., Smagin A.I., Tetkin G.A. (1992). Voprosy ikhtiologii, 32 (1), 79-87. (in Russian)
29. Dubinin N.P., Shevchenko V.A., Kalchenko V.A. et al. (1980). In the book: Mutagenesis under Action of Physical Factors. Nauka, Moscow. 3-44. (in Russian)

276

30. Fetisov A.N., Smagin A.I., Rubanovich A.V. (1993). Radiobiologya, 33 (1), 160-165. (in Russian)
31. Shevchenko V.A., Fetisov A.N. (1989). Doklady AN SSSR, 307 (5), 1249-1252. (in Russian)
32. Kryshev I.I., Alexakhin R.M., Ryabov I.N. et al. (1991). Radioecological Consequences of the Chernobyl Accident. Nuclear Soc. of USSR. 190 p. (in Russian)
33. Sorochisky B.V., Grodzinsky D.M. (1993). Radiobiol. Congress, Kiev, 20-25 Sept. 1993. Absracts. Part III. Pushchino. 944-945. (in Russian)
34 Skotnikova O.G., Kazimirko Y,V,, Kolokolova N.V. et al. (1990). In the book: Biol. and Radioecol. Aspects of Consequences of the Chernobyl NPP Accidents. Ist Internat.Conf.(Abstracts). "Zeleny Mys", 10-18 Sept., 1990. Moscow 1990. p.110. (in Russian)
35. Shevstova N.A. (1993). Radiobiol. Congress. Kiev, 20-25 Sept. 1993. (Abstracts). Part III. Pushchino. 1993. 1134-1135. (in Russian)
36. Klokov V.M., Pankov I.V., Volkova E.N. et al. (1993). Ibid. Part II. Pushchino. 456-457. (in Russian)
37. Polikarpov G.G., Tsytsugina V.G., Zherko N.V., Baranova O.K. (1992). Doklady AN Ukrainy. N 4. 147-150. (in Russian)
38. Tsytsugina V.G. Polikarpov G.G. Radiobiol. Congress. Kiev, 20-25 Sept. 1993. (Abstracts). Part III. Pushchino. 1089-1090. (in Russian)
39. Ryabov I.N. (1992). Radiobiologiya. 32 (5). 662-666. (in Russian)
40. Fetisov A.N., Rubanovich A.V., Slipchenko T.S. (1990). [See: Ref. 34], p.141. (in Russian)
41. Khmeleva N.N., Golubev A.P., Plenkin A.E. et al.(1990). [See: Ref. 34], p.125. (in Russian)
42. Starobogatov Y.I. (1990). [See: Ref. 34], p.118. (in Russian)
43. Molotkov D.V., Morozov A.M. (1991). All-Union Conf. "Radiobiol. Consequences of the Chernobyl NPP Accident". (Abstracts). 30 Oct. - 1 Nov., 1991. Minsk. p.92. (in Russian)
44. Petrova N.A. (1990). [See: Ref. 34}, p.159. (in Russian)
45. Shuvalikov V.B. (1990). Ibid., p.161. (in Russian)
46. Vyatchinkina V.I. (1993). [See: Ref. 33], Part I, p. 195. (in Russian)
47. Izopeskov E.E., Davydov A.I. (1990). [See: Ref. 34], p.111. (in Russian)
48. Kokhnenko O.S., Petchikov A.M., Voronovich A.I. (1991). [See: Ref. 43], p. 60. (in Russian)
49. Goncharova R.I., Slukvin A.M. (1993). [See: Ref. 33], p. 259. (in Russian)
50. Goncharova R.I., Slukvin A.M., Gurnovsky A.S. (1991). [See: Ref. 43], p. 28. (in Russian)
51. Goncharova R.I., Slukvin A.M. (1993). [See: Ref. 33], 240-241. (in Russian)
52. Gorodilov Y.N., Dukina N.A., Chmilevsky D.A. et al. (1990). [See: Ref. 34], p.113. (in Russian)
53. Kokhenko O.S., Petrikov A.M. (1993). [See: Ref. 33], Part I, 513-514. (in Russian)
54. Pechkurenkov V.L. (1990). [See: Ref.34], 143. (in Russian)
55. Pechkurenkov V.L. (1991). Radiobiologiya. 31 (5).704-707. (in Russian)
56. Smirnov S.A., Gorodilov Y.N., Markov K.P. (1990). [See: Ref. 34], 109. (in Russian)
57. Khadachi O., Kotelevtsev S.K., Stepanova L.I., Ryabov I.N. (1993). [See: Ref. 33], Part III, 1063-1064. (in Russian)
58. Belova N.V. (1990). [See: Ref. 34], p. 112. (in Russian)
59. Verigin B.V., Mekeeva A.P., Emelyanova N.G. et al. (1990). Ibid., p.119. (in Russian)
60. Makeeva A.P., Emelyanova N.G., Belova N.V., Ryabov I.N. (1990). Ibid., p.120. (in Russian)
61. Krysanov E.Y., Krysanova I.A. (1990). Ibid., p. 140. (in Russian)
62. Ryabstev I.A., Krysanov E.Y. (1989). Ist All-Union Radiobiol. Congress. Moscow, 21-27 Aug.1989. (Abstracts). Vol. II.Pushchino. 526-527. (in Russian)
63. Vinogradov A.E., Rozanov Y.M., Borkin L.Y. (1990). [See: Ref. 34]. p.124. (in Russian)
64. Eliseeva K.G., Voytovich A.M., Ploskaya M.V. (1989). [See: Ref. 62], p.443. (in Russian)
65. Eliseeva K.G., Voytovich A.M., Ploskaya M.V. et al. (1991). [See: Ref. 43]. p. 41-42. (in Russian)
66. Eliseeva K.G., Voytovich A.M., Trusova V.D. Serzhanin Y.I. (1993). [See: Fef. 33], Part I, 339-340. (in Russian)
67. Kornilova M.B. (1993). Ibid., Part II, p.502. (in Russian)
68. Mikityuk A.Y. (1993). Ibid., 679-680. (in Russian)
69. Mikityuk A.Y. (1993). Ibid., 680-681. (in Russian)
70. Mikituik A.Y. (1993). Ibid. p.682. (in Russian)
71. Hydroecological Consequences of the Chernobyl NPP Accident. (1992). (Ed. D.M.Grodzinsky). Naukova dumka, Kiev. 267 p. (in Russian)
72. Radioactive and Chemical Pollution of Dnieper and its Reservoirs after the Chernobyl NPP Accident. (1992). (Ed. D.M.Grodzinsky). Naukova dumka, Kiev, 194 p. (in Russian)

73. Polikarpov G.G., Tsytsugina V.G. (1993). Radiobiologiya.33(2). 205-213. (in Russian)
74. Polikarpov G.G.(1978). Effects of ionizing radiation upon aquatic organisms (Chronic irradiation). In: XX Congr. Nazionalle (AIFSPR). Atti della giornata sultema "Alcuni Aspetti di Radioecologia". (Bologna 1977.) Poligrafici Parma. 25-46.
75. Polikarpov G.G. (1987). Gidrobiologicheskiy Zhurnal. 23 (6), 29-38. (in Russian)
76. Woodhead D.S. (1993). Dosimetry and the assessment of environmental effects of radiation exposure. In: Radioecology after Chernobyl. Biogeochemical Pathways of Artificial Radionuclides. SCOPE 50.
    Ed. F. Warner, R.M. Harrison. John Wiley and Sons, Chichester, N.Y., Brisbane, Toronto, Singapore. 291-306.
77. Belova N.V., Verigin B.V. (1993). Voprosy ikhtiologii, 33 (6), 814-828. (in Russian)
78. Lackey J., Bennett C.F. (1963). In : Radioecology. Proc. of the 1st Nat. Symp., Sept. 10-15, 1961. (Ed. V.Schultz and A.W.Klement, Jr.). Reinhold Publ. Co., N.Y., Amer. Inst. Biol. Sci., Wash. D.C. 175-177.
79. Whicker F.W., Schultz V. (1982). Radioecology: Nuclear Energy and the Environment. Vol.2. CC Press, Inc., Boa Raton, Florida. 228p.
80. Polikarpov G.G. (1979). In : Methodology for Assessing Impacts of Radioactivity on Aquatic Ecosystems. Report of an Advisory Group Meeting, IAEA, Vienna, 21-25 Nov., 1977. STI/DOC/10/190.IAEA, Vienna. 173-194.
81. Tsytsugina V.G. (1979). Ibid. 369-380.

LIST OF CONTRIBUTORS

ASKER AARKROG
  Risø National Laboratory
  P.O. Box 49,
  DK-4000 Roskilde,
  Denmark.

O. BURTON
  Radioecology Laboratory,
  Physics Department,
  Faculty of Agronomic Sciences,
  B-5030 Gembloux (Belgium).

ARRIGO A. CIGNA
  President of the UIR
  Fraz Tuffo
  I-14023 Cocconato AT, Italy

P.J. COUGHTREY
  International Union of
  Radioecologists

M. CRICK
  IAEA P.O. Box 100.
  A-1400 Vienna,
  Austria

Ye.G. DROZHKO
  "Mayak"
  Experimental Research Station
  CIS-454060 Chelyabinsk-60,
  Russia

E. HOLM
  Radiation Physics Department.
  Lund University Hospital.
  S-221 85 Lund,
  Sweden

M.J. FRISSEL
  Senior Officer
  Torenlaan 3,
  6866-BS Heelsum,
  Netherlands

E.N. KARAVAEVA
  Biophysical station of the Institute
  of Plant and Animal Ecology
  Urals Division,
  Russian Academy of Sciences
  ulitsa 8 Marta, 202
  CIS-620219 Yekatarinburg,
  Russia

R. KIRCHMANN
  Radioecology Laboratory,
  Botany Department,
  University of Liège, Sart-Tilman,
  B-4000 Liège
  Belgium

I. LINKOV
  Department of Environmental
  and Occupational Health
  Graduate School of Public
  Health
  University of Pittsburgh,
  Pittsburgh, PA 15261,
  USA

I.M. MOLCHANOVA
  Biophysical station of the Institute
  of Plant and Animal Ecology
  Urals Division,
  Russian Academy of Sciences
  ulitsa 8 Marta, 202
  CIS-620219 Yekatarinburg,
  Russia

M.G. NIFONTOVA

Biophysical station of the
Institute of Plant and
Animal Ecology
Urals Division,
Russian Academy of Sciences
ulitsa 8 Marta, 202
CIS-620219 Yekatarinburg,
Russia

G.G. POLIKARPOV

A.O. Kovalevsky Institute of
Biology of the Southern Seas
Prospekt Nakhimova 2,
CIS 335011 Sevastopol, Crimea,
Ukraine

V. POZOLOTINA

Institute of Plant and Animal
Ecology
Urals Division,
Russian Academy of Sciences
ulitsa 8 Marta, 202
CIS-620219 Yekatarinburg,
Russia

V. P. REMEZ

JU "Compomet Cantec"
8th March street 5, Suite 414,
Ekaterinburg 620014,
Russia

C. ROBINSON

NRPB Chilton, Didcot.
Oxfordshire OX11 0RQ
UK

G.N. ROMANOV

"Mayak"
Experimental Research Station
CIS-454060 Chelyabinsk-60,
Russia

G. SHAW

Centre for Analytical Research in
the Environment.
Imperial College
at Silwood Park, Ascot Berkshire.
SL5 7TE
UK

W.R. SCHELL

Department of Environmental
and Occupational Health
Graduate School of Public
Health
University of Pittsburgh,
Pittsburgh, PA 15261,
USA

D.A. SPIRIN

'Mayak'
Experimental Research Station
(ONIS)
CIS-454060 Cheliabinsk-60,
Russia

B.V. TESTOV

Natural Sciences Institute
Perm State University
ulitsa Genkelya, 4
CIS-614005 Perm,
Russia

A. TRAPEZNIKOV

Biophysical station of the
Institute of Plant and Animal
Ecology
Urals Division,
Russian Academy of Sciences
ulitsa 8 Marta, 202
CIS-620219 Ekatarinburg,
Russia

V.G. TSYTSUGINA

A.O. Kovalevsky Institute of Biology of the Southern Seas
Prospekt Nakhimova 2,
CIS 335011 Sevastopol, Crimea,
Ukraine

O. VOITSEKHOVITCH

Ukrainian Hydrometeorological
Institute
Prospect Nauki, 37
CIS-252028 Kiev,
Ukraine

P.I. YUSHKOV

Biophysical station of the
Institute of Plant and Animal
Ecology
Urals Division,
Russian Academy of Sciences
ulitsa 8 Marta, 202
CIS-620219 Yekatarinburg,
Russia

LIST OF PARTICIPANTS

VANDENHOVE Hildegarde
Grees, 35,
B-2490 Balen, Belgium

WAUTERS Jan
J. Ceulemansstraat, 8,
B-3128 Tremelo,
Belgium

ZHU Hiu-Chao
Rue de Houdain, 9,
B-7000 Mons,
Belgium

PLENIN Alexandr
Institute of Zoology,
Academy of Sciences of Belarus,
ulitsa, CIS-220000 Minsk,
Belarus

RIMKEVITCH Vitaly
International Saharov Institute
on Radioecology.
Minsk,
Belarus

MAVRODIEV Vasile, Miltchev
Sofia University,
Department of Chemistry.
Radiochemical Laboratory,
1 James Bouchier Av.
1126 Sofia,
Bulgaria

NIZAMSKA Marina, Vladova
Committee on the use of Atomic
Energy for Peacefull Purpose,
67, Shipchensky prohod Klvd
1574 Sofia,
Bulgaria

CHAMNEY Larry
P.O.Box 1046, Station B,
Slater Street 280, Ottawa,
Ontario,
Canada KIP 559

GARBA Alois
Slezka 9,
12029 Prague 2,
Czech Republic

REALO Enn
Ann 77-15,
EE-2400 Tarty,
Estonia

BLEHER Martin
Institut fur Strahlenschutz GSF,
Neuherberg, Postfach 1129,
D-85758 Oberschleisheim,
Germany

NALEZINSKI Sibylle
Bundesamt for Strahlensch1itz,
Ingolstdter, Landstr.1,
85764 Oberschleisheim
(Neuherberg)
Germany

TAESCHNER Michael, Hans
Frerkingweg 32,
D-30455 Hannover,
Germany

KARAMANIS Dimitrios
71 Nikopoleos,
Gr-45221 Ioannina,
Greece

TRABIDOU Georgia
Erechtiou 4,
11742, Gargaretta, Athens,
Greece

283

KOBLINGER-BOKORI Edith
Arany J.u. 41 1/5,
H-1221 Budapest,
Hungary

NAGY Arpad Z.
Godollo University of Agricul-
tural Sciences,
Faculty of Agricultural
Engineering,
P.O.Box 303,
H-2103 Godollo,
Hungary

SZERBIN Pavel
Adria s. 4/b,
H-1149 Budapest,
Hungary

PALSSON Sigurdur, Emil
Videmelur 39,
Is-107 ReykJavik,
Iceland

CARINI Franca
Via Cervini, 67,
I-29100 Piacenza,
Italy

GIOVANI Concettina
Via pre'Zaneto 28,
I-33050 Pavia di Udine (UD),
Italy

KONDO Kunio
Misawe-City,
Aomori Pref.033,
Japan

YOSHIDA Satoshi
1-103 Atagojutaku, 1-1 Bunkyo,
Mito-shi,
Ibaraki, 310
Japan

JANSSEN Martien
P.O.Box 9035,
NL-3720 BA Bilthoven ,
Netherlands

TIMMERMANS Cor W.M.
KEMA, P.O.Box 9035,
NL-6800 ET Arnhem
Netherlands

SELNES Tone, Dorothea
Veker veien 218,
N-0751 Oslo,
Norway

KRAJEWSKI Pawel
ul. Konwaliowa 7,
03-194 Warsaw,
Poland

WACTAWEK Zofia,
KazimieraHawajska 1/30,
OE-776 Warsaw,
Poland

BARCA Cataline, Vasile
Boul. Libertatii, 12,
Bucharest-6,
Romania

ILIESCU Nicolae
Alea Sandulesti nr.2, Bl. OD7,
sc.D et 9 ap.157,
CP. 77394, Sector 6, Bucharest,
Romania

POPESCU Ion I.
P.O.Box 78, Pitesti,
Romania

KOBZOVA Darina
Hlboka 2,
012235 Bratislava,
Slovak Republic

VRABCEK Peter
Ul. Oslobodenia 21/292,
90101 Malacky,
Slovakia

MILLAN Rocio
Real Vieja 36,
S.Sebastian de los Reyes,
E-28700 -Madrid,
Spain

ANDREEV Alexey
Institute of Hydrobiology,
National Academy of Sciences of
Ukraine,
Prospekt Geroyev
Stalingrada,12,
CIS-252210 Kiev,
Ukraine

KOROTKOV Andrey
IBSS, Prospekt Nakhimova, 2,
CIS-335011 Sevastopol, Crimea,
Ukraine

LAZAREV Nikolay
Institute of Agricultural
Radiology,
ulitsa Mashinostroiteley,7,
CIS-255205 Chabany,
Kievskaya Oblast,
Ukraine

MIKHEEV Alexandr
ICBGE, ulitsa Lebedeva,l,
CIS-252022 Kiev,
Ukraine

SIKORENKO-GUSAR Varvara
Ukrainian Hygenic Centre,
Ministry of Health of Ukraine,
Prospekt Popudrenko, 50
CIS-252600 Kiev,
Ukraine

VEMBER Valeria
Institute of Microbiology and
Virology,
ulitsa Zabolotnogo, 154,
CIS-252143 Kiev,
Ukrain

MURDOCK Robert
157 Avondale, Ash Vale,
Aldershot, Hampshire,
GU12 5SN,
United Kingdom

WEBBE-WOOD David
35, Kingsand Rd Lee, London
SE12 OLE,
United Kingdom

LINKOV Igor
Department of Environmental
and Occupational Health
Graduate School of Public
Health
130, DeSoto Street, Pittsburgh,
PA 15261,
USA

EKIDIN Alexey
Institute of Industrial Ecology,
Urals Division,
Russian Academy of Sciences,
ulitsa Permvomayskaya, 91,
CIS-620219 Yekaterinburg
GSP-169,
Russia

KONONOVITCH Alexandr
All-Russian Scientific Research
Institute on Exploitation of
Nuclear Power Plants,
ulits Ferganskaya, 25,
CIS-109507 Moscow,
Russia

KOZELTSEV Andrei
The Institute Druzba Narodov,
Moscou.
Russia

KRUZALOV A exander
Urals Technical University
UPI-UGTU,
CIS-620002, Yekatarinburg,
Russia

LOGUNTSEV Andrai
The Tomek State University,
Tomsk,
Russia

PANTELEEV Vladimir
The Department on Remediation
of Radioactive Areas
The Urals Region,
Chelayabinsk,
Russia.

PAVLOTSKAYA Fanni
The V.I.Vernadsky'Institute of
Geochemistry and Analytical
Chemistry,
Russian Academy of Sciences,
ulitsa Kosygina, 19,
CIS-111975 Moscow,
Russia

POKARZEVSKY Andrey
Institute of Evoluational Mor-
phology and Ecology,
Moscow StateUniversity school,
Russian Academy of Sciences,
Leninsky Propspekt, 33,
CIS-117071 Moscov,
Russia

SHATILIN Sergei
The territoria Inspection of
Environmental Control.
Glasov,
Russia

SPIRIDONOV Sergei
The Russian Institute of Agricul-
tural Radiology and
Agroecoloay.
Obninsk,
Russia

TARASOV Oleg
ONIS,
CIS-454060 Chelyabinsk-60,
Russia

TASKEV Anatoly
Instituta of Biology,
Komi Scientific Centre,
Urals Division,
Russian Academy of Sciences,
ulitsa Kommunisticheskaya, 24,
CIS-167610 Syktyvkar, Komi,
Russia

VINOGRAD0V Vladimir
The Ministry for Science And
Technical Polivy of Russian
Federation.
Moscow,
Russia

VOITSCICKY Adward
Direction of Gosprogra mme on
rehabilitation of the territory of
VURS at Administration Of
Sverdlovsk Oblast,
Oktyabrskaya Ploshchad, 3,
CIS-620000 Yekaterinburg,
Russia

ZHEBEL Viktor
Urals Scientific Practical Centre
of Radiation Medicine
Medgorodok,
CIS-454076 Chelyabinsk,
Russia

ZHIGALSKY Oleg
IEPA, ulitsa 8 Marta, 202,
CIS-620219 Yekaterinburg
GSP-511,
Russia

ZIRAYANOV Aleksei
Research and Development
Institute of Power Engineering
Sverdlovsk Branch (SFNIKIET)
Zarechny,
Russia